Image Recognition and Classification

Image Recognition and Classification

Algorithms, Systems, and Applications

edited by

Bahram Javidi

University of Connecticut
Storrs, Connecticut

MARCEL DEKKER, INC.   NEW YORK · BASEL

ISBN: 0-8247-0783-4

This book is printed on acid-free paper.

Headquarters
Marcel Dekker, Inc.
270 Madison Avenue, New York, NY 10016
tel: 212-696-9000; fax: 212-685-4540

Eastern Hemisphere Distribution
Marcel Dekker AG
Hutgasse 4, Postfach 812, CH-4001 Basel, Switzerland
tel: 41-61-261-8482; fax: 41-61-261-8896

World Wide Web
http://www.dekker.com

The publisher offers discounts on this book when ordered in bulk quantities. For more information, write to Special Sales/Professional Marketing at the headquarters address above.

Current printing (last digit):
10 9 8 7 6 5 4 3 2 1

PRINTED IN THE UNITED STATES OF AMERICA

For my Aunt Matin

Preface

Image recognition and classification is one of the most actively pursued areas in the broad field of imaging sciences and engineering. The reason is evident: the ability to replace human visual capabilities with a machine is very important and there are diverse applications. The main idea is to inspect an image scene by processing data obtained from sensors. Such machines can substantially reduce the workload and improve accuracy of making decisions by human operators in diverse fields including the military and defense, biomedical engineering systems, health monitoring, surgery, intelligent transportation systems, manufacturing, robotics, entertainment, and security systems.

Image recognition and classification is a multidisciplinary field. It requires contributions from diverse technologies and expertise in sensors, imaging systems, signal/image processing algorithms, VLSI, hardware and software, and packaging/integration systems.

In the military, substantial efforts and resources have been placed in this area. The main applications are in autonomous or aided target detection and recognition, also known as automatic target recognition (ATR). In addition, a variety of sensors have been developed, including high-speed video, low-light-level TV, forward-looking infrared (FLIR), synthetic aperture radar (SAR), inverse synthetic aperture radar (ISAR), laser radar (LADAR), multispectral and hyperspectral sensors, and three-dimensional sensors. Image recognition and classification is considered an extremely useful and important resource available to military personnel and operations in the areas of surveillance and targeting.

In the past, most image recognition and classification applications have been for military hardware because of high cost and performance demands. With recent advances in optoelectronic devices, sensors, electronic hardware, computers, and software, image recognition and classification systems have become available with many commercial applications.

While there have been significant advances in image recognition and classification technologies, major technical problems and challenges face this field. These include large variations in the inspected object signature due to environmental conditions, geometric variations, aging, and target/sensor behavior (e.g., IR thermal signature fluctuations, reflection angles, etc.). In addition, in many applications the target or object of interest is a small part of a very complex scene under inspection; that is, the distorted target signature is embedded in background noise such as clutter, sensor noise, environmental degradations, occlusion, foliage masking, and camouflage. Sometimes the algorithms are developed with a limited available training data set, which may not accurately represent the actual fluctuations of the objects or the actual scene representation, and other distortions are encountered in realistic applications. Under these adverse conditions, a reliable system must perform recognition and classification in real time and with high detection probability and low false alarm rates. Therefore, progress is needed in the advancement of sensors and algorithms and compact systems that integrate sensors, hardware, and software algorithms to provide new and improved capabilities for high-speed accurate image recognition and classification.

This book presents important recent advances in sensors, image processing algorithms, and systems for image recognition and classification with diverse applications in military, aerospace, security, image tracking, radar, biomedical, and intelligent transportation. The book includes contributions by some of the leading researchers in the field to present an overview of advances in image recognition and classification over the past decade. It provides both theoretical and practical information on advances in the field.

The book illustrates some of the state-of-the-art approaches to the field of image recognition using image processing, nonlinear image filtering, statistical theory, Bayesian detection theory, neural networks, and 3D imaging. Currently, there is no single winning technique that can solve all classes of recognition and classification problems. In most cases, the solutions appear to be application-dependent and may combine a number of these approaches to acquire the desired results.

Image Recognition and Classification provides examples, tests, and experiments on real world applications to clarify theoretical concepts. A bibliography for each topic is also included to aid the reader. It is a practical book, in which the systems and algorithms have commercial applications and can be implemented with commercially available computers, sensors, and processors. The book assumes some elementary background in signal/image processing. It is intended for electrical or computer engineers with interests in signal/image processing, optical engineers, computer scientists, imaging scientists, biomedical engineers, applied physicists, applied mathematicians, defense technologists, and graduate students and researchers in these disciplines.

I would like to thank the contributors, most of whom I have known for many years and are my friends, for their fine contributions and hard work. I also thank Russell Dekker for his encouragement and support, and Eric Stannard for his assistance. I hope that this book will be a useful tool to increase appreciation and understanding of a very important field.

Bahram Javidi

Contents

Contributors

Bir Bhanu   Center for Research in Intelligent Systems, University of California, Riverside, California

Francis Chan   Naval Undersea Warfare Center, Newport, Rhode Island

Lipchen Alex Chan   U.S. Army Research Laboratory, Adelphi, Maryland

Rama Chellappa   University of Maryland, College Park, Maryland

Sandor Z. Der   U.S. Army Research Laboratory, Adelphi, Maryland

Frank Dubois   Université Libre de Bruxelles, Bruxelles, Belgium

Olivier Germain   Ecole National Supérieure de Physique de Marseille, Domaine Universitaire de Saint-Jérôme, Marseille, France

Seung Hyun Hong   University of Connecticut, Storrs, Connecticut

Bahram Javidi   University of Connecticut, Storrs, Connecticut

Grinnell Jones III   Center for Research in Intelligent Systems, University of California, Riverside, California

Sherif Kishk   University of Connecticut, Storrs, Connecticut

Abhijit Mahalanobis   Lockheed Martin, Orlando, Florida

Hesham Mahmoud   Wireless Facilities, Inc., San Diego, California

Osamu Matoba   Institute of Industrial Science, University of Tokyo, Tokyo, Japan

Christophe Minetti   Université Libre de Bruxelles, Bruxelles, Belgium

Nasser M. Nasrabadi   U.S. Army Research Laboratory, Adelphi, Maryland

Luting Pan   University of Connecticut, Storrs, Connecticut

Elisabet Pérez   Polytechnic University of Catalunya, Terrassa, Spain

Brian Redman   Arete Associates, Tucson, Arizona

Philippe Réfrégier   Ecole National Supérieure de Physique de Marseille, Domaine Universitaire de Saint-Jérôme, Marseille, France

Joseph Rosen   Ben-Gurion University of the Negev, Beer-Sheva, Israel

Firooz Sadjadi   Lockheed Martin, Saint Anthony, Minnesota

Enrique Tajahuerce   Universitat Jaume I, Castellon, Spain

Wen-Yi Zhao   Sarnoff Corporation, Princeton, New Jersey

Qinfen Zheng   University of Maryland, College Park, Maryland

Image Recognition
and Classification

1

Neural-Based Target Detectors for Multiband Infrared Imagery

Lipchen Alex Chan, Sandor Z. Der, and Nasser M. Nasrabadi
U.S. Army Research Laboratory, Adelphi, Maryland

1.1 INTRODUCTION

Human visual performance greatly exceeds computer capabilities, probably because of superior high-level image understanding, contextual knowledge, and massively parallel processing. Human capabilities deteriorate drastically in a low-visibility environment or after an extended period of surveillance, and certain working environments are either inaccessible or too hazardous for human beings. For these reasons, automatic recognition systems are developed for various military and civilian applications. Driven by advances in computing capability and image processing technology, computer mimicry of human vision has recently gained ground in a number of practical applications. Specialized recognition systems are becoming more likely to satisfy stringent constraints in accuracy and speed, as well as the cost of development and maintenance.

The development of robust automatic target recognition (ATR) systems must still overcome a number of well-known challenges: for example, the large number of target classes and aspects, long viewing range, obscured targets, high-clutter background, different geographic and weather conditions, sensor noise, and variations caused by translation, rotation, and scaling of the targets. Inconsistencies in the signature of targets, similarities between the signatures of different targets, limited training and testing data, camouflaged targets, nonrepeatability of target signatures, and difficulty using available contextual information makes the recognition problem even more challenging.

A complete ATR system typically consists of several algorithmic components, such as preprocessing, detection, segmentation, feature extraction, classification, prioritization, tracking, and aimpoint selection [1]. Among these components, we are particularly interested in the detection-classification modules, which are shown in Fig. 1. To lower the likelihood of omitting targets of interest, a detector must accept a nonzero false-alarm rate. Figure 1 shows the output of a detector on a typical image. The detector has found the target but has also selected a number of background regions as potential targets. To enhance the performance of the system, an explicit clutter rejector may be added to reject most of the false alarms produced by the detector while eliminating only a few of the targets. Clutter rejectors tend to be much more complex than the detector, giving better performance at the cost of greater computational complexity. The computational cost is often unimportant because the clutter rejector needs to operate only on the small subset of the image that is indicated by the detector.

The ATR learning environment, in which the training data are collected, exerts a powerful influence on the design and performance of an ATR system. Dasarathy [2] described these environments in an increasing order of difficulty, namely the supervised, imperfectly supervised, unfamiliar, vicissitudinous, unsupervised, and partially exposed environments. In this chapter, we assume that our training data come from an unfamiliar environment, where the labels of the training data might be unreliable to a level that is not known a priori. For the experimentation presented in this chapter, the input images were obtained by forward-looking infrared

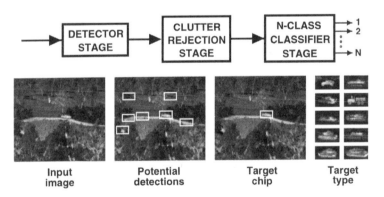

Figure 1   A typical ATR system.

(FLIR) sensors. For these sensors, the signatures of the targets within the scene are severely affected by rain, fog, and foliage [3]. Clark et al. [4] used an information-theoretic approach to evaluate the information bound of FLIR images in order to estimate the best possible performance of any ATR algorithm that uses the given FLIR images as inputs. On the other hand, some FLIR enhancement techniques may be used to preprocess the FLIR input images. Lo [5] examined six of these techniques and found that a variable threshold zonal filtering technique performed most satisfactorily.

The major goal of this research is to examine the benefits of using two passive infrared images, sensitive to different portions of the spectrum, as inputs to a target detector and clutter rejector. The two frequency bands that we use are normally described as mid-wave (MW, 3–5 μm) and long-wave (LW, 8–12 μm) infrared. Two such images are shown in Fig. 2. Although these images look roughly similar, there are places where different intensities can be noted. The difference tends to be more significant during the day, because reflected solar energy is significant in the mid-wave band, but not in the long-wave band. These differences have indeed affected the detection results of an automatic target detector. As shown in Fig. 3, different regions of interest were identified by the same target detector on these two images. Because a different performance is obtained using either the MW or the LW imagery, our first question is which band alone provides better performance in target detection and clutter rejection? The second question is whether combining the bands results in better performance than using either band alone, and if so, what are the best methods of combining these two bands.

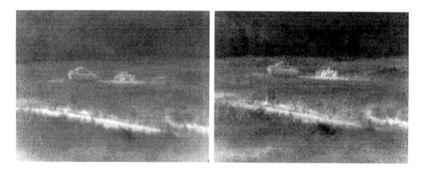

Figure 2 Typical FLIR images for the mid-wave (left) and long-wave (right) bands, with an M2 tank and a HMMWV around the image center. Different degree of radiation, as shown by the windshield of the HMMWV, is quite apparent.

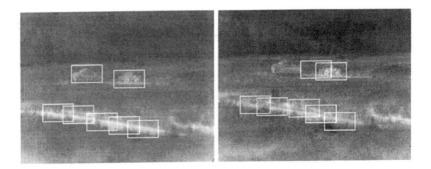

Figure 3 The first seven regions of interest detected on the mid-wave (left) and the long-wave (right) bands. Note that the M2 tank is missed in the case of the long-wave image but detected in the mid-wave image.

To answers these questions, we developed a set of eigen-neural-based modules and use them as either a target detector or clutter rejector in our experiments. As shown in Fig. 4, our typical detector/rejector module consists of an eigenspace transformation and a multilayer perceptron (MLP). The input to the module is the region of interest (target chip) extracted either from an individual band or from both of the MW and LW bands simultaneously. An eigen transformation is used for feature extraction and dimensionality reduction. The transformations considered in this chapter are principal component analysis (PCA) [6], the eigenspace separation transform (EST) [7], and their variants that were jointly optimized with the MLP. These transformations differ in their capability to enhance class separability and to extract component features from a training set. When both bands are input together, the two input chips are transformed through either a set of jointly obtained eigenvectors or two sets of band-specific eigenvectors. The result of the eigenspace transformation is then fed to the MLP that predicts the identity of the input, which is either a target or clutter. Further descriptions about the eigenspace transformation and the MLP are provided in the next two sections. Experimental results are presented in Section 4. Some conclusions are given in the final section of this chapter.

1.2 EIGENTARGETS

We used two methods to obtain the eigentargets from a given set of training chips. PCA is the most basic method, from which the more complicated EST method is derived.

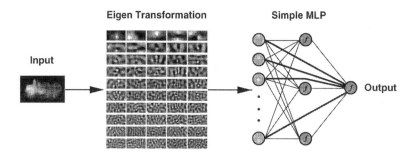

Figure 4   Schematic diagram of our detector/rejector module.

1.2.1   Principal Component Analysis

Also referred to as the Hotelling transform or the discrete Karhunen–Loève transform, PCA is based on statistical properties of vector representations. PCA is an important tool for image processing because it has several useful properties, such as decorrelation of data and compaction of information (energy) [8]. Here, we provide a summary of the basic theory of PCA.

Assume a population of random vectors of the form

$$\mathbf{x} = \begin{bmatrix} x_1 \\ x_2 \\ \vdots \\ x_n \end{bmatrix} \tag{1}$$

The *mean vector* and the *covariance matrix* of the vector population $\mathbf{x}$ are defined as

$$\mathbf{m_x} = E\{\mathbf{x}\} \tag{2}$$

$$\mathbf{C_x} = E\{(\mathbf{x} - \mathbf{m_x})(\mathbf{x} - \mathbf{m_x})^T\} \tag{3}$$

respectively, where $E\{arg\}$ is the expected value of the argument and T indicates vector transposition. Because $\mathbf{x}$ is n dimensional $\mathbf{C_x}$ is a matrix of order $n \times n$. Element c_{ii} of $\mathbf{C_x}$ is the variance of x_i (the ith component of the $\mathbf{x}$ vectors in the population) and element c_{ij} of $\mathbf{C_x}$ is the covariance between elements x_i and x_j of these vectors. The matrix $\mathbf{C_x}$ is real and symmetric. If elements x_i and x_j are uncorrelated, their covariance is zero and, therefore, $c_{ij} = c_{ji} = 0$. For N vector samples from a random population, the mean vector and covariance matrix can be approximated respectively from the samples by

$$\mathbf{m_x} = \frac{1}{N} \sum_{p=1}^{N} \mathbf{x}_p \tag{4}$$

$$\mathbf{C_x} = \frac{1}{N} \sum_{p=1}^{N} (\mathbf{x}_p \mathbf{x}_p^T - \mathbf{m_x} \mathbf{m_x}^T) \tag{5}$$

Because $\mathbf{C_x}$ is real and symmetric, we can always find a set of n orthonormal eigenvectors for this covariance matrix. A simple but sound algorithm to find these orthonormal eigenvectors for all really symmetric matrices is the Jacobi method [9]. The Jacobi algorithm consists of a sequence of orthogonal similarity transformations. Each transformation is just a plane rotation designed to annihilate one of the off-diagonal matrix elements. Successive transformations undo previously set zeros, but the off-diagonal elements get smaller and smaller, until the matrix is effectively diagonal (to the precision of the computer). The eigenvectors are obtained by accumulating the product of transformations during the process, and the main diagonal elements of the final diagonal matrix are the eigenvalues. Alternatively, a more complicated method based on the QR algorithm for real Hessenberg matrices can be used [9]. This is a more general method because it can extract eigenvectors from a nonsymmetric real matrix. It becomes increasingly more efficient than the Jacobi method as the size of the matrix increases. Because we are dealing with large matrices, we used the QR method for all experiments described in this chapter. Figure 5 shows the first 100 (out of the 800 possible in this case) most dominant PCA eigentargets and eigenclutters, which were extracted from the target and clutter chips in the training set, respectively. Having the largest eigenvalues, these eigenvectors capture the greatest variance or energy as well as the most meaningful features among the training data.

Let $\mathbf{e}_i$ and λ_i, $i = 1, 2, \ldots, n$, be the eigenvectors and the corresponding eigenvalues, respectively, of $\mathbf{C_x}$, sorted in a descending order so that $\lambda_j \geq \lambda_{j+1}$ for $j = 1, 2, \ldots, n-1$. Let $\mathbf{A}$ be a matrix whose rows are formed from the eigenvectors of $\mathbf{C_x}$, such that

$$\mathbf{A} = \begin{bmatrix} \mathbf{e}_1 \\ \mathbf{e}_2 \\ \vdots \\ \mathbf{e}_n \end{bmatrix} \tag{6}$$

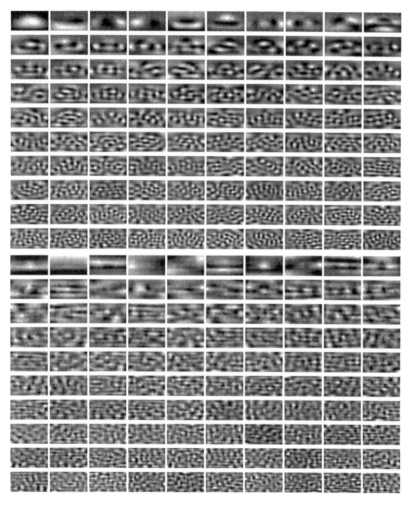

Figure 5 First 100 most dominant PCA eigenvectors extracted from the target (top) and clutter (bottom) chips.

This **A** matrix can be used as a linear transformation matrix that maps the **x**'s into vectors, denoted by **y**'s, as follows:

$$\mathbf{y} = \mathbf{A}(\mathbf{x} - \mathbf{m_x}) \tag{7}$$

The **y** vectors resulting from this transformation have a zero mean vector; that is, $\mathbf{m_y} = 0$. The covariance matrix of the **y**'s can be computed from **A** and $\mathbf{C_x}$ by

$$\mathbf{C_y} = \mathbf{A}\mathbf{C_x}\mathbf{A}^T \tag{8}$$

Furthermore, $\mathbf{C_y}$ is a diagonal matrix whose elements along the main diagonal are the eigenvalues of $\mathbf{C_x}$; that is,

$$\mathbf{C_y} = \begin{bmatrix} \lambda_1 & & & 0 \\ & \lambda_2 & & \\ & & \cdot & \\ & & & \cdot \\ 0 & & & \lambda_n \end{bmatrix} \tag{9}$$

Because the off-diagonal elements of $\mathbf{C_y}$ are zero, the elements of the **y** vectors are uncorrelated. Because the elements along the main diagonal of a diagonal matrix are its eigenvalues, $\mathbf{C_x}$ and $\mathbf{C_y}$ have the same eigenvalues and eigenvectors.

On the other hand, we may want to reconstruct vector **x** from vector **y**. Because the rows of **A** are orthonormal vectors, $\mathbf{A}^{-1} = \mathbf{A}^T$. Therefore, any vector **x** can be reconstructed from its corresponding **y** by the relation

$$\mathbf{x} = \mathbf{A}^T\mathbf{y} + \mathbf{m_x} \tag{10}$$

Instead of using all the eigenvectors of $\mathbf{C_x}$, we may pick only k eigenvectors corresponding to the k largest eigenvalues and form a new transformation matrix $\mathbf{A}_k$ of order $k \times n$. In this case, the resulting **y** vectors would be k dimensional, and the reconstruction given in Eq. (10) would no longer be exact. The reconstructed vector using $\mathbf{A}_k$ is

$$\hat{\mathbf{x}} = \mathbf{A}_k^T\mathbf{y} + \mathbf{m_x} \tag{11}$$

The mean square error (MSE) between **x** and $\hat{\mathbf{x}}$ can be computed by the expression

$$\epsilon = \sum_{j=1}^{n} \lambda_j - \sum_{j=1}^{k} \lambda_j = \sum_{j=k+1}^{n} \lambda_j \tag{12}$$

Because the λ_j's decrease monotonically, Eq. (12) shows that we can minimize the error by selecting the k eigenvectors associated with the k largest eigenvalues. Thus, the PCA transform is optimal in the sense that it minimizes the MSE between the vectors $\mathbf{x}$ and their approximations $\hat{\mathbf{x}}$.

1.2.2 Eigenspace Separation Transform

The EST has been proposed by Torrieri as a preprocessor to a neural binary classifier [10]. The goal of the EST is to transform the input patterns into a set of projection values such that the size of a neural classifier is reduced and its generalization capability is increased. The size of the neural network is reduced, because the EST projects an input pattern into an orthogonal subspace of smaller dimensionality. The EST also tends to produce projections with different average lengths for different classes of input and, hence, improves the discriminability between the targets. In short, the EST preserves and enhances the classification information needed by the subsequent classifier. It has been used in a mine-detection task with some success [11].

The transformation matrix $\mathbf{S}$ of the EST can be obtained as follows:

1. Computer the $n \times n$ correlation difference matrix

$$\hat{\mathbf{M}} = \frac{1}{N_1} \sum_{p=1}^{N_1} \mathbf{x}_{1p} \mathbf{x}_{1p}^T - \frac{1}{N_2} \sum_{q=1}^{N_2} \mathbf{x}_{2q} \mathbf{x}_{2q}^T \tag{13}$$

where N_1 and $\mathbf{x}_{1p}$ are the number of patterns and the pth training pattern of Class 1, respectively. N_2 and $\mathbf{x}_{2q}$ are similarly related to Class 2 (which is the complement of Class 1).

2. Calculate the eigenvalues of $\hat{\mathbf{M}}$ $\{\lambda_i | i = 1, 2, \dots, n\}$.

3. Calculate the sum of the positive eigenvalues

$$E_+ = \sum_{i=1}^{n} \lambda_i \quad \text{if } \lambda_i > 0 \tag{14}$$

and the sum of the absolute values of the negative eigenvalues

$$E_- = \sum_{i=1}^{n} |\lambda_i| \quad \text{if } \lambda_i < 0 \tag{15}$$

(a) If $E_+ > E_-$, then take all the k eigenvectors of $\hat{\mathbf{M}}$ that have positive eigenvalues and form the $n \times k$ matrix $\mathbf{S}$.

(b) If $E_+ < E_-$, then take all the k eigenvectors of $\hat{\mathbf{M}}$ that have negative eigenvalues and form the $n \times k$ matrix $\mathbf{S}$.

(c) If $E_+ = E_-$, then use either subset of eigenvectors to form the matrix $\mathbf{S}$, preferably the smaller subset.

Given the $\mathbf{S}$ transformation matrix, the projection $\mathbf{y}_p$ of an input pattern $\mathbf{x}_p$ is computed as $\mathbf{y}_p = \mathbf{S}^T\mathbf{x}_p$. The $\mathbf{y}_p$, with a smaller dimension (because $k \leq n$) and presumably larger separability between the classes, can then be sent to a neural classifier. Figure 6 shows the eigenvectors associated with the positive and negative eigenvalues of the $\hat{\mathbf{M}}$ matrix that was computed with the target chips as Class 1 and the clutter chips as Class

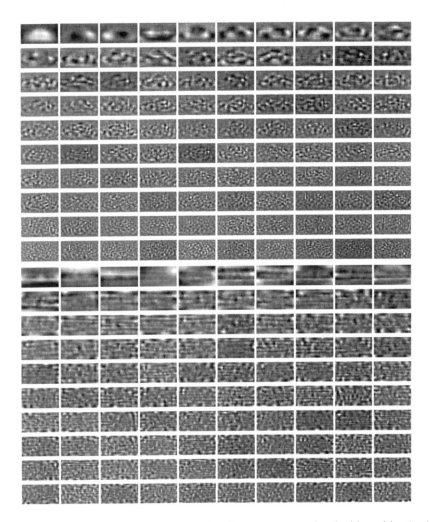

Figure 6 First 100 most dominant EST eigenvectors associated with positive (top) and negative (bottom) eigenvalues.

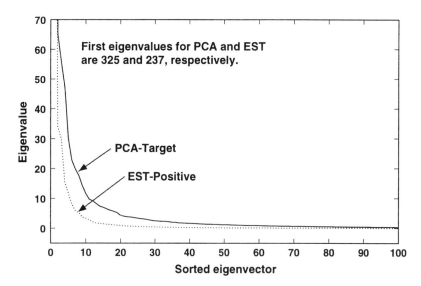

Figure 7 Rapid attenuation of eigenvalues.

2. From the upper part of Fig. 6, the signature of targets can be clearly seen. On the other hand, the lower part represents all the features of clutters.

As we can see from Figs. 5 and 6, only the first few dozens of the eigentargets contain consistent and structurally significant information pertaining to the training data. These eigentargets exhibit a reduction in information content as their associated eigenvalues rapidly decrease, which is depicted in Fig. 7. For the less meaningful eigentargets, say the 50th and all the way up to the 800th, only high-frequency information is present. In other words, by choosing $k = 50$ in Eq. (12) when $n = 800$, the resulting distortion error, ϵ, would be small. Although the distortion is negligible, there is a 16-fold reduction in input dimensionality.

1.3 MULTILAYER PERCEPTRON

After projecting an input chip to a chosen set of k eigentargets, the resulting k projection values are fed to an MLP classifier, where they are combined nonlinearly. A typical MLP used in our experiments, as shown on the right-hand side in Fig. 4, has $k + 1$ input nodes (with an extra bias input), several layers of hidden nodes, and one output node. In addition to full connections between consecutive layers, there are also shortcut connections directly from one layer to all other layers, which may speed up the learning process. The MLP classifier is trained to perform a two-class problem, with training output values of ± 1. Its sole task is to decide whether a given input pattern is a target (indicated by a high output value of around $+1$) or clutter (indicated by a low output value of around -1). The MLP is trained in batch mode using Qprop [12], a modified backpropagation algorithm, for a faster but stable learning course.

Alternatively, the eigenspace transformation can be implemented as an additional linear layer that attaches to the input layer of the simple MLP above. As shown in Fig. 8, the resulting augmented MLP classifier, which is collectively referred to as a PCAMLP network in this chapter, consists of a transformation layer and a back-end MLP (BMLP). When the weights connecting the new input nodes to the kth output node of the transformation layer are initialized with the kth PCA or EST eigenvector, the linear summation at the kth transformation output node is equivalent to the kth projection value. The advantage of this augmented structure is to enable a joint optimization between the transformation (feature extraction) layer and the BMLP classifier, which is achieved by adjusting the corresponding weights of the transformation layer based on the error signals backpropagated from the BMLP classifier.

The purpose of joint optimization is to incorporate class information in the design of the transformation layer. This enhancement is especially

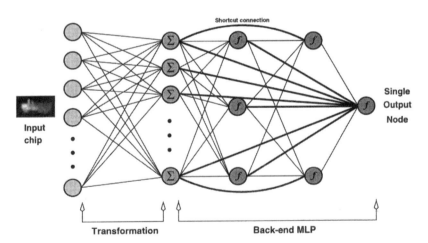

Figure 8   An augmented MLP (or PCAMLP) that consists of a transformation layer and a back-end MLP.

useful to the PCA eigenvectors, because the class-separation issue has never been considered during their derivation. During the joint operation process, the transformation weights are gradually adjusted, suing a variety of gradient descent-based algorithms, so that the overall error is reduced at the output node of the back-end MLP. Although the discriminability of the transformation layer is enhanced, it may lose some of its energy compaction capability in exchange. These changes are exhibited in Fig. 9, where the structural characteristics of the PCA eigenvectors are gradually given away to local emphases that distinguish the targets from clutter. After a prolonged joint optimization process, the succinct PCA structures could be completely replaced by incomprehensible patterns that have overfitted the training samples. Care should be taken to avoid overtraining the transformation layer.

It is interesting to observe that similar evolutions also occur when we initialize the transformation layer with random weights, instead of initializing with the PCA or EST eigenvectors. Adjusted through a supervised gradient descent algorithm, these random weights connected to each output node of the transformation layer gradually evolve into certain features that try to maximize the class separation for the BMLP classifier. A typical evolution of a five-node supervised transformation matrix is shown in Fig. 10, after it had been trained for 689, 3690, 4994, and 9987 epochs, respectively. Note that the random weights at the early stage evolved into more structural features that resemble those of the PCA eigenvectors shown

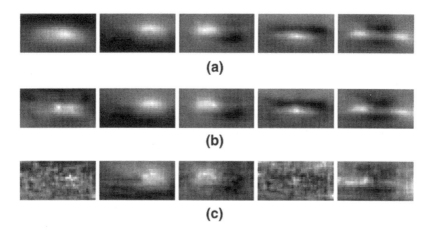

(a)

(b)

(c)

Figure 9 Changes in PCA eigenvectors after (a) 0, (b) 4752, and (c) 15751 epochs of backpropagation training to enhance their discriminability.

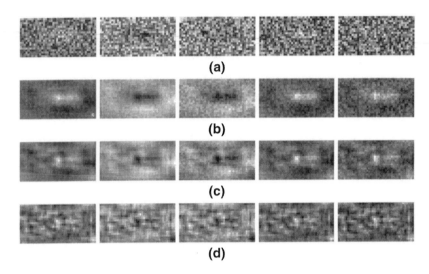

Figure 10   The evolution of transformation vectors that were initialized with random weights and trained with a gradient descent algorithm, after (a) 689, (b) 3690, (c) 4994, and (d) 9987 epochs of training.

in Fig. 9a. Nonetheless, these features became incomprehensible and less structural again when the training session was extended.

In contrast to the PCA transformation, the above supervised transformation does not attempt to optimize the energy compaction on the training data. In addition, the gradient descent algorithm is very likely to be trapped at a local minimum in the treacherous weight space of $p \times m$ dimensions or in its attempts to overfit the training data with strange and spurious solutions. A better approach would be using a more sophisticated training algorithm that is capable of optimizing both the interclass discriminability and energy compaction simultaneously.

Let us first consider the issue of energy compaction during joint discrimination-compression optimization training. Instead of extracting the PCA eigenvectors from the covariance matrix C_x, we can compute them directly from the x input vectors via a single-layer self-organized neural network [13]. An example of such a neural network, with predefined p input nodes and m linear output nodes, is shown in Fig. 11. If the network is trained with the generalized Hebbian algorithm (GHA) proposed by Sanger [14], the activation value of the kth output neuron, y_k, converges to the kth most dominant eigenvalue associated with the input data. At the same time, the p weights leading to the kth output neuron, $w_{ki}, i = 1, \ldots, p$,

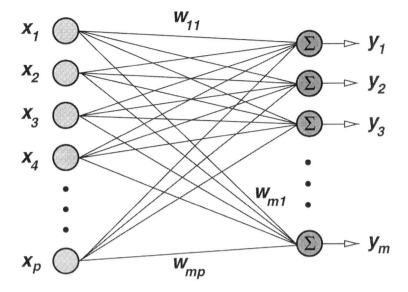

Figure 11 A single-layer self-organized neural network.

become the eigenvector associated with the kth dominant eigenvalue. Suppose we want to find the m most dominant eigenvalues and their associated eigenvectors based on S input samples of size p, namely x_i^s, $s = 1, \ldots, S$, $i = 1, \ldots, p$. The corresponding GHA network can be trained through the following steps:

1. At iteration $t = 1$, initialize all the adjustable weights, w_{ji}, $j = 1, \ldots, m$, $i = 1, \ldots, p$, to small random values. Choose a small positive value for the learning rate parameter η.

2. Compute the output value $y_j^s(t)$ and weight adjustment $\Delta w_{ji}^s(t)$ for $s = 1, \ldots, S$, $j = 1, \ldots, m$, $i = 1, \ldots, p$, as follows:

$$y_j^s(t) = \sum_{i=1}^{p} w_{ji}(t) x_i^s \tag{16}$$

$$\Delta w_{ji}^s(t) = \eta y_j^s(t) \left(x_i^s - \sum_{k=1}^{j} w_{ki}(t) y_k^s(t) \right) \tag{17}$$

3. Modify the weights, w_{ji}, $j = 1, \ldots, m$, $i = 1, \ldots, p$ for this iteration:

$$w_{ji}(t+1) = w_{ji}(t) + \frac{1}{S}\sum_{s=1}^{S} \Delta w_{ji}^s(t) \tag{18}$$

4. Increment t by 1 and go back to Step 2. Repeat Steps 2–4 until all the weights reach their steady-state values.

We combine the unsupervised GHA with a supervised gradient descent algorithm (such as the Qprop algorithm) to perform a joint discrimination–compression optimization. Note that the GHA network in Fig. 11 structurally and functionally resembles the transformation layer of the PCAMLP shown in Fig. 8. Therefore, we may adjust the weights of the transformation layer in Fig. 8 as follows:

$$w_{ji}(t+1) = w_{ji}(t) + \alpha \text{ [PCA contribution]} + \beta \text{ [BMLP contribution]}$$

$$= w_{ji}(t) + \alpha\left(\frac{1}{S}\sum_{s=1}^{S}\Delta w_{ji}^s(t)\right) - \beta\left(\frac{1}{S}\sum_{s=1}^{S}x_i^s\delta_j^s(t)\right) \tag{19}$$

$$= w_{ji}(t) + \frac{1}{S}\sum_{s=1}^{S}\left[\alpha\Delta w_{ji}^s(t) - \beta x_i^s\delta_j^s(t)\right] \tag{20}$$

The PCA contribution in Eq. (19) is defined earlier as the second term on the right-hand side of Eq. (18). The $\delta_j^s(t)$ in Eq. (20) is the error signal back-propagated from the BMLP to the jth output neuron of the transformation layer for training sample s at iteration t, whereas the x_i^s is the same input vector defined in Eq. (16). The strength of the PCA contribution on the joint transformation is controlled by α, whereas β controls the contribution of gradient descent learning. If $\alpha = 0$, a regular supervised transformation is performed. Setting $\beta = 0$ results in a standard PCA transformation, provided that the η in Eq. (17) is small enough [14].

For the joint transformation to acquire PCA-like characteristics, the η in Eq. (17) and α in Eq. (20) must be small. To prevent the gradient descent effect from dominating the joint transformation, the β has to be small also. As a result, the training process is slow. To speed up the process, we first obtain the standard PCA eigenvectors using the much more efficient QR algorithm [9] and initialize the transformation layer in Fig. 8 with these eigenvectors. Equation (20) is then used to jointly optimize the transformation layer and the classifier together. It is easier to observe performance changes in this way, as the joint transformation attempts to maximize its discriminative power while maintaining its energy compression capability simultaneously.

The effect of this joint discrimination–compression optimization can be clearly seen in Fig. 12. Figure 12a shows the first five most dominant

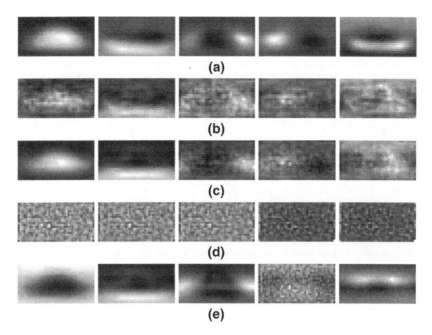

Figure 12 The effect of joint discrimination–compression optimization. The five transformation vectors show as standard PCA eigenvectors (a), after 12519 epochs of Qprop (b), or after 12217 epochs of Qprop + GHA training (c). With randomly initialized values, they appear after 17654 epochs of Qprop (d) or 34788 epochs of Qprop + GHA training (e).

PCA eigenvectors obtained with the standard QR algorithm. If we initialize the transformation layer of the PCAMLP with these standard PCA eigenvectors and adjust them based on the supervised Qprop algorithm only, the resulting weight vectors, as shown in Fig. 12b and similarly in Fig. 9c, would gradually lose all of their succinct structures to quasirandom patterns. However, if Eq. (20) with small nonzero α and β are used, the most important structures of the PCA eigenvectors are always preserved, as we can see in Fig. 12c. If we initialize the transformation vectors with random weights rather than PCA eigenvectors, the Qprop algorithm alone could only forge them into incomprehensible features, as shown in Fig. 12d as well as Fig. 10d, after an extended period of training. With the joint discrimination–compression optimization, even the random weights evolve into the mostly understandable features as shown in Fig. 12e. Out of the five feature vectors displayed in Fig. 12e, only the fourth one fails to exhibit a clear structure. Comparing the other four vectors of Fig. 12e to the corresponding vectors in Fig. 12a, a clear relationship can be established. Reverse-video of the first vector and fifth vector might be caused by an α value that is too large or might be an anomaly of the GHA algorithm when initialized with random weights. The sign of both $w_{ki}(t)$ and $y_k^s(t)$ can flip without affecting the convergence of the algorithm, as can be seen in Eq. (17). The only effect on the back end of the MLP is to flip the signs of the weights that are connected to the $y_k^s(t)$. The other minor differences in these vector pairs are probably the work of the Qprop algorithm.

1.4 EXPERIMENTAL RESULTS

A series of experiments was used to examine the performance of the PCAMLP, either as a target detector or clutter rejector. We also investigated the usefulness of a dual-band FLIR input dataset and the best way to combine the two bands in order to improve the PCAMLP target detector or clutter rejector. We used 12-bit gray-scale FLIR input frames similar to those shown in Fig. 2, each of which measured 500×300 pixels in size. There were 461 pairs of LW–MW matching frames, with 572 legitimate targets posed between 1 and 4 km in each band. First, we trained and tested the PCAMLP as a clutter rejector that processed the output of an automatic target detector called NVDET (developed at the U.S. Army Research Laboratory). Then, we used the trained PCAMLP as a target detector on its own and compared its detection performance to that of NVDET on the same dataset.

1.4.1 PCAMLP as a Clutter Rejector

In order to find the answers for the three questions raised in Section 1.1, we have designed four different clutter rejection setups. As shown in Fig. 13, the first two setups use an individual MW or LW band alone as input. Based on the results from these two setups, we should be able to answer the first question, namely which band alone may perform better in our clutter rejection task? For setup c, we stack the MW and LW chips extracted at the same location before the eigenspace transformations. In this case, the size of each eigenvector is doubled, but not the number of projection values fed to the MLP. If the performance of setup c is better than both setups a and b, then we may say that there is an advantage to using dual band simultaneously. Finally, setup d is almost the same as combining setups a and b, except the

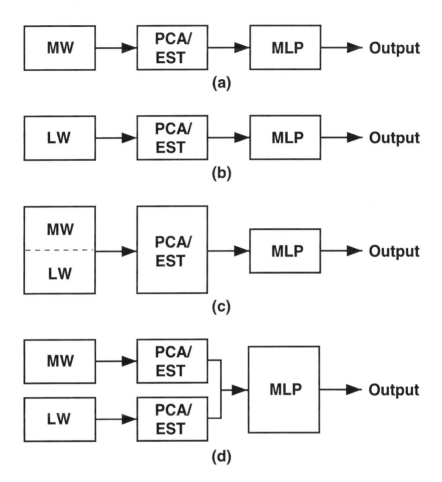

Figure 13 Four different setups for our clutter rejection experiments.

projection values resulting from each eigenspace transformation are now combined before feeding to an MLP with twice as many input nodes. Comparing the performance of setups c and d, we can find out if it is better to combine the two bands before or after the eigenspace transformation.

The chips extracted from each band has a fixed size of 75 × 40 pixels. Because the range to the targets varies from 1 to 4 km, the size of the targets varies considerably. For the first dataset, the chips were extracted from the location suggested by the NVDET. As shown in Fig. 14, many of these so-

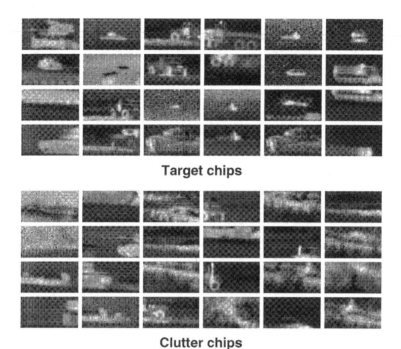

Target chips

Clutter chips

Figure 14   Examples of detector-centered chips.

called detector-centered chips end up with the target lying off-center within the chip. This is a very challenging problem, because the chips of a particular target, posed at the same viewing distance and aspect, may appear different. Furthermore, any detection point would be declared as a miss when its distance from the ground-truth location of a target is greater than a predefined threshold. Hence, a clutter chip extracted around a miss point may contain a significant part of a target, which is very similar to an off-centered target chip. Therefore, it is difficult to find an unequivocal class boundary between the targets and the clutter. The same numbers of chips were created for the MW and LW in all experiments.

We have also created ground-truth-centered chips, which were extracted around the ground truth location of a detected target, as our second dataset. The extraction process of this dataset is almost the same as in the previous dataset, except that whenever a detection suggested by the target detector is declared as an acceptable hit, we move the center of chip extraction from the detected location to the ground-truth center of the corresponding target. In this case, all the target silhouettes were properly centered within the chips, so that a class boundary between the targets and the clutter becomes more feasible. However, some partial targets still appeared on some of the clutter chips, undermining the notion of clear-cut class boundaries. Also, the size of targets continue to fluctuate considerably at different viewing ranges, which complicates the culmination of target distinction. Examples of good-truth-centered chips are given in Fig. 15.

The third dataset consists of chips that were properly centered and zoomed based on ground-truth location and range. The target appears at the center of each chip with a relatively consistent silhouette size. Nonetheless, the signatures of the same target may still exhibit a wide scope of appearances due to differences in zoomed resolution, viewing aspect, operational and weather conditions, environmental effects, and many other factors. Figure 16 shows a few chips from the third dataset.

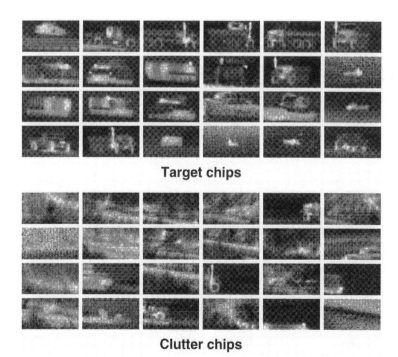

Target chips

Clutter chips

Figure 15   Examples of ground-truth-centered chips.

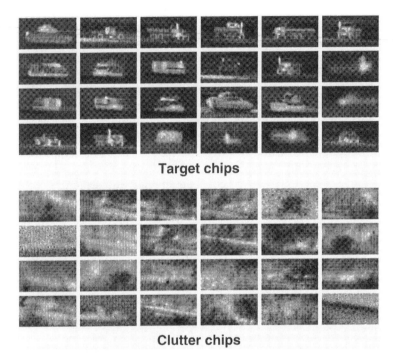

Target chips

Clutter chips

Figure 16   Examples of ground-truth-centered and zoomed chips.

To reduce the computational complexity while retaining enough information embedded in the chips, we down-sampled the input image chip from 75×40 pixels to 40×20 pixels. As shown in Fig. 7, the eigenvalues diminish rapidly for both the PCA and EST methods, but those of the EST decrease even faster. In other words, the EST may produce a higher compaction in information. The eigenvalues approach zero after the 40th or so eigentarget, so we were interested in no more than the 40 most dominant eigentargets, instead of all 800 eigentargets. For setups a, b, and c, we used the 1, 5, 10, 20, 30, and 40 most dominant eigentargets of each transformation to produce the projection values for the MLP. For setup d, we used the 1, 5, 10, 20, and 25 projection values of each band to feed the corresponding MLPs with 2, 10, 20, 30, 40, and 50 input nodes, respectively. In each case, five independent training processes were tried with different initial MLP weights. The average hit rates of each setup for detector-centered chips, at a controlled false-alarm rate of 3%, are tabulated in Table 1. The bold numbers in the table indicate the best PCA and EST performance achieved for each setup with this dataset.

Table 1 Performance on Detector-Centered Chips at 3% False-Alarm Rate

No. of MLP inputs[a]	Data type	Average hit rates of five runs (%)							
		a		b		c		d	
		PCA	EST	PCA	EST	PCA	EST	PCA	EST
1/2	Train	21.08	46.31	26.31	43.96	25.41	48.83	25.55	50.05
	Test	20.07	42.68	23.69	40.87	22.78	45.21	23.98	45.28
5/10	Train	78.02	79.10	72.93	78.02	82.84	85.05	87.14	85.48
	Test	70.27	70.78	61.05	65.50	74.40	77.21	76.20	74.07
10/20	Train	79.93	81.69	76.40	79.86	88.25	**90.59**	**88.22**	**90.20**
	Test	73.24	72.88	63.00	67.05	79.49	**81.88**	**78.66**	**74.14**
20/30	Train	**83.35**	**85.01**	**79.06**	**85.30**	89.69	89.04	85.66	87.57
	Test	**74.50**	**74.47**	**66.91**	**69.26**	81.66	76.17	77.87	74.29
30/40	Train	79.17	80.29	78.81	76.72	**91.78**	85.55	80.94	88.32
	Test	66.91	64.34	66.76	61.05	**80.25**	71.86	73.27	72.19
40/50	Train	68.18	57.48	70.09	62.25	88.50	82.63	74.67	76.14
	Test	62.82	48.35	62.17	51.97	78.70	68.54	70.38	65.06

[a] First number is for setups a, b, and c. Second number is for setup d.

Comparing setup a and b in Table 1, we can see that the MW band performed better than the LW band when a moderate number of 5–30 projection values were fed to the MLP. For both setups, the peak performance was achieved with 20 MLP inputs. Although their peak hit rates for the training set are somewhat comparable, the MW leads in the testing performance by 5–8%. Therefore, the MW sensor seems to be the better candidate than the LW, if we have to choose only one of them for our clutter rejector. It should be noted that this conclusion may apply only to the specific sensors used for this study. If we compare setup a with setup c, we note significant improvement achieved by the stacked dual-band input in both training and testing sets, which ranges from 5% to 8% again. In other words, processing the MW and LW jointly is better than using either one of them alone. The way we merge the two bands also affects the clutter rejection performance. Although the performances of setups c and d are similar, setup c is the clear winner when it comes to the peak performance and in the cases where 20 or more MLP inputs were used. Therefore, combining the dual band before the eigenspace transformation, rather than after, is the better way to utilize the MW and LW jointly.

In order to examine the effect on the clutter rejector of accurate centering of the targets within the input chips, we repeated the above experiments with the second dataset. Once again, we tabulated the average hit rates achieved by each setup in Table 2 and marked with bold numbers the best performance of all setups. When we look at the best performance in Table 2, the relationships among the four setups are similar to those exhibited in Table 1. Due to the distinctly improved target chips in this case, performance of all setups have dramatically improved. Emerging from much lower hit rates on the first dataset, the single-band setups have made a greater gain than the dual-band setups with the improved target centering offered by the second dataset. As a result, the performance edge of the dual-band clutter rejectors has shrunk to about 5%. In other words, the usefulness of dual-band input would be reduced if the prior target detector could detect the ground-truth target center more accurately.

Finally, we repeated the same set of experiments on the third dataset, in which the target chips were centered and zoomed correctly using the ground-truth information. We give the average hit rates of each setup in Table 3. With a quick glance on the bold numbers in Table 3, one can see that near-perfect hit rates were achieved by almost every setup for the training set, even at a demanding 3% false-alarm rate. The performance on the testing set are not far behind either, with those of the setup a tailing at around 94%. In other words, accurate zooming of the target has helped every setup, especially the weaker single-band clutter rejectors.

Table 2   Performance on Ground-Truth-Centered Chips at 3% False-Alarm Rate

No. of MLP inputs[a]	Data type	Average hit rates of five runs (%)							
		a		b		c		d	
		PCA	EST	PCA	EST	PCA	EST	PCA	EST
1/2	Train	26.50	45.16	35.24	48.64	31.01	50.72	34.14	56.63
	Test	27.37	47.26	35.57	48.26	29.95	51.74	34.18	56.86
5/10	Train	89.92	89.93	87.44	85.41	92.31	94.34	94.00	**95.38**
	Test	85.92	83.83	85.42	85.42	90.25	90.85	88.71	**91.14**
10/20	Train	**92.11**	**93.40**	**91.02**	**88.88**	94.84	96.58	**97.87**	93.60
	Test	**85.27**	**85.07**	**86.81**	**86.37**	88.26	89.35	**89.40**	87.21
20/30	Train	90.47	88.69	83.47	80.00	**97.47**	**97.37**	95.43	95.39
	Test	86.97	79.31	80.20	73.73	**91.29**	**90.94**	89.80	87.31
30/40	Train	71.96	67.10	77.02	66.70	97.96	92.11	87.84	89.83
	Test	71.69	62.84	71.14	60.60	89.65	86.82	84.83	81.15
40/50	Train	77.92	70.67	79.30	69.53	82.08	84.96	87.59	73.10
	Test	75.57	62.64	73.58	64.93	81.14	80.65	84.83	66.07

[a] First number is for setups a, b, and c. Second number is for setup d.

Table 3 Performance on Ground-Truth-Centered-Zoomed Chips at 3% False-Alarm Rate

No. of MLP inputs[a]	Data type	Average hit rates of five runs (%)							
		a		b		c		d	
		PCA	EST	PCA	EST	PCA	EST	PCA	EST
1/2	Train	68.98	77.42	79.40	82.88	80.40	86.10	80.55	86.85
	Test	70.15	78.86	75.87	82.09	78.11	83.33	78.76	83.83
5/10	Train	70.22	97.17	78.86	**99.01**	77.32	**100.00**	80.35	**100.00**
	Test	71.49	95.62	80.55	**96.51**	79.15	**98.46**	82.59	**97.11**
10/20	Train	83.97	**99.95**	87.10	96.43	92.36	99.55	93.20	90.08
	Test	88.65	**94.73**	90.55	94.48	95.82	96.92	96.32	89.16
20/30	Train	88.29	92.70	90.57	92.61	94.64	98.96	96.43	95.09
	Test	91.99	85.97	93.63	87.56	96.12	92.78	97.51	89.55
30/40	Train	90.42	93.50	93.45	84.77	99.06	92.31	99.20	99.01
	Test	92.19	82.04	95.52	86.47	95.07	89.50	96.37	89.55
40/50	Train	**96.77**	93.30	**100.00**	87.74	**100.00**	98.51	**99.30**	99.35
	Test	**94.58**	83.93	**96.47**	85.82	**98.36**	89.50	**97.66**	89.60

[a] First number is for setups a, b, and c. Second number is for setup d.

In Table 4, we show the average value of the bold numbers in Tables 1–3 for the single-band (columns 3–6) and dual-band (columns 7–10) setups, respectively. The benefit of dual-band data decreases gradually as more ground-truth information is added to the process of chip extraction. It should be noted that as the performance improves, the performance estimates become relatively less accurate because of reduced number of samples.

The average recognition rates usually increase with the number of eigenvectors used for feature extraction, but they approach saturation at around 20 projection values. Theoretically, the more eigenvectors employed in the transformation, the larger the amount of information that should be preserved in the transformed data. However, using more transformed inputs increases the complexity of the MLP, prolongs the training cycle, results in an overfitted MLP with reduced generalization capability, and increases the chance of getting stuck in a nonoptimal solution. In our experiments, many clutter rejectors with a large number of projection values have shown a steady decrease in their peak performance, mainly because of the weakening in their generalization capability to recognize the targets in the testing set. When fewer projection values are used, a higher performance is achieved by the EST. This improvement can be attributed to the better compaction of

Table 4 Performance Improvement (%) by Dual-Band Data at 3% False-
Alarm Rate

Data type	Single band	Dual band	Improvement
Detector centered			
Train	83.43	90.20	6.77
Test	71.29	78.73	7.44
Ground-truth centered			
Train	91.35	97.02	5.67
Test	85.88	90.69	4.81
Ground-truth-centered zoomed			
Train	98.93	99.83	0.90
Test	95.57	97.90	2.33

information associated with EST. However, the PCA performed as good or even better when more projection values were used, which may indicate that some minor information might have been lost in the EST method. Nonetheless, the EST should be a better transformation when only a small number of projection values can be processed, because of speed or memory constraints.

We also investigated the effect on the performance of clutter rejectors of jointly optimizing the transformation layer with the BMLP. Consider the room for potential improvement at a 3% false-alarm rate; we chose the best PCA setups with 5 (10 for setup d) MLP inputs that were trained with the third dataset. First, we tried to minimize the overall output error of the PCAMLP by modifying the PCA eigenvectors, based on the errors back-propagated from the BMLP, using the supervised Qprop algorithm only. The clutter rejection rates of these four PCAMLPs for the first 4000 epochs of joint Qprop optimization are shown in Fig. 17. Due to the increased discriminability at the PCA transformation layer, their hit rates were improved by 15–25%. The improvements achieved by single-band setups were especially significant and, therefore, further diminished the dwindling advantage held by dual-band setups for this dataset. The best testing performance of setups a–d were achieved at epoch 5862, 5037, 1888, and 5942 of training, with corresponding hit rates of 99.78%, 100.00%, 97.99%, and 100.00% for the training set and 98.22%, 98.66%, 96.44%, and 99.78% for the testing set, respectively.

We also attempted to modify the PCA transformation layer with Eq. (20), where the Qprop and GHA were applied simultaneously. The resulting improvements of the same PCA setups are shown in Fig. 18. Comparing the corresponding curves in Figs. 17 and 18, we found that the GHA appeared

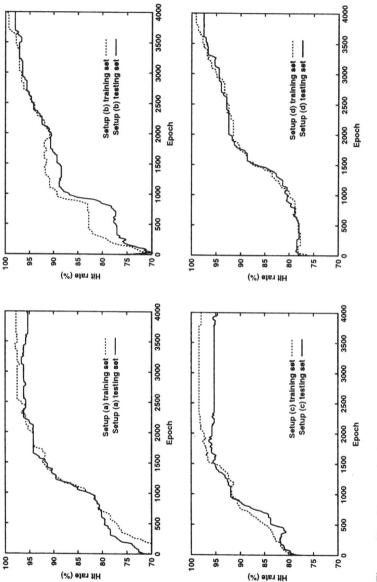

Figure 17 Clutter rejection performance of PCAMLP were enhanced by optimizing the PCA layer using the Qprop algorithm only.

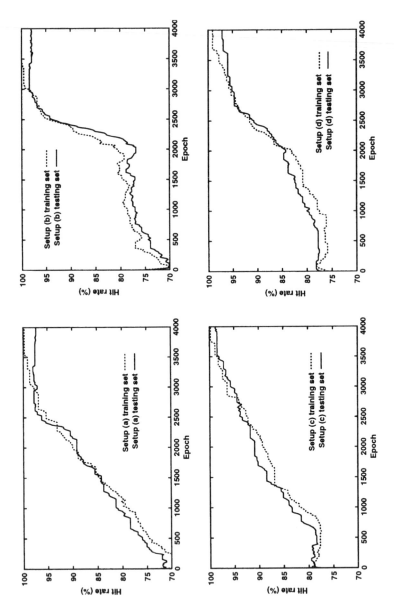

Figure 18   Clutter rejection performance of PCAMLP were enhanced by optimizing the PCA layer using Qprop and GHA algorithms simultaneously.

to slow down the improvement during the early stage of training, but then accelerated at the later stage to performance peaks that rival or beat those in Fig. 17. In this case, their best testing performance were achieved at epoch 3293, 3413, 3952, and 4531 of training, with corresponding hit rates of 99.11%, 100.00%, 99.78%, and 99.78% for the training set, and 98.00%, 98.44%, 99.33%, and 98.44% for the testing set, respectively. The early damping in learning curves indicates conflicting roles played by GHA and Qprop. The GHA tried to preserve the compaction characteristics of the transformation layer by maintaining the structures of those standard PCA eigenvectors, whereas the Qprop attempted to modify them in order to minimize the overall errors at the BMLP output node. The result of this struggle is a transformation layer that maintained most of its structure while emphasized some key areas, as exemplified by Fig. 12e.

Although the GHA did help the curves in Fig. 18 to reach their peaks sooner or higher, these differences in performance are statistically questionable because of the extremely small sample size. (The number of additional targets that are rejected by a system with 98.44% performance, versus 98.66% performance, is 1.) A larger or more difficult dataset is required to adequately measure the performance of this algorithm.

The added cost of computing the GHA is quite significant. Therefore, the usefulness of Eq. (20) is not proven by these experiments, where the transformation layer was initialized with standard PCA eigenvectors rather than random weights. In situations where the PCAMLP setups were equipped with the EST transformation layer, the effect of either joint optimization above was insignificant. The main reasons are thought to be associated with the integrated class separation formulation of the EST, as well as their near-perfect performance with merely five projection values.

1.4.2 PCAMLP as a Target Detector

The PCAMLP structure can be used as a target detector instead of a clutter rejector. As shown in Fig. 19, successive and overlapping chips can be extracted from the input frames and fed to the PCAMLP. For single-band detection, each chip is evaluated by the PCAMLP and the resulting output value indicates the likelihood of having a target situated at the location where the center of that chip is extracted. For setups c and d, a pair of chips must be extracted from the corresponding locations on the two bands for each evaluation. After the whole frame is evaluated, a number of locations with high PCAMLP scores are selected as potential target areas. High scores within a small neighborhood are combined and represented by the highest-scoring pixel among them. Any detection that lies sufficiently close to the ground-truth location is declared a hit, and if not, it is declared a

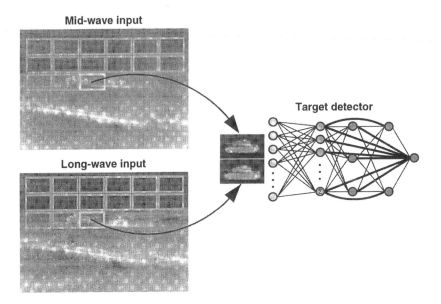

Figure 19 PCAMLP as a target detector.

false alarm. The numbers of hits and false alarms per frame could be changed by considering a different number of top detections from each frame.

We split the 461 pairs of LW–MW matching frames into two near-equal sets, each containing 286 targets of interest. We used the half with 231 frames as a training set, from which we extracted the training chips that were used in the previous clutter rejection experiments. In other words, the trained PCAMLP clutter rejections had "seen" parts of these frames, the parts where the NVDET detector declared as potential target areas. The other 230 served as a testing set, from which we extracted the testing chips for the clutter rejectors.

The same PCA setups chosen for the joint optimization experiments in Sections 1.4.1 were used as target detectors on these frames. With the standard PCA eigenvectors as their transformation layer, the detection performance of all four setups are presented as receiver operating characteristics (ROC) curves. The ROC curves obtained from the training and testing frames are shown at the upper and lower parts of Fig. 20 respectively. For the purpose of comparison, the ROC curves of the NVDET detector for MW and LW frames are also provided. Clearly, the single-band PCAMLPs outperformed the NVDET in both MW and LW cases at lower false-alarm rates, and the dual-band PCAMLPs excelled over the

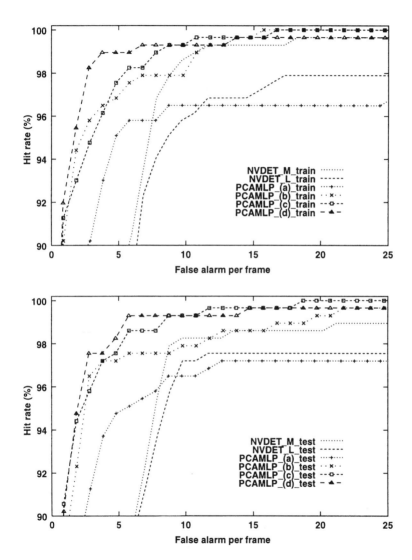

Figure 20 Detection performance of PCAMLP and NVDET.

single-band PCAMLPs. Reversing their order of achievement as clutter rejectors, setup d performed slightly better than setup c as a target detector, especially at low false-alarm rate operation. The MW detectors scored lower than their LW counterparts, even though the MW clutter rejectors were preferred in Tables 1 and 2.

We also applied the Qprop-optimized PCAMLPs, whose clutter rejection improvements were characterized in Fig. 17, as target detectors. Their detection performance is shown by the ROC cures in Fig. 21. Compared to Fig. 20, the performance of setup c was significantly improved at any false-alarm rate on both the training and testing sets and reached the peaks much sooner than other setups. Meanwhile, setup d showed solid gains and continued its leading edge at the regions of five false alarm per frame or lower, but nearly unchanged at higher false-alarm rates. In contrast, setup b showed substantial improvement in the testing performance at five false alarms per frame or higher, but did not change a lot at lower false-alarm rates. Finally, setup a performed significantly worse after this joint optimization on both training and testing sets at any false-alarm rates.

To examine the performance of GHA–Qprop-optimized PCAMLPs at a target detection task, we used the PCAMLPs whose clutter rejection characteristics were charted in Fig. 18. The detection performance of these PCAMLPs for the training and testing sets are presented as ROC curves shown in Fig. 22. Compared to Figs. 20 and 21, the ROC curves in Fig. 22 were changed significantly from those in Fig. 21. Only setup a appeared to benefit from this GHA–Qprop-optimization, whereas all the other setups suffered deteriorations in performance. For the first time, setup a outperformed other setups as a target detector, even though it was limited to low false-alarm regions of the ROC curves.

1.5 CONCLUSIONS

Developing learning algorithms requires overcoming a number of common, challenging design issues. The size of the training dataset requires that the architecture of the learning algorithm does not use too many trainable parameters. Dimensionality reduction on the raw data is a feasible way of reducing the number of weights in a neural network. Dimensionality reduction must be done with an eye toward maintaining much of the information that is useful for discrimination. This implies that the dimensionality reduction should be done with a specific discrimination task in mind.

Because of the limited number of training samples available, our benchmark BMLP classifier consisted of only 5 projection inputs, 1 output, and 111 adjustable weights. In order to represent an extracted area of 75 ×

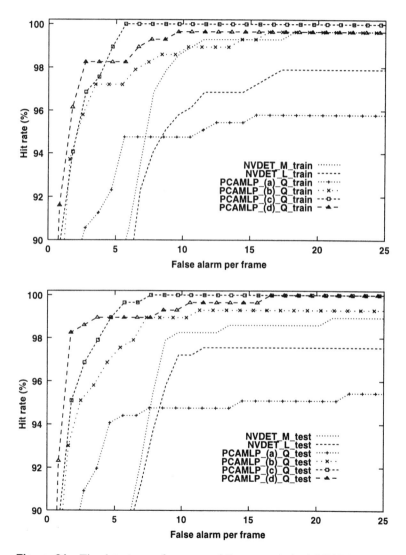

Figure 21 The detector performance of Qprop-optimized PCAMLPs.

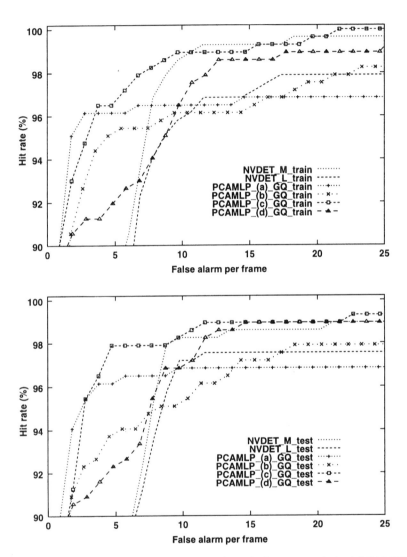

Figure 22   The detector performance of GHA–Qprop = optimized PCAMLPs.

40 pixels with only 5 coefficients, we first shrank the input chips to 40×20 pixels and then projected them to the first five eigenvectors of PCA and EST, resulting in a dimensionality reduction ratio of 600. The feature extraction capability of the transformation layer was enhanced through a joint optimization with the BMLP, so that the class membership and task-specific information were fused into the feature extraction process.

Based on the experiments in which the PCAMLPs were used as clutter rejectors, we conclude that the performance of clutter rejection can be improved by using dual-band FLIR inputs, instead of MW or LW band alone. However, the margin of improvement is inversely affected by the consistency of the input chips. When the targets were not properly centered and scaled within the chips, the dual-band clutter rejectors were able to increase the testing performance by over 7%. This improvement shrank to slightly over 2% when all the targets were properly centered and scaled within the chips. In other words, finding the correct center and range of the targets is a very crucial step in the boosting of clutter rejection performance.

From the results of target detection experiments, we again conclude that the dual-band inputs provide increased performance over the single-band PCAMLP target detectors. Although the MW detectors alone performed considerably worse than the LW detectors, critical information from MW inputs still enabled the dual-band detectors to achieve hit rates superior to those of LW detectors. Even though the PCAMLPs were trained only for cluster rejection tasks, with limited training on the background, constrained to those views that were prescreened by the NVDET detector, they still performed well as target detectors. Jointly optimizing the transformation layer with the BMLP enabled the dual-band PCAMLP detectors to reach the best possible results in these experiments. Given the opportunity to further improve these PCAMLP detectors via direct and complete training on the input scenes, they should be able to perform even better.

ACKNOWLEDGMENT

This research was sponsored by the Army Research Laboratory (ARL) and was accomplished under the ARL/ASEE postdoctoral fellowship program, contract DAAL01-96-C-0038. The views and conclusions contained in this document are those of the authors and should not be interpreted as representing the official policies, either expressed or implied, of the Army Research Laboratory or the U.S. Government. The U.S. Government is authorized to reproduce and distribute reprints for government purposes not withstanding any copyright notation hereon.

REFERENCES

1. B Bhanu. Automatic target recognition: State of the art survey. IEEE Trans Aerospace electron Syst 22(4):364–379, 1986.
2. B Dasarathy. Information processing for target recognition from autonomous vehicles. Proc SPIE Electro-Opt Tech Autonomous Vehicles 219:86–93, 1980.
3. B Bhanu, T Jones. Image understanding research for automatic target recognition. IEEE Aerospace Electron Syst Mag 15–22, Oct. 1993.
4. L Clark, L Perlovsky, W. Schoendorf, C. Plum, T Keller. Evaluation of forward-looking infrared sensors for automatic target recognition using an information-theoretic approach. Opt Eng 31(12):2618–2627, 1992.
5. CM Lo. Forward-looking infrared (FLIR) image enhancement for the automatic target cuer system. Proc SPIE Image Process Missile Guidance 238:91–102, 1980.
6. IT Jolliffe. Principal Component Analysis. New York: Springer-Verlag, 1986.
7. D Torrieri. The eigenspace separation transform for neural-network classifiers. Neural Networks 12:419–427, 1999.
8. RC Gonzalez, RE Woods. Digital Image Processing. New York: Addison-Wesley, 1992.
9. WH Press, SA Teukolsky, WT Vetterling, BP Flannery. Numerical Recipes in C, 2nd ed. New York: Cambridge University Press, 1992.
10. D Torrieri. A linear transform that simplifies and improves neural network classifiers. Proc Int Conf Neural Networks 3:1738–1743, 1996.
11. GL Plett, T Doi, D Torrieri. Mine detection using scattering parameters and an artificial neural network. IEEE Trans Neural Networks 8(6):1456–1467, 1997.
12. S Fahlman. Faster learning variations on back-propagation: An empirical study. Proceedings of the 1988 Connectionist Models Summer School. San Mateo, CA: Morgan Kaufmann, pp 38–51, 1988.
13. S Haykin. Neural Networks: A Comprehensive Foundation. New York: Macmillan College, 1994.
14. TD Sanger. Optimal unsupervised learning in a single-layer linear feedforward neural network. Neural Networks 2:459–473, 1989.

2

Passive Infrared Automatic Target Discrimination

Firooz Sadjadi
Lockheed Martin, Saint Anthony, Minnesota

2.1 INTRODUCTION

Automatic target recognition (ATR) technology is entering a critical phase of the technology cycle. ATR functions are now being incorporated as a requirement in several defense systems like Joint Strike Fighter aircraft (JSF) and the unmanned air vehicles (UAV). These systems will impose increasingly demanding requirements on ATR performance.

During the course of ATR evolution, numerous algorithms have been developed and tested. However, the algorithms are limited to a very small scope of applications. Outside this limited scenario, many ATRs fail their performance requirements. For future systems, the ATR algorithms should be able to maintain optimal performance under a variety of applications, such as scenarios, environments, countermeasures, clutter density, and others.

Our technical approach addresses two technology areas:

- Target segmentation algorithm development
- Evaluation technology

In the algorithm development area, we present a target discrimination/segmentation technique based on hypothesis testing of the statistical decision theory. This approach relies on the assumption that the target segments in infrared imagery have different conditional probability density functions (PDFs) than those of the background. Based on this assumption, we present a technique for target segmentation by estimating the differences of these PDFs, creating a new image based on their statistical differences and then classifying this distance image into two distinct classes based on another hypothesis testing approach.

In the second part of the approach, we present a methodology for the scientific evaluation of the segmentation algorithm and use it for testing the behavior of the algorithm on a large amount of target data from a number of different databases.

This chapter is structured in the following way: In Section 2.2, a description of the segmentation algorithm in the context of the statistical decision theory is given. Section 2.3 provides the details of the experimental design methodology used in this study for the evaluation of the performance of the target segmentor as functions of the changes in the image content and algorithms internal parameters. Section 2.4 documents the results of the performance analysis on a number of real infrared (IR) imagery containing thousands of tactical targets. Finally, in Section 2.5 we provide a summary of the results of the study and its main conclusions.

2.2 TARGET SEGMENTATION

Image segmentation is one of the most error-prone steps in any autonomous image exploitation system [1–5] and has been the subject of research and development for the past decades [1–18]. These studies can be grouped into several categories. Heuristic approaches [6] and the more recent statistical techniques that make use of the Bayesian cost minimization and also techniques based on Markov models [7–18] are the major trends in the field.

Our presented approach uses a heuristically defined distance metric and multilevel Bayesian hypothesis testing techniques for separating IR targets from their backgrounds.

Target segmentation is viewed in this study as the process of separating two statistically distributed population of pixels, each originating from the targets and backgrounds.

In the sense of minimizing the total probability of error, the optimal separation between these two populations can be achieved by the Bayes rule. Each population can be described statistically by means of its conditional PDFs, conditioned on the observations that are denoted by their corresponding image pixel intensity values. Each PDF can itself be represented in a one-to-one correspondence by the set of its statistical moments. In general, an infinite number of moments are needed for this representation. However, for some special distributions such as Gaussian, a few of the lower-order moments can uniquely describe the PDF. The lower-order moments such as means and variance have been used in the definitions of a number of bounds on the Bayesian total probability of errors. The Fisher distance is an example of such a bound.

In the following, we describe a technique for first estimating the target and background PDFs in terms of their first two moments. Then, a distance metric [5,19,20] is defined that is a function of these moments for the target and background conditional PDFs. This distance function is then used in hypothesis testing to separate targets from backgrounds.

2.2.1 A Brief Description of the Segmentor

Define the conditional PDF for the target originated IR pixels as the $P(I|T)$ and the conditional PDF for the background originated IR pixels as $P(I|B)$. The qth-order moment of PDF is defined as

$$m_q = \int x^q P(I|\cdot)\, dx \tag{1}$$

It is well known in *probability theory* [21–25] that any PDF can be uniquely represented by the set of all moments. For several cases of special interest, this uniqueness moment representation can be achieved for low values of q (e.g., for $q = 2$ for the case of Gaussian PDF). In the target segmentation, we limit ourselves to only finding the difference of the statistical PDF of the target and background. This leads, in an alternate way, to the issue of obtaining a measure of separability or distance in the moments of the two populations.

In the literature, there exists a number of distance functions for measuring the statistical dissimilarity of two populations. In our study, we used a new distance function with a particular form that has been heuristically motivated.

Consider d as a distance that is a function the first- and second-order moments of the target and background originated PDFs:

$$d = d(m_T, \sigma_T, m_B, \sigma_B) \tag{2}$$

The particular form of $d(\cdot)$ will be discussed later in this section.

Using d, we consider two other functions: $P(E|d)$, the probability density of the having a target boundary pixel conditioned on the observation of the distance pixel value d and the $P(\bar{E}|d)$ as the probability density function of having a nontarget boundary conditioned on the observation pixel d. These two functions are obtained by the use of the Bayes theorem from the $P(d|E)$ and $P(d|\bar{E})$, the probability densities for distance d conditioned on being from target and nontarget boundaries, respectively. In this context, the optimal decision about which pixels belong to the targets boundaries and which pixels belong to the background is reduced to a hypothesis testing problem in the context of the statistical decision-making. The solution is shown to reduce to a comparing of the ratio of the two conditional PDFs with a threshold value p:

$$l = \frac{P(E|d)}{P(\bar{E}|d)} \begin{cases} > p \Rightarrow \text{target boundary} \\ \leq p \Rightarrow \text{background} \end{cases} \tag{3}$$

The $P(\cdot|d)$ is related to the $P(d|\cdot)$ through the Bayes theorem:

$$P(\cdot|d) = \frac{P(d|\cdot)P(\cdot)}{\sum_{i=1}^{2} P(d|E_i)P(E_i)} \tag{4}$$

where E_i for $i = 1$ stands for E, and $i = 2$, it stands for $\bar{E} \cdot P(E)$ and $P(\bar{E})$ and the a priori probabilities of E and $\bar{E}$, respectively.

2.2.2 Estimating the *d* Function

For this study, we assume that the sample moments are equivalent to the spatial moments (ergodicity assumption). The means and variances of the target and background generated IR pixel PDEs are estimated on a $n \times n$ moving window centered at any particular pixel in the IR image of the scene. The values of n are determined empirically and are adjusted for changing scene conditions. Thus, we associate a set of moments with each pixel in the IR imagery. Using the estimated values for the moments, the d function is estimated as the maximum of the local distance d values obtained on another $k \times k$ moving window that is centered at any given pixel in the IR image.

Defining d_{ij} as

$$d_{ij} = \left\{ \left(m_i - m_j\right)^2 + \left(\sigma_i - \sigma_j\right)^2 \right\}^{1/2} \tag{5}$$

where i and j indices refer to two distinct pixels in the $k \times k$ region. Finally, d is defined as

$$d = \max_{i, j \in \{k \times k \text{ region}\}} \{d_{ij}\} \tag{6}$$

In this way, we associate a distance d with each pixel in the IR image. This d value is then used to generate a two-dimensional array, referred to as the distance image.

The final decision regarding the presence or lack of a target boundary edge pixel is determined using this distance image. This task is achieved by thresholding the distance image systematically by comparing its pixel values with a threshold p, as was described in Eq. (2).

The final target segments are obtained from their corresponding target boundaries through the applications of a connected component algorithm that establishes the different target regions as distinct labeled domains and the observation that pixels within a distinct labeled region have their x values limited from below and above by their corresponding values on the labeled region boundary pixel sets.

It is to be noted that there are mainly three critical internal parameters (n, k, p) for the operation of this algorithm.

2.3 EXPERIMENTAL DESIGN METHODOLOGY

To evaluate the performance of the target segmentor, we followed the rules of the experimental design methodology. This procedure is generic and applies to systems independent of the sensor types, the systems output, the goals of the evaluation, and performance measures that are used. The methodology of experimental design has been developing since 1920 and has been successfully used in numerous real-world applications such as industrial quality control. However, the use of the tools developed in experimental design has not been seen until recently in the field of signal and image processing [26–29].

2.3.1 Data Characterization

The concept of quantifying data to account for meaningful variations in imagery forms the goals of the Image Metric Theory. Computation of image metrics is akin to low-level image processing functions. It requires both global and local measurements to be made. The most useful metrics require some "ground-truth" information. In our study, ground truthing was achieved by associating each IR frame of data with another frame where at the locations of the targets, geometrical wire-frame silhouettes of the targets are synthetically generated. These models are then used in the computation of a set of image metrics. In these models, no consideration is given to the thermal signature of the targets [30–32].

The target metrics that we used were Target Interference Ratio Squared (TIR^2), Target Background Interference Ratio Squared (TBIR^2), Edge Strength Ratio (ESR), and Resolution Cells on Target (RNO). TIR^2 is defined as

$$\text{TIR}^2 = \frac{\left(1/N_0 \sum_{i=1}^{N_0} x_{01} - 1/N_b \sum_{i=1}^{N_0} x_{bi}\right)^2}{1/N_0 \sum_{i=1}^{N_0} x_{bi}^2 - \left(1/N_b \sum_{i=1}^{N_0} x_{bi}\right)^2} = \frac{(\bar{x}_0 - \bar{x}_b)^2}{\sigma_b^2} \qquad (7)$$

where N_0 is the number of pixels in the object, x_{0i} is the intensity level of the ith pixel in the object, N_b is the number of pixels in the background, x_{bi} is the intensity level of the ith pixel in the background, $\bar{x}_0$ is the mean target intensity value, $\bar{x}_b$ is the mean background intensity value, and σ_b is the background standard deviation. The background used in this computation is a local window around the minimum bounding rectangle of the target.

The TBIR^2 is defined similarly:

$$\text{TBIR}^2 = \frac{(\bar{x}_0 - \bar{x}_b)^2}{\sigma_b \sigma_0} \qquad (8)$$

where σ_0 is the standard deviation of the target.

The edge strength ratio is defined as the filtered, range-compensated Sobel operator output divided by local background variance.

2.3.2 Data Bases

As part of our study, we had access to relatively large databases, each exhibiting a wide spectrum of targets at various ranges, clutter condition, and viewing angles. The databases that were used in our experiments were the following:

1. 29-Palm Database. This database consisted of two different sets, each containing 32 frames. There were about 184 targets in this database.

2. Aberdeen Proving Ground (APG) database: This database consisted of three different sets. Two of them contained 45 frames and the remaining database contained 46 frames. There were a total of 378 targets in this database.

Both of these databases have ground-truth data available, which consisted of the wire-frame model of the targets at their appropriate locations on separate frames for every frame.

3. Aberdeen Proving Ground Test database: This database consisted of 197 frames of data; this set was used only to perform detection and segmentation test experiment.

Figures 1–8 show selected frames from databases and some of the results of the segmentation algorithm for these frames.

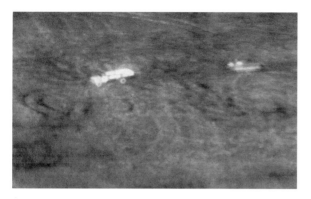

Figure 1 A sample of IR imagery.

2.3.3 Performance Measures

The relevant performance metrics for evaluating the target segmentation algorithms were as follows:

Probability of detection (PD):

$$PD = \frac{\text{Number of ground-truth targets detected}}{\text{Number of ground truth targets}} \tag{9}$$

False-alarm rate (FAR) per frame performance:

$$FAR = \frac{\text{Total number of clutter regions declared target}}{\text{Total number of frames}} \tag{10}$$

Segmentation accuracy (SA):

$$SA = \frac{\text{Number of pixels of (segmented region)} \cap \text{(Ground-truth region)}}{\text{Number of pixels of ground truth} \cup \text{(segmented region)}} \tag{11}$$

2.3.4 Internal Algorithms Parameters

It is always the case that a particular algorithm depends on a set of parameters for its execution. In general, each performance measure (PM) depends on a set of metrics (M) and a set of algorithms parameters (P): $PM = PM(P, M)$. Our particular target segmentation algorithm, as was discussed earlier, had three major internal parameters: n, k, and p.

Figure 2   Another sample of IR data showing several targets at close range.

2.4   THE SEGMENTOR PERFORMANCE ANALYSIS EXPERIMENTS

The objective of this experiment was to investigate the variations of the segmentation accuracy as the three parameters were varied. The outcome of this investigation that provided insight into the sensitivities of the segmentation accuracy with respect to the (n, k, p) set was used for the appropriate selection of the parameters for a particular database.

In our first experiment, we used the 29-Palm database. The first 28 images in this database were divided into two major groups based on the similarities of the member of each group. From each group, a single typical image was selected.

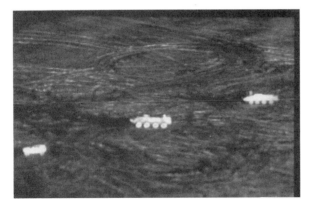

Figure 3   IR image of a scene showing a number of armored personnel carriers.

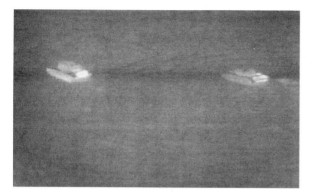

Figure 4   A sample of IR imagery showing two tanks.

The parameters n and k were varied by three different values. The parameter p varied by four different values because past experience had suggested that it was the most sensitive of the three parameters. Thus, a $3 \times 3 \times 4$ factorial design was performed and a total of 36 runs on each frame were made. Figure 9 shows the variations of mean segmentation accuracy versus the percent parameter, p, for three different values of n and k for the two frames. As can be seen in this figure, for the first frame, SA achieved a maximum at $p = 0.25$ for $n = 4$ and $k = 6$. As we increased the value of n from 4 to 5, the maximum SA reached its peak at $n = 5$, $k = 7$, and $p = 0.35$. Moreover, the value of maximum increased also from about

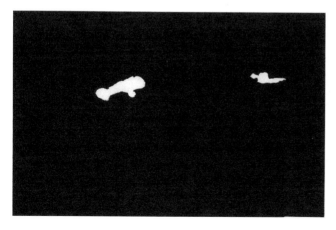

Figure 5   The result of applying the segmentation algorithm on Fig. 1.

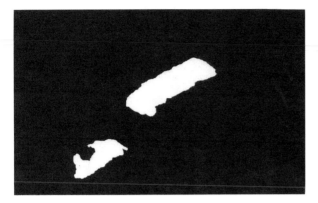

Figure 6   The result of applying the segmentation algorithm on Fig. 2.

0.31 to about 0.36. Figure 9 also shows the effect of increasing n to 6. In this case, SA achieved its highest value of 0.41 at $n = 6$, $k = 6$, and $p = 0.35$. Consequently, the best result was achieved at $n = 6$, $k = 6$, and $p = 0.35$. Figure 9 also shows the variations of the SA for the second frame as functions of n, k, and p. SA achieved its highest value of 0.48 at $n = 6$, $k = 7$ or 8, and $p = 0.15$ as seen in this figure.

Based on the results of this experiment, an average appropriate parameter set was derived. This set was then used for the 28 frames form the 29-Palm database.

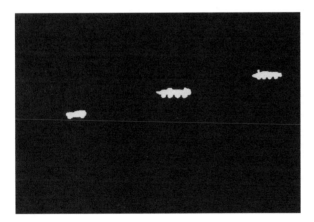

Figure 7   The result of applying the segmentation algorithm on Fig. 3.

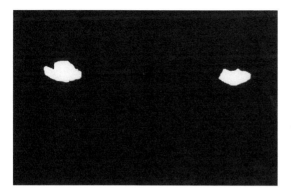

Figure 8 The result of applying the segmentation algorithm on Fig. 4.

2.4.1 Performance on the 29-Palm Database

Twenty-eight frames from the 29-Palm database were segmented by the target segmentor and the segmentation accuracy performance measures were derived.

Tables 1–3 show the variations of the SA values versus TIR^2, $TBIR^2$, and ESR metrics for the 29-palm dataset.

SA Versus TIR^2

As can be seen from Figs. 4–10, for the TIR^2 values in the first bin $(0.00185 < TIR^2 < 8.70)$, the SA has a mean value of 0.30. In the second bin $(8.7 < TIR^2 < 7.4)$, the SA average was 0.43. In the third bin $(17.4 < TIR^2 < 26.1)$, the SA average was 0.49. Finally, in the fourth bin $(26.1 < TIR^2 < 34.0)$, the SA average was 0.01, due to only one sample that produced this SA value.

SA Versus $TBIR^2$

Table 2 shows the variations of SA versus $TBIR^2$ metric. In the first bin $(0.00141 < TBIR^2 < 1.58)$, the SA average was 0.29. In the second bin $(1.58 < TBIR^2 < 3.15)$, the SA average was 0.37. In the third bin $(3.15 < TBIR^2 < 4.73)$, the SA average was at 0.63. Finally, in the fourth bin $(4.73 < TBIR^2 < 6.3)$, the SA average was 0.31.

SA Versus ESR

Table 3 shows the variations of the SA versus ESR metric. In the first bin $(0.944 < ESR < 103)$, the SA average was 0.30. In the second bin $(103 < ESR < 205)$, the SA average was 0.41. In the third bin

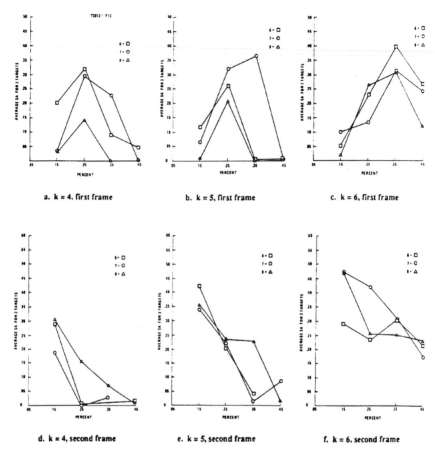

a. k = 4, first frame   b. k = 5, first frame   c. k = 6, first frame

d. k = 4, second frame   e. k = 5, second frame   f. k = 6, second frame

Figure 9   Variation of SA as a function of range.

(205 < ESR < 307), the SA average was 0.30. Finally, in the fourth bin (307 < ESR < 409), there was only one sample due to one target that was not detected.

In the above analysis we have excluded the samples that produced SA values of zero in the computation of average SA values.

2.4.2   Aberdeen Proving Ground Database Results

In this experiment, a total of 136 frames containing 372 targets were used. There were three different sets of data in the this database that we refer to as T3002 and T3004. In the following, the results for each set is reported separately.

Table 1 Variation of SA as a Function of TIR^2

Database	Bin	Bin range Min.	Bin range Max.	SA average	Comments	SA / TIR^2
29-Palms	1	0	8.7	0.30		
	2	8.71	17.4	0.43		
	3	17.41	26.1	0.49		
	4	26.11	34.8	0.01	Only one sample	
T3002	1	0	37.5	0.33		
(expanding	2	37.51	75.0	0.47		
the first bin)	3	75.10	113.0	0.49		
	4			0	One undetected target	
T3003	1	0	32.5	0.36		
(expanding	2	32.51	65.0	0.43		
the first bin)	3	65.01	97.5	0.36		
	4	97.51	130.0	0.60		
T3004	1	0	76.5	0.38		
	2	76.51	153.0	0.18		
	3	153.01	230.0	0.39	Only one sample	
	4	230.01	306.0		Only one sample	
T3004	1	0	37.5	0.33		
(expanding first	2	37.51	75.0	0.55		
two bins)	3	75.00	113.0	0.33		(No plot)
	4	113.01	150	0.03		
Combined	1	0	139	0.35	0.20 SD[a]	
databases	2	139.01	277	0.35	0.18 SD	
	3	277.01	416	0.40	0.09 SD	
	4	416.01	554	0.60	0 SD (only one sample)	
Combined	1	0	37.5	0.31	0.24 SD	
databases (for	2	37.51	75.0	0.49	0.25 SD	
0–150 only)	3	75.01	113.0	0.44	0.21 SD	
	4	113.01	150.0	0.03	0.02 SD (two undetected targets)	

[a]Standard deviation

Table 2   Variation of SA as a Function of TBIR^2

Database	Bin	Bin range Min.	Max.	SA average	Comments	SA / TBIR^2
29-Palms	1	0	1.58	0.29		
	2	1.59	3.15	0.37		
	3	3.16	4.73	0.63		
	4	4.74	6.30	0.31		
T3002	1	0	7.17	0.33		
	2	7.18	14.30	0.29		
	3	14.31	21.50	0.69		
	4	21.51	28.70	0.36		
T3003	1	0	25.50	0.36		
	2	25.51	51.10	0.60		
	3	51.11			Only one sample	
	4		100.00		(target not detected)	
T3004	1	0	8.50	0.34		
	2	8.51	17.00	0.42		
	3	17.01	25.50	0.57		
	4	25.51	34.00		Only one sample (target not detected)	
Combined databases	1	0	25.50	0.35	0.24 SD[a]	
	2	25.51	51.10	0.60	0 SD	
	3	51.11	76.60		No sample available	
	4	76.61	102.00		One target not detected	
Combined (expanding TBIR^2 range to between 0 and 25)	1	0	6.25	0.32	0.22 SD	
	2	6.26	12.50	0.38	0.28 SD	
	3	12.51	18.80	0.48	0.28 SD	
	4	18.81	25.00	0.45	0.08 SD	

[a]Standard deviation

Table 3   Variation of SA as a Function of ESR

Database	Bin	Bin range Min.	Bin range Max.	SA average	Comments	
29-Palms	1	0	103	0.30	⎫ Excludes samples	
	2	104	205	0.41	⎬ that produced	
	3	206	307	0.30	⎭ SA values of 0	
	4	308	409		Only one sample (target not detected)	
T3002	1	0	750	0.33		
(expanding	2	751	1500	0.61		
the first bin)	3	1501	2250	0.49	Only one sample	
	4	2251	3000	0	One sample not detected	
T3003	1	4	1770	0.36		
	2	1771	3530	0.33		
	3	3531	5290	0.28	One sample	
	4	5291	7060	0.52		
T3003	1	0	475	0.32		(No plot)
(expanding	2	476	950	0.45		
the first bin)	3	951	1430	0.56		
	4	1431	1900	0.43		
T3004	1	3	436	0.34		
	2	437	868	0.43		
	3	869	1300	0.01	Mainly due to	
	4	1301	1730		one target	
Combined	1	1	1810	0.35	0.24 SD[a]	
databases	2	1811	3620	0.33	0 SD (only one sample)	
	3	3621	5420	0.39	0.11 SD	
	4	5421	7230	0.52	0.08 SD	
Combined	1	0	375	0.32	0.23 SD	
(expanding	2	376	750	0.40	0.28 SD	
range to	3	751	1130	0.52	0.20 SD	
between 0	4	1131	1500	0.56	0.19 SD	
and 1500)						

[a]Standard deviation

T3002 Set Results

Table 1 shows the variations of SA as function of TIR^2. In the first bin $(0 < TIR^2 < 37.5)$, the average segmentation accuracy was 0.33. In the second bin $(37.5 < TIR^2 < 75)$, the average SA was 0.47. In the third bin $(75 < TIR^2 < 113)$, the average value of SA was 0.49. Finally, in the fourth bin, the average SA value was zero due to the one target that was not detected.

Variation of SA Versus $TBIR^2$

Table 2 shows the variations as a function of $TBIR^2$. In the first bin $(1.43 \times 10^{-3} < TBIR^2 < 7.17)$, the average SA value was 0.33. In the second bin $(7.17 < TBIR^2 < 14.3)$, the average SA value was 0.29. In the third bin $(14.3 < TBIR^2 < 21.5)$, the average SA was 0.69. Finally, in the fourth bin $(21.5 < TBIR^2 < 28.7)$, the average SA was down to 0.36.

Variation of SA Versus ESR

Table 3 shows the variations of SA as a function of ESR. In the first bin $(0 < ESR < 750)$, the average SA was 0.33. In the second bin $(750 < ESR < 1500)$, the average SA was 0.61. In the third bin $(1500 < ESR < 2250)$, the average SA was 0.49 due to only one sample. Finally, in the fourth bin $(2250 < ESR < 3000)$, the average value of SA was zero due to one sample that was not detected.

T3003 Set Results

Variation of SA Versus TIR^2

Table 1 shows the variations of the SA as a function of TIR^2. In the first bin $(0 < TIR^2 < 32.5)$, the average SA value was 0.36. In the second bin $(32.5 < TIR^2 < 65)$, the average SA value was 0.43. In the third bin $(65 < TIR^2 < 97.5)$, the average value of SA was 0.36. In the fourth bin $(97.5 < TIR^2 < 130)$, the average SA value was 0.6.

Variation of SA Versus $TBIR^2$

Table 2 shows the variations of SA as a function of $TBIR^2$. In the first bin $(3 \times 19^{-6} < TBIR^2 < 25.5)$, the average SA value was 0.36. In the second bin $(25.5 < TBIR^2 < 51.1)$, the average SA value was 0.60. In the third and fourth bins $(51.1 < TBIR^2 < 100)$, there was only one sample due to a target that was not detected.

Variations of SA Versus ESR

Table 3 shows the variation of SA as a function of ESR. In the first bin $(4.06 < ESR < 1770)$, the average SA value was 0.36. In the second bin $(1770 < ESR < 3520)$, the average SA value was 0.33. In the third bin

(3520 < ESR < 5250), the average SA was 0.28 due to one sample. Finally, for (5250 < ESR < 7000), the average SA value was 0.52.

Further expanding the first bin into four subbins, one obtains the following results: In the first subbin (0 < ESR < 475), the average SA value was 0.32. In the second subbin (47 < ESR < 950), the average SA value was 0.45. In the third subbin (960 < ESR < 1430), the average SA value was 0.56. Finally, in the fourth bin (1430 < ESR < 1900), the average SA value was 0.43.

T3004 Dataset

Variation of SA Versus TIR^2

Table 1 shows the variation of SA as a function of TIR^2. In the first bin $(3.74 \times 10^{-3} < TIR^2 < 76.5)$, the average SA value was 0.38. In the second bin $(76.5 < TIR^2 < 153)$, the average SA value was 0.18. In the third bin $(153 < TIR^2 < 230)$, the average SA value was 0.39 due to only one sample. In the fourth bin $(230 < TIR^2 < 306)$, only one sample was located, which was due to one target that was not detected. Further expanding the first two bins into four equal-sized bins, one obtains the following results. In the first subbin $(0 < TIR^2 < 37.5)$, the average SA was 0.33. In the second subbin $(37.5 < TIR^2 < 75.0)$, the average SA was 0.55. In the third subbin $(75.0 < TIR^2 < 113)$, the average value of SA was 0.33, and, finally, in the fourth subbin $(113 < TIR^2 < 150)$, the average SA value was 0.03.

Variation of SA Versus $TBIR^2$

Table 2 shows the variations of the segmentation accuracy as a function of $TBIR^2$. In the first bin $(4.10 \times 10^{-3} < TBIR^2 < 8.50)$, the average SA value was 0.34. In the second bin $(8.50 < TBIR^2 < 17.0)$, the average SA value was 0.42. In the third bin $(17.0 < TBIR^2 < 25.5)$, the average SA value was 0.57. In the fourth bin $(25.5 < TBIR^2 < 34.0)$, there was only one sample due to one target that was not detected.

Variation of SA Versus ESR

Table 3 shows the variation of SA as a function of ESR. In the first bin (3.03 < ESR < 436), the average SA was 0.34. In the second bin (436 < ESR < 860), the average SA value was 0.43. In the third bin (860 < ESR < 1300), the average value of SA was 0.01, mainly due to one target.

Variation of SA Versus Range for the APG and 29-Palm Databases

Figure 10a shows the variation of SA versus range for the first 28 frames from the 29-Palms database. There are three distinct ranges: 800,

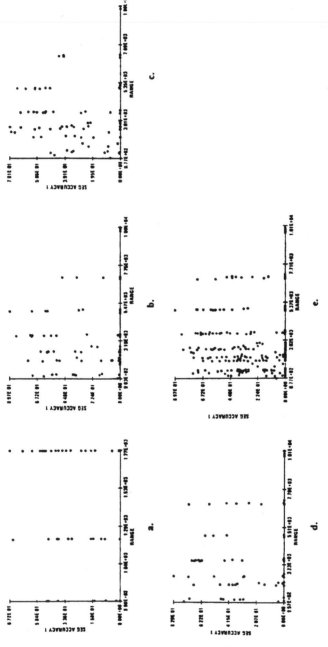

Figure 10 Variation of SA as a function of range. (a) 29-Palms database; (b) T3002 database; (c) T3003 database; (d) T3004 database; and (e) combined databases.

1210, and 1770. In the highest, SA occurs at 1770. At this range, the highest SA attains a value of 0.672.

Figure 10b shows the variation of SA as a function of range for the T3002 dataset. As can be seen, SA attains its highest values at ranges between 3000 and 5470 and then decreases. The highest SA is 0.897.

Figure 10c shows the variation of SA versus range for the T3003 dataset. In this dataset, SA attains its highest values at ranges between 3010 and 5350 and then decreases. The highest value of SA is 0.781.

Figure 10d shows the variation of SA versus range for the T3004 dataset. The highest SA value decreases at 0.6 at 3230. The highest SA value decreases to 0.6 at 5510 and then increases again to 0.72 at around 7000.

2.4.3 Performance of Target Segmentation Algorithm on the Combined Databases of Aberdeen Proving Ground and 29-Palms

SA Versus TIR^2

Table 1 shows the variation of SA as a function of TIR^2. In the first bin $(4.0 \times 10^{-6} < TIR^2 < 139)$, the SA has an average value of 0.35 and a standard deviation of 0.25. In the second bin $(139 < TIR^2 < 277)$, the SA has an average value of 0.35 and a standard deviation of 0.18. In the third bin $(277 < TIR^2 < 416)$, the SA has an average value of 0.40 and a standard deviation of 0.09. In the fourth bin $(416 < TIR^2 < 554)$, the SA has an average value of 0.60 and a standard deviation of zero due to the only sample detected in this bin.

Table 1 shows the variation of SA versus TIR^2 for $(0 < TIR^2 < 150)$ only. Dividing this metric space into four equal and consecutive bins, one obtains the following results. In the first subbin $(0 < TIR^2 < 37.5)$, the SA has an average value of 0.31 and a standard deviation of 0.24. In the second subbin $(37.5 < TIR^2 < 75)$, the SA has an average of 0.49 and a standard deviation of 0.25. In the third subbin $(75 < TIR^2 < 113)$, the SA has an average value of 0.44 and a standard deviation of 0.21. In the fourth bin $(113 < TIR^2 < 150)$, the SA has an average of 0.03 and a standard deviation of 0.02 due to two detected targets.

SA Versus $TBIR^2$

Table 2 shows the variation of SA as a function of $TBIR^2$. In the first bin $(3 \times 10^{-6} < TBIR^2 < 25.5)$, the SA has an average value of 0.35, and a standard deviation of 0.24. In the second bin $(25.5 < TBIR^2 < 51.5)$, the SA has an average value of 0.60 and a standard deviation of zero due to the presence of only one sample in this bin. In the third bin (51.1 < $TBIR^2$ < 76.7), no sample was available. In the fourth bin (76.6 < $TBIR^2$ < 102), there was one target that was not detected. Expanding the values of $TBIR^2$ between 0 and 25 further as shown in Fig. 10, the following results are obtained. In the first subbin (0 < $TBIR^2$ < 6.25), the SA has an average value of 0.32 and a standard deviation of 0.22. In the second subbin (6.25 < $TBIR^2$ < 12.5), the SA has an average value of 0.38 and a standard deviation of 0.28. In the third subbin (6.25 < $TBIR^2$ < 18.8), the SA has an average value of 0.46 and a standard deviation of 0.28. In the fourth bin (18.8 < $TBIR^2$ < 25), the SA attains an average value of 0.45 and a standard deviation of 0.08.

SA Versus ESR

Table 3 shows the variation of SA as a function of ESR. In the first bin (0.944 < ESR < 1810), the SA has an average value of 0.35 and a standard deviation of 0.24. In the second bin (1810 < ESR < 3620), the SA has an average of 0.33 and a standard deviation of zero due to the presence of a single sample. In the third bin (3620 < ESR < 5420), the SA has an average value of 0.39 and a standard deviation of 0.11. In the fourth bin (5420 < ESR < 7230), the SA attains an average value of 0.52 and a standard deviation of 0.08. Expanding the ESR range between 0 and 1500 further, as seen in Fig. 10, one obtains the following. In the first subbin (0 < ESR < 375), the SA has an average value of 0.32 and a standard deviation of 0.23. In the second subbin (375 < ESR < 750), the SA has an average of 0.40 and a standard deviation of 0.28. In the third subbin (750 < ESR < 1130), the SA attains an average value of 0.52 and a standard deviation of 0.20. Finally, in the fourth bin (1130 < ESR < 1500), the SA has an average range of 0.56 and a standard deviation of 0.19.

SA Versus Range

Figure 10e shows the variation of SA as a function of range. SA attains its highest value of 0.897 at 5370 and then falls to around 0.7 at around 7000.

Target Segmentor Probability of Detection Versus Metrics on the Combined APG and 29-Palm Databases

PD Versus TIR^2

Figure 11a shows the variation of PD as a function of TIR^2. As TIR^2 increases, PD tends to increase and attain the value of 1 for most TIR^2 values higher than 350.

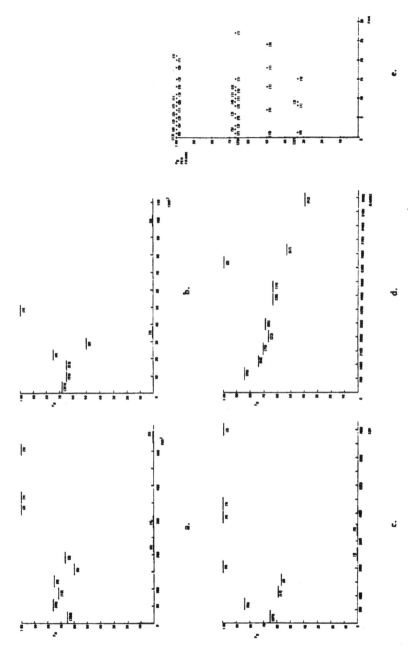

Figure 11 Variation of PD for the combined databases: (a) as a function of TIR^2; (b) as a function of TBIR^2; (c) as a function of ESR; (d) as a function of range; and (e) as a function of FAR per frame.

PD Versus TBIR²

Figure 11b shows the variations of PD as a function of $TBIR^2$. As $TBIR^2$ increases, PD tends to increase. In the range $0 < TBIR^2 < 10$, the PD average is 0.70 with a standard deviation of 0.21. For $10 < TBIR^2 < 20$ PD has an average value of 0.56 and a standard deviation of 0.35. For $20 < TBIR^2 < 30$, the PD has an average value of 0.53 and a standard deviation of 0.50, for $30 < TBIR^2 < 110$, the PD has an average value of 0.5 and a standard deviation of 0.577.

PD Versus ESR

Figure 11c shows the variation of PD as a function of ESR. For values of $0 < ESR < 50$, the average value of PD was 0.61 and standard deviation of 0.32. For values of $50 < ESR < 100$, the average value of PD was 0.91 and a standard deviation was 0.12. For $100 < ESR < 1500$, the average value of PD was 0.7 with a standard deviation of 0.4. For values of $500 < ESR < 7200$, the average PD was 0.6 and the standard deviation was 0.459.

PD Versus Range

Figure 11d shows the variations of PD as a function of range. The worst PDs occur at the 9800 range value. The PD has maximum value of 0.53 at ranges between 1400 and 2100 and the PD has a maximum of 1 and a minimum of around 0.66.

PD Versus False-Alarm Range

Figure 11e shows the variation of PD as a function of false-alarm rate (FAR) performance. The numbers inside the parentheses show the number of samples having the same position on the graph. As can be seen there is a large number of samples; 53 have their PD equal to 1. Fifty-four samples have a PD value of around 0.67. The rest of the 55 samples have PD values of 0.5 or smaller.

2.5 CONCLUSION AND SUMMARY OF THE RESULTS

In this study, we presented a statistical technique for the segmentation of targets in infrared imagery. We then used elements of experimental design methodology to systematically evaluate the performance of the algorithm on large and varied numbers of IR targets under a variety of backgrounds and scene conditions. The measures of performance that we used were probability of detection, false-alarm rate, and segmentation accuracy. The imagery that were used were characterized in terms of a set of image metrics, among them were TIR^2, $TBIR^2$, ESR, and range. All of the performance measures were tabulated in terms of the image metrics for a total of 163 frames and 448 targets for the single-frame evaluations. The study showed that generally as TIR^2, $TBIR^2$, and ESR increase, at least in the lower domains of metric space, performance measures increase. However, for higher domains of metric space, the behavior of algorithms were different. This is partly due to the lack of sufficient data in this domain and, consequently, relating to very few samples.

As part of this study, we gained significant insight into the algorithm parameter selection. The study showed that the algorithm performance metrics were very sensitive to the setting of the parameters. Moreover, the study showed that there existed optimum sets of parameters for the algorithm as functions of the image metrics. Thus, one can embark on an adaptive parameter selection based on the image metrics and, consequently, based on scene parameters.

REFERENCES

1. EL Hall. Computer Image Processing and Recognition. New York: Academic Press, 1979.
2. BKP Horn. Robot Vision. Cambridge, MA: MIT Press, 1986.
3. DH Ballard, CM Brown. Computer Vision. Englewood Cliffs, NJ: Prentice-Hall, 1982.
4. RC Gonzales, P Wintz. Digital Image Processing. Reading, MA: Addison-Wesley, 1977.
5. RO Duda, PH Hart. Pattern Classification and Scene Analysis. New York: Wiley, 1973.
6. CR Brice, CL Fennema. Scene analysis using regions. Artif Intell 1, pp. 205–226, 1970.
7. W Oh, WB Lindquist. Image thresholding by indicator kriging. IEEE Trans Pattern Anal Machine Intell 21(7), 1999.
8. NR Pal, SK Pal. A review on image segmentation techniques. Pattern Recogn 26(9), 1993.
9. JS Weszka, A Rosenfeld. Threshold evaluation techniques. IEEE Trans Syst Man Cybern 8, 1978.
10. AD Lanteman, U Grenander, M Miller. Bayesian segmentation via asymptotic partition functions. IEEE Trans Pattern Anal Machine Intell 22(8), 2000.
11. J-F Bonnet, D Duchos, G Stamon, R Samy. Ground target classification using robust active contour segmentation. Automatic Target Recognition. SPIE, 1999, Vol. 3718, Bellingham, Washington.
12. S-C Zhu. Embedding gestault laws in Markov random fields. IEEE Trans Pattern Anal Machine Intell 21(11), 1170–1187, 1999.

13. P Andrey, P Tarroux. Unsupervised segmentation of Markov random field modeled textured images using selectionist relaxation. IEEE Trans Pattern Anal Machine Intell 20(3), 1998.

14. TCM Lee. Segmenting images corrupted by correlated noise. IEEE Trans Pattern Anal Machine Intell 20(5), 1998.

15. J-P Wang. Stochastic Relaxation on partitions with connected components and its approach to image segmentations. IEEE Trans Pattern Anal Machine Intell 20(6), 1998.

16. T Hofman, J Puzicha, JM Bushmann. Unsupervised texture segmentation in a deterministic annealing framework. IEEE Trans Pattern Anal Machine Intell 20(8), 1998.

17. J Shi, J Malik. Normalized cuts and image segmentation. IEEE Trans Pattern Anal Machine Intell 22(8), 1988.

18. S Geman, D Geman. Stochastix relaxation, Gibbs distribution, and the Bayesian restoration of imagery. IEEE Trans Pattern Anal Machine Intell 6, pp. 721–741, November 1984.

19. PA Devijver, J Kittler. Pattern Recognition. A Statistical Approach. London: Prentice-Hall, 1982.

20. K Fukunaga. Introduction to Statistical Pattern Recognition. New York: Academic Press, 1972.

21. W Feller. An Introduction to Probability Theory and its Applications. New York: Wiley, 1971, Vol. 2.

22. SM Kay. Fundamentals of Statistical Signal Processing—Estimation Theory. Englewood Cliffs, NJ: Prentice-Hall, 1993, Vol. 1.

23. D Middleton. Introduction to Statistical Communication Theory. Los Altos, CA: Peninsula Publishing, 1987.

24. A Papoulis. Probability, Random Variables, and Stochastic Processing. New York: McGraw-Hill, 1965.

25. WB Davenport, WL Root. An Introduction to the Theory of Random Signals and Noise. New York: IEEE Press, 1987.

26. GEP Box, WG Hunter, JS Hunter. Statistics for Experimenters, An Introduction to Design, Data Analysis, and Model Building. New York: Wiley, 1978.

27. WG Cochran, GM Cox. Experimental Designs. 2nd ed. New York: Wiley, 1957.

28. H Scheffe. The Analysis of Variance. New York: Wiley, 1959.

29. FA Sadjadi. Experimental Design Methodology, the Scientific Tool for Performance Evaluation. In: *Signal and Image Processing Systems Perf. Evaluation.* SPIE, 1990, Vol. 1310.

30. FA Sadjadi, H Nasr, H Amehdi, M Bazakos. Knowledge and model based automatic target recognition algorithm adaptation. Opt Eng, 30(2), 183–188, 1991.

31. FA Sadjadi. Application of image metrics in dynamic scene analysis. Adaptive and Learning Systems. SPIE, 1992, Vol. 1706, Bellingham, Washington.

32. ATR Definitions and Performance Measures, Automatic Target Recognizers Working Group (ATRWG) Publications No. 86-001, 1986, Dayton, Ohio.

3
Recognizing Objects in SAR Images

Bir Bhanu and Grinnell Jones III
Center for Research in Intelligent Systems, University of California, Riverside, California

3.1 INTRODUCTION

3.1.1 Problem Definition and Scope

Automated object recognition in synthetic aperture radar (SAR) imagery is a significant problem because recent developments in image collection platforms will soon produce far more imagery (terabytes per day per aircraft) than the declining ranks of image analysts are capable of handling. In this chapter, the problem scope is the recognition subsystem itself, starting with chips of military target vehicles from real SAR images at 1-ft resolution and ending with vehicle identification. The specific challenges for the recognition system are the need for automated recognition of vehicles with articulated parts (like the turret of a tank), or that have significant external configuration variants (like fuel barrels, searchlights, etc.), or that can be partially hidden. Previous recognition methods involving detection theory [1,2], pattern recognition [3–5], and neural networks [6,7] are not useful in these cases because articulation or occlusion changes global features like the object outline and major axis [8]. In order to characterize the performance of the recognition subsystem, we approach the problem scientifically from fundamentals. We characterize SAR azimuth variance to determine the number of models required, we utilize the invariance of the targets, and based on these invariants, we develop a SAR-specific recognition system. We characterize the performance of this system in terms of invariance of features, number of features, and amount of occlusion for recognition of articulated objects, occluded objects, and configuration-variant objects. All of the experimental data are based on 1-ft resolution real SAR images of actual vehicles from the MSTAR (Public) targets dataset [9].

3.1.2   Overview of Approach for Object Identification

Our approach to object identification is specifically designed for SAR. The peaks (local maxima) in radar return are related to the physical geometry of the object. The relative locations of these scattering centers are independent of translation and serve as distinguishing features. The specular radar return varies greatly with the uncontrolled target orientation (azimuth) and this is captured by using models of the objects at $1°$ azimuth increments. The radar depression angle to the target is controllable, or known, and it is fixed at $15°$ for this research (unless otherwise noted).

We demonstrate that quasi-invariant scattering center *locations* exist and that their *magnitudes* are also quasi-invariant for articulation and configuration variants. These invariants permit building standard nonarticulated recognition models and using them to successfully recognize nonstandard and articulated targets. This avoids the combinatorial problem of modeling 360 turret angles times 360 aspect views as well as numerous configuration differences. The SAR recognition system has an off-line model construction phase and an on-line recognition process. The recognition model is a look-up table that basically relates the relative distances (in the radar range and cross-range directions) among the scattering centers to object type and azimuth (as well as recording scatterer magnitudes). The recognition process is an efficient search for *positive evidence*, using relative locations of test scattering centers to access the look-up table and (if the model and test scatterer magnitudes are similar, within limits) to generate votes for the appropriate object (and azimuth).

3.1.3   Key Contributions

The major contributions of this chapter are as follows:

1. Quantifies the azimuthal variance of scattering center locations in real SAR data
2. Demonstrates that quasi-invariant scattering center *locations* exist and that their *magnitudes* are also quasi-invariant for (a) articulation and (b) configuration variants for actual vehicles in real SAR data

3. Develops a new recognition system based on scattering center location, and magnitude that achieves significant vehicle recognition performance for articulation, configuration variants, and large amounts of occlusion with real SAR data.

3.2 BACKGROUND AND RELATED WORK

3.2.1 Background

Automatic target recognition (ATR) is the use of computer processing to detect and recognize object (target) signatures in sensor data. General reviews of ATR concepts and technologies can be found in Refs. 10–13. ATR systems generally have separate detection and recognition stages. The goal of the detection stage is to eliminate most of the sensor data from further consideration and find small regions potentially containing the targets of interest. The goal of the recognition stage is to classify or identify the targets. The term *classification* is generally used for a coarse categorization (e.g., the target is a tank, not a truck) and *identification* is used for a fine categorization (e.g., a specific type of tank). Detecting targets in SAR imagery typically involves a prescreen stage {e.g., a constant false alarm rate (CFAR) thresholding technique [14]} and a discriminator stage to separate the targets from the background clutter (a detailed description of features used for this segmentation can be found in Ref. 15). Other methods for target detection include filters [16], using the variations in return from man-made objects and natural clutter with changes in image resolution [17], likelihood images with image relational graphs [18], and neural networks [7]. There are several different categories of algorithms used for recognition in ATR systems: detection theory, pattern recognition, artificial neural networks, and model-based recognition [13]. Several of these can be used for detection as well as recognition and several ATR systems use combinations of approaches.

Detection Theory

The detection theory approach uses filters to separate the distributions of target and clutter signatures so they can be distinguished by a simple statistical test. This approach has been applied to classification of different objects in SAR images using Minimum Noise and Correlation Energy (MINACE) filters [2,16] and Optimal Trade-off Distance Classifier Correlation Filters (OTDCCFs) [1] (which use a distance measure for classification instead of just the output correlation peak). Because of the azimuth variation of SAR signatures, multiple filters are used for each object, each covering a range of target azimuth angles. Limited tests showed that the correlation peak drops proportionally to the percent occlusion and the MINACE filters performed well at up to 20% occlusion [2].

Pattern Recognition

The pattern recognition approach uses SAR image templates [5] or feature vectors {such as Topographical Primal Sketch (TPS) categories, based on zero crossings of directional derivatives [3]} and determines the best match between the target image and an exemplar database. In contrast to detection theory algorithms that are derived by using statistical models of raw data, pattern recognition uses more ad hoc approaches to the definition and extraction of the features used to characterize targets. In contrast to the later model-based approach, the pattern recognition approaches usually use global measures to represent the target. These global features are very susceptible to occlusion and articulation effects. Many of the pattern recognition approaches suffer from an exhaustive search of a database of exemplars, although a hierarchical index of distance transforms with composite models had been used [4] to convert the problem to a tree search.

Neural Networks

The neural network approach uses learning by example to discover and use signature differences that distinguish different types of targets. Adaptive Resonance Theory networks [7] have been used to categorize SAR target aspects and learn SAR target templates. Feature Space Trajectory neural networks have been used for SAR detection and classification [6]. One problem for neural network approaches is to achieve good performance with a range of target signatures (with occlusion, articulation, and nonstandard configurations) and varying background clutter, given limited amounts of training data.

Model-Based Recognition

The model-based recognition approach typically uses multiple local features (involving object parts and the relationships between the parts) and matching of sensor data to predictions based on hypotheses about the target type and pose. A current state-of-the-art example is the Moving and Stationary Target Acquisition and Recognition (MSTAR) program that uses a search module for hypothesis generation and refinement [19], a feature prediction module that captures the target signature phenomenology [20] and a matching module [21]. Most model-based systems are optimized for unobscured targets and, therefore, degrade rapidly for increasing occlusion. One model-based approach, Partial Evidence Reconstruction from Object Restricted Measures (PERFORM) [22] uses a linear signal decomposition, direction of arrival pose estimation technique, and attempts to overcome the difficulties of recognizing an obscured target by breaking the problem into smaller parts (e.g., separately recognize the front, center, and rear regions of the vehicle and fuse the results). Test results with the PERFORM system show significant reductions in performance with 5–25% occlusion.

3.2.2 Related Work

The detection theory, pattern recognition, and neural network approaches to SAR recognition, as discussed in the previous subsection, all tend to use global features that are optimized for standard, nonarticulated, non-occluded configurations. Approaches that rely on global features are not appropriate for recognizing articulated or occluded objects because these conditions change global features like the object outline and major axis [8]. Template matching (whether the templates are built by hand, as in Ref. 5, or developed by neural networks, as in Ref. 7) has been successfully used for object recognition with SAR images, but this approach is not suitable for recognizing articulated objects because there will be a combinatorial explosion of the number of templates with varying articulations. Some of the SAR recognition techniques (e.g., MINACE filters [2], PERFORM [22], mean squared error template matching [23], stochastic models [47], and invariant histograms [24]) have reported limited test results for small amounts of occlusion, typically 25% or less, which would also indicate some potential for articulated object recognition. In addition, the developers of the MSTAR search engine reported [25] using a shadow inferencing technique to hypothesize targets with up to 30% occlusion in the cross-range direction. The standard approaches used for articulated object recognition in optical images (such as recognition by components [26], constrained models of parts [27], and joint articulation [28] are not appropriate for the relatively low-resolution, nonliteral nature, and complex part interactions of SAR images.

Table 1 compares the recognition results based on different approaches, all using real SAR images from the MSTAR public data [9]. This is an active area of current research; new approaches are evolving and this comparison should be viewed as a snapshot in time. The results are presented in terms of probability of correct identification (PCI) for cases with target articulation, depression angle change, and target configuration variants. Many of the results are for forced recognition, where the ATR system is forced to make a target decision (i.e., there is no

Table 1 Related Work Comparison for Target Recognition Using Real MSTAR SAR Data

Approach	Ref	Probability of correct identification			Remarks
		Art.	Depr.	Config	
MSTAR predict models, mean square error matching	29	—	—	0.74–0.78[a]	Forced average of 11 T72s
Template matching, correlation/mean sq. error	30	—	0.99	0.93[b]	Forced recognition
	31	—	—	0.40–0.75	11 T72s at 0.10 Pfa
	23	0.35	0.93	0.73	Forced recognition
	9	—	—	0.79	0.10 Pmiss
Radon transform, neural net	32	—	0.93	—	Forced recognition
Radon transform, hidden Markov models	33	—	0.94	—	Forced recognition
This chapter		1.00	0.99	0.95	Forced recognition
		0.97	0.82	0.55–0.95	at 0.10 Pfa

[a]Estimates.
[b]Configuration and depression.

"unknown" class). Other results with an "unknown" or "reject" class are presented at a given probability of false alarm (Pfa) or probability of miss (Pmiss). The MSTAR predict approach [29] uses CAD object models as the basis for generating synthetic SAR signature model predictions that are matched to the real SAR data; all of the other approaches build models from real SAR images. The template-matching approaches [9,23,30,31] use (scaled) intensity values of the target chip as features, but they require either an accurate estimate of the target pose or an exhaustive search of a large database of model templates. Others [32,33] use the Radon transform as a way to reduce the number of models required to handle the SAR signature variations with target azimuth rotation. None of these approaches is specifically designed to accommodate articulated objects and the one attempt to recognize articulated objects [23] confirms that the template-matching approach is not well suited for articulated objects. In contrast, this chapter presents an approach to SAR target recognition, specifically designed to accommodate articulated and occluded targets, that achieves excellent recognition results for articulated objects and for highly occluded data with over 50% target occlusion.

3.3 SAR TARGET CHARACTERISTICS

The typical detailed edge and straight-line features of man-made objects in the visual world do not have good counterparts in SAR images for sub-components of vehicle-sized objects at 1-ft resolution; however, there is a wealth of peaks corresponding to scattering centers. The relative locations of SAR scattering centers, determined from local peaks in the radar return, are related to the aspect and physical geometry of the object, independent of translation and serve as distinguishing features. Target regions of interest (ROIs) are found in the MSTAR SAR chips by reducing speckle noise using the Crimmins algorithm in Khoros [34], thresholding at the mean plus two standard deviations, dilating to fill small gaps among regions, eroding to have one large ROI and small regions, discarding the small regions with a size filter, and dilating to expand the extracted ROI. The scattering centers are extracted from the SAR magnitude data (within the boundary contour of the ROI) by finding local eight-neighbor maxima. The parameters used in extracting ROIs are held constant for all of the results reported. Objects from the MSTAR public data used in this research include BMP2 armored personnel carriers (APCs), a BTR70 APC, T72 tanks, a ZSU23/4 antiair-craft gun, and a BRDM2 APC. Photo images of the MSTAR articulated objects used in this chapter, T72 tank serial number (#) a64 and ZSU 23/4 antiaircraft gun #d08, are shown in Figs. 1 and 2. Example SAR images and the ROIs with the locations of the scattering centers superimposed, are shown in Fig. 3 for baseline and articulated version of the T72 and ZSU (at 30° radar depression angle, 66° target azimuth).

(a) (b)

Figure 1 T72 tank #a64: (a) turret straight; (b) turret articulated.

(a)   (b)

Figure 2   ZSU 23/4 antiaircraft gun #d08: (a) turret straight; (b) turret articulated.

3.3.1   Azimuthal Variance of Scatterer Locations

The typical rigid-body rotational transformations for viewing objects in the visual world do not apply much for the specular radar reflections of SAR images. This is because a *significant* number of features *do not* typically persist over a few degrees of rotation. Because the radar depression angle is generally known, the significant unknown target rotation is 360° in azimuth. Azimuth persistence or invariance can be expressed in terms of the

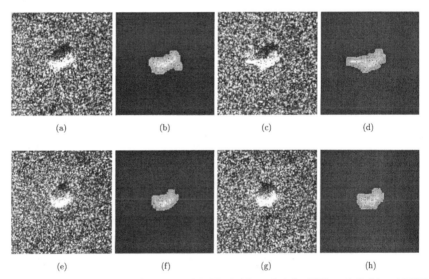

(a)   (b)   (c)   (d)

(e)   (f)   (g)   (h)

Figure 3   MSTAR SAR images and ROIs (with peaks) for T72 tank #a64 and ZSU 23/4 #d08 at 66° azimuth. (a) T72 image; (b) ROI; (c) articulated image; (d) articulated ROI; (e) ZSU image; (f) ROI; (g) articulated image; and (h) articulated ROI.

percentage of scattering center locations that are unchanged over a certain span of azimuth angles. It can be measured (for some base azimuth θ_0) by rotating the pixel locations of the scattering centers from an image at azimuth θ_0 by an angle $\Delta\theta$ and comparing the resulting range and cross-range locations with the scatterer locations from an image of the same object at azimuth $\theta_0 + \Delta\theta$. More precisely, because the images are in the radar slant plane, we actually project from the slant plane to the ground plane, rotate in the ground plane, and project back to the slant plane. Because the objects in the chips are not registered, we calculate the azimuth invariance as the maximum number of corresponding scattering centers (whose locations match within a given tolerance) for the optimum integer pixel translation. This method of registration by finding the translation that yields the maximum number of correspondences has the limitation that for very small or no actual invariance, it may find some false correspondences and report a slightly higher invariance than in fact exists. To determine scattering center locations that persist over a span of angles, there is an additional constraint that for a matching scattering center to "persist" at the kth span $\Delta\theta_k$, it must have been a persistent scattering center at all smaller spans $\Delta\theta_j$, where $0 \leq j < k$. Averaging the results of these persistent scattering center locations over 360 base azimuths gives the mean azimuth invariance of the object.

Figure 4 shows an example of the mean scatterer location invariance (for the 40 strongest scatterers) as a function of azimuth angle span using T72 tank #132, with various definitions of persistence. In the "exact match"

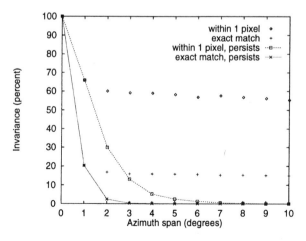

Figure 4 Scatterer location persistence, T72 #132.

cases, the center of the rotated scatterer pixel from the image at θ_0 azimuth is within the pixel boundaries of a corresponding scatterer in the image at $\theta_0 + \Delta\theta$. In the "within 1 pixel" cases, the scatterer location is allowed to move into one of the eight adjacent pixel locations. Note that for a $1°$ azimuth span, although only 20% of the scatterer locations are invariant for an "exact match," 65% of the scatterer locations are invariant "within 1 pixel." The cases labeled "persists" in Fig. 4 enforce the constraint that the scatterer exist for the entire span of angles and very few scatterers continuously persist for even $5°$. In the upper two cases (not labeled "persists"), scintillation is allowed and the location invariance declines slowly with azimuth span. The "within 1 pixel" results (which allow scintillation) are consistent with the 1-ft ISAR results of Dudgeon et al. [35], whose definition of persistence allowed scintillation. Because of the higher scatterer location invariance with a $1°$ azimuth span, in this research we use azimuth models at $1°$ increments for each target, in contrast to others who have used $5°$ [36], $10°$ [24], and 12 models [5].

3.3.2 Scatterer Location Invariance

Many of the scatterer locations are invariant to target conditions such as articulation or configuration variants. Because the object and ROI are not registered, we express the scatterer center location invariance with respect to articulation or configuration differences as the maximum number of corresponding scattering centers (whose locations match within a stated tolerance) for the optimum integer pixel translation. Given an original version of a SAR target image with n scattering centers, represented by points at pixel locations $P_i = (x_i, y_i)$ for $1 \leq i \leq n$ and a translated, distorted version $P'_j = (x'_j, y'_j)$ ($1 \leq j \leq n$) at a translation $t = (t_x, t_y)$, we define a *match* between points P'_j and P_i as

$$M_{ij}(t) = \begin{cases} 1 & \text{if } |x'_j - t_x - x_i| \leq l \text{ and } |y'_j - t_y - y_i| \leq l \\ 0 & \text{otherwise} \end{cases}$$

where $l = 0$ for an "exact" match and $l = 1$ for a match "within 1 pixel." The scatterer location invariance, L_n of n scatterers, expressed as a percentage of matching points, is given by

$$L_n = \frac{\max}{t} \left\{ \frac{100}{n} \sum_{j=1}^{n} \min\left(\left(\sum_{i=1}^{n} M_{ij}(t) \right), 1 \right) \right\}$$

where each point P'_j is restricted to at most one match.

Figure 5 shows the location invariance, L_{40}, of the strongest 40 scattering centers with articulation for T72 tank #a64 and ZSU 23/4 antiaircraft

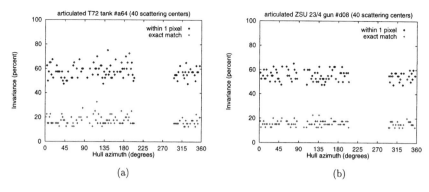

Figure 5   Scatterer location invariance with articulation: (a) T72 tank; (b) ZSU 23/4.

gun #d08 (at a 30° depression angle) as a function of the hull azimuth. The combined average invariance for both vehicles is 16.5% for an exact match of scattering centers and 56.5% for a location match within one pixel (3×3 neighborhood) tolerance. Similarly, Fig. 6 shows the percent of the strongest 40 scattering center locations that are invariant for configuration variants, T72 #812 versus #132 and BMP2 vehicle #C21 versus #9563, at a 15° depression angle. The mean and standard deviation for percent location invariance (for 40 scatterers and depression angle ϕ) are shown in Table 2 for articulated versions of the T72 and ZSU23/4, for configuration variants of the T72 and BMP2.

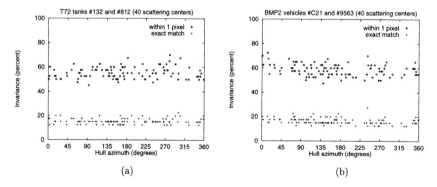

Figure 6   Scatterer location invariance with configuration: (a) T72 tank; (b) BMP2.

Table 2 Scatterer Percent Location Invariance for Targets with
Articulation and Configuration Variants

	Depression angle	Exact match invariance Mean	s.d.	Within 1 pixel invariance Mean	s.d.
Articulation					
T72 #a64	30	17.17	1.47	57.83	2.23
ZSU #d08	30	15.69	0.91	55.05	1.72
Average		16.45		56.47	
Configuration variants					
T72					
#812 vs. #132	15	15.34	0.89	55.34	1.91
#s7 vs. #132	15	15.40	0.83	56.68	1.95
BMP2					
#9563 vs. #c21	15	16.34	0.84	58.52	1.97
#9566 vs. #c21	15	16.17	0.99	57.93	1.97
Average		15.83		57.15	

3.3.3 Scatterer Magnitude Invariance

Using a scaled scatterer amplitude (S), expressed as a radar cross section in
square meters, given by $S = 100 + 10\log_{10}(i^2 + q^2)$, where i and q are the
components of the complex radar return, we define a percent amplitude
change (A_{jk}) as $A_{jk} = 100(S_j - S_k)/S_j$. (This form allows a larger variation
for the stronger signal returns.) A location and magnitude match $Q_{jk}(t)$ is
given by

$$Q_{jk}(t) = \begin{cases} 1 & \text{if } M_{jk}(t) = 1 \text{ and } |A_{jk}| \le l_A \\ 0 & \text{otherwise} \end{cases}$$

where l_A is the percent amplitude change tolerance. The scatterer magnitude
and location invariance (I_n), expressed as a percentage of n scatterers, is
given by

$$I_n = \frac{\max}{t} \left\{ \frac{100}{n} \sum_{k=1}^{n} \min\left(\left(\sum_{j=1}^{n} Q_{jk}(t)\right), 1\right) \right\}$$

Figure 7 shows the probability mass functions (PMFs) for percent
amplitude change for the strongest 40 articulated versus nonarticulated
scattering centers of T72 tank #a64 and ZSU 23/4 gun #d08. Curves are

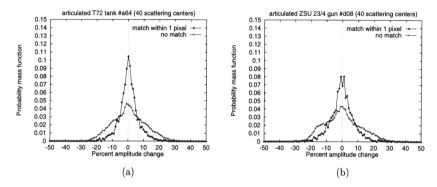

Figure 7   Scatterer magnitude invariance with articulation: (a) T72 tank; (b) ZSU 23/4.

shown both for the cases where the scattering center locations correspond within one pixel tolerance and for all the combinations of scatterers whose locations do not match. Similarly, Figure 8 shows the PMFs for percent amplitude change for the strongest 40 scattering centers with configuration variants, T72 #812 versus #132 and BMP2 #c21 versus #9563, at a 15° depression angle. The mean and standard deviation for these matching and nonmatching scatterers and the crossover points for the PMFs are given in Table 3. Table 4 shows the mean and standard deviation for the percent location and magnitude invariance (within one-pixel location tolerance and an amplitude change tolerance of l_A) of the strongest 40 scatterers for these same articulation and configuration difference cases. Similar findings for scatterer location and magnitude invariance are presented in Ref. 37 for a 2° depression angle change.

3.3.4   Target Occlusion

No real SAR data with occluded objects is available to the general public (limited data on vehicles in revetments [23] and partially hidden behind walls [25] have been reported to exist, but they have not yet been released for unrestricted use). In addition, there is no standard, accepted method for characterizing or simulating occluded targets. Typically, occlusion occurs when a tank backs up into a tree line, for example, so that the back end is covered by trees and only the front portion of the tank is visible to the radar. Thus, the "bright target" becomes a much smaller-sized object to the ATR. In addition, the treetops can produce "bright" peaks that are of similar strength to target peaks at many azimuths.

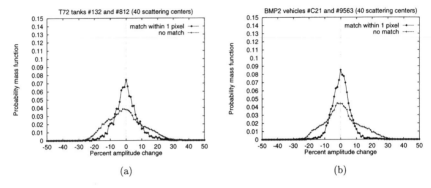

(a)   (b)

Figure 8 Scatterer magnitude invariance with configuration: (a) T72 tank; (b) BMP2.

The occluded test data in this chapter are simulated by starting with a given number of the strongest scattering centers and then removing the appropriate number of scattering centers encountered in order, starting in one of four perpendicular directions d_i (where d_1 and d_3 are the cross-range directions, along and opposite the flight path, respectively, and d_2 and d_4 are the up range and down range directions, respectively). Then, the same number of scattering centers (with random magnitudes) are added back at *random locations* within the original bounding box of the chip. This keeps the number of scatterers constant and acts as a surrogate for some potential occluding object. Our approach, using simulated occlusion, provides an

Table 3 Scatterer Percent Amplitude Change

	Within 1 pixel		No match		
	Mean	s.d.	Mean	s.d.	Crossover
Articulation					
T72 #a64	0.51	5.91	0.75	10.44	−5/+6
ZSU #d08	0.06	7.44	0.08	11.37	±9
Configuration variants					
T72					
#812 vs. #132	0.15	7.29	−0.38	11.12	±8
#s7 vs. #132	0.48	6.69	2.20	11.15	±9
BMP2					
#9563 vs. #c21	0.35	5.72	0.94	10.88	−8/ + 9
#9566 vs. #c21	0.48	6.20	0.56	10.68	−7/ + 8

Table 4 Scatterer Percent Location and
Magnitude Invariance for Locations Within
One Pixel and Amplitude Tolerance l_A (in
%)

	l_A	Mean	s.d.
Articulation			
T72 #a64	±9	53.47	2.63
ZSU #d08	±9	47.98	2.22
Average		50.78	
Configuration variants			
T72			
#812 vs. #132	±9	48.40	2.42
#s7 vs. #132	±9	50.69	2.44
BMP2			
#9563 vs. #c21	±9	54.38	2.34
#9566 vs. #c21	±9	53.00	2.51
Average		51.68	

enormous amount of data with varying known amounts of occlusion for
carefully controlled experiments.

3.4 SAR RECOGNITION SYSTEM

The basic SAR recognition algorithm is an off-line model construction
process and a similar on-line recognition process. The approach is designed
for SAR and is specifically intended to accommodate recognition of
occluded and articulated objects. Standard nonarticulated models of the
objects are used to recognize these same objects in nonstandard, articulated,
and occluded configurations. The models are a look-up table and the recognition process is an efficient search for *positive evidence*, using relative locations of the scattering centers in the test image to access the look-up table
and generate votes for the appropriate object (and azimuth pose).

Establishing an appropriate local coordinate reference frame is critical
for reliably identifying objects (based on locations of features) in SAR images
of articulated and occluded objects. These problems require the use of a local
coordinate system; global coordinates and global constraints do not work, as
illustrated in Fig. 3, where the center of mass and the principal axes change
with articulation. In the geometry of a SAR sensor, the "squint angle," the
angle between the flight path (cross-range direction) and the radar beam

(range direction), can be known and fixed at 90°. Given the SAR squint angle, the image range and cross-range directions are known and any local reference point chosen, such as a scattering center location, establishes a reference coordinate system. (The scattering centers are local maxima in the radar return signal.) The relative distance and direction of the other scattering centers can be expressed in radar range and cross-range coordinates and naturally tessellated into integer buckets that correspond to the radar range/cross-range bins. The recognition system takes advantage of this natural system for SAR, where a single basis point performs the translational transformation and fixes the coordinate system to a "local" origin.

The model construction algorithm for the recognition system is outlined in Fig. 9 (where the "origin" is the stronger of a pair of scatterers and a "point" is another weaker scatterer). Because of the specular radar reflections in SAR images, a significant number of features do not typically persist over a few degrees of rotation (as shown in Fig. 4). Consequently, we model each object at 1° azimuth increments. The relative locations and magnitudes of the N strongest SAR scattering centers are used as characteristic features (where N, the number of scattering centers used, is a design parameter). Any local reference point, such as a scattering center location, could be chosen as a basis point ("origin") to establish a reference coordinate system for building a model of an object at a specific azimuth angle pose. For ideal data, selecting the location of the strongest scattering center as the origin is sufficient. However, for potentially corrupted data where any scattering center could be spurious or missing (due to the effects of noise, target articulation, occlusion, nonstandard target configurations, etc.), we use all N strongest scattering centers in turn as origins to ensure that a valid origin is obtained Thus, to handle occlusion and articulation, the size of the look-up table models (and also the number of relative distances that are considered in the test image during recognition) are increased from N to $N(N-1)/2$.

---

1. For each model Object do 2

2. For each model Azimuth do 3, 4, 5

3. Obtain the location (R, C) and magnitude (S) of the strongest N scatterers.

4. Order (R, C, S) triples by descending S.

5. For each origin O from 1 to N do 6

6. For each point P from O+1 to N do 7, 8

7. $dR = R_P - R_O$; $dC = C_P - C_O$.

8. At look-up table location dR, dC append to list entry with: Object, Azimuth, R_O, C_O, S_O, S_P.

Figure 9   Model construction algorithm.

Using a technique like geometric hashing [38], the models are constructed using the relative positions of the scattering centers in the range (R) and cross-range (C) directions as the initial indices to a look-up table of labels that give the associated target type, target pose, "origin" range and cross-range positions, and the magnitudes (S) of the two scatterers. Because the relative distances are not unique, there can be several of these labels (with difference target, pose, etc. values) at each look-up table entry.

The recognition algorithm is outlined in Fig. 10. The recognition process uses the relative locations of the N strongest scattering centers in the test image to access the look-up table and generate votes for the appropriate object, azimuth, range and cross-range translation. (In contrast to many model-based approaches to recognition [39], we are not "searching" all of the models.) Further comparison of each test data pair of scatterers with the model look-up table result(s) provides information on the magnitude changes (between the data and the model) for the two scatterers. Limits on allowable values for translations and magnitude changes are used as constraints to reduce the number of false matches. The number of scattering centers used and the various constraint limits are design parameters that are

---

1. Obtain from test image the location (R, C) and magnitude (S) of N strongest scatterers.

2. Order (R, C, S) triples by descending S.

3. For each origin O from 1 to N do 4

4. For each point P from O+1 to N do 5, 6

5. $dR = R_P - R_O$; $dC = C_P - C_O$.

6. For DR from dR-1 to dR+1 do 7

7. For DC from dC-1 to dC+1 do 8, 9, 10

8. weighted_vote = |DR| + |DC|.

9. Look up list of model entries at DR, DC.

10. For each model entry E in the list do 11

11. IF $|tr = R_O - R_E|$ < translation_limit and $|tc = C_O - C_E|$ < translation_limit

and $|1 - S_{EO}/S_O|$ < magnitude_limit and $|1 - S_{EP}/S_P|$ < magnitude_limit

THEN increment accumulator array [Object, Azimuth, tr, tc] by weighted_vote.

12. Query accumulator array for each Object, Azimuth, tr and tc, summing the votes in a 3x3

neighborhood in translation subspace about tr, tc; record the maximum vote_sum and the

corresponding Object.

13. IF maximum vote_sum > threshold

THEN result is Object ELSE result is "unknown".

Figure 10 Recognition algorithm.

optimized, based on experiments, to produce the best recognition results. (Another approach to optimizing these tuning parameters, based on reinforcement learning, is presented in Ref. 40.) Given that MSTAR targets are "centered" in the chips, a ± 5 pixel limit on allowable translations is imposed for computational efficiency. As will be shown later, the experimentally determined optimum limit on the allowable percent difference in the magnitudes of the data and model scattering centers was $\pm 9\%$, which is consistent with the measured probability mass functions of scatterer magnitude invariance with target configuration variants and articulations (previously shown in Figs. 8 and 7, respectively). To accommodate some uncertainty in the scattering center locations, the eight neighbors of the nominal range and cross-range relative location are also probed in the look-up table and the translation results are accumulated for a 3×3 neighborhood in the translation subspace. A (city-block) weighted voting method is used to reduce the impact of the more common small relative distances. The recognition process is repeated with different scattering centers as basis points, providing multiple "looks" at the model database to handle spurious scatterers that arise due to articulation, occlusion, or configuration differences. The recognition algorithm actually makes a total of $9N(N - 1)/2$ queries of the look-up table to accumulate evidence for the appropriate target type, azimuth angle, and translation. The models (labels with object, azimuth, etc.) associated with a specific look-up table entry are the "real" model and other models that happen, by coincidence, to have a scatterer pair with the same (range, cross-range) relative distance. The constraints on magnitude differences filter out many of these false matches. In addition, although these collisions may occur at one relative location, the same random object–azimuth pair does not keep showing up at other relative locations with appropriate scatterer magnitudes and mapping to a consistent 3×3 neighborhood in translation space, whereas the "correct" object does.

The basic decision rule used in the recognition is to select the object–azimuth pair (and associated "best" translation) with the highest accumulated vote total. To handle identification with "unknown" objects, we introduce a criteria for the quality of the recognition result that the votes for the potential winning object exceed some minimum threshold v_{min}. By varying the decision rule threshold, we obtain a form of the receiver operating characteristic (ROC) curve with probability of correct identification, PCI = $P\{$decide correct object|object is true$\}$, versus probability of false alarm, P_f = $\{$decide any object|unknown is true$\}$. We call the algorithm a six dimensional (6D) recognition algorithm because, in effect, we use the range and cross-range positions and the magnitudes of pairs of scattering centers. (When using 40 scatterers, this 6D algorithm takes an average of 2.5 s to process a test chip on a Sun Ultra2 without any optimizations.)

3.5 RECOGNITION RESULTS

3.5.1 Configuration Experiments

In the configuration-variant experiments, a single configuration of the T72 tank (#132) and BMP2 (#C21) APC are used as the models and the test data are two other variants of each vehicle type (T72 #812, #s7 and BMP2 #9563, #9566) and different "unknown" confuser test vehicles (all at 15° depression angle). Although more extensive T72 configuration-variant data are available, only two configurations are used so that the amount of test data for the T72 and BMP2 is comparable and the results are not artificially biased toward recognizing the T72. In "forced recognition" experiments, the test data only include variants of the modeled classes and the system is forced to choose an alternative among the modeled classes. The forced recognition confusion matrix for these configuration variants is shown in Table 5, with an overall recognition rate of 94.7%. (The 94.7% rate with this 6D recognition system is a great improvement over the directly comparable 68.4% rate for an earlier 2D version of the recognition system given in Ref. 41.) These results were obtained with the 6D system using 36 scattering centers, a translation limit of ±5 pixels, and a percent magnitude change of less than ±9%. These parameter settings were optimum for the configuration-variant experiments, the most difficult case, and the same settings were also used with the 6D system in the articulation and depression angle change results given in subsequent sections. The effect on the forced recognition PCI of the number of scattering centers used is shown in Fig. 11, and Fig. 12 shows the effect of varying the amplitude change limit.

Figure 13 illustrates the pose accuracy of the forced recognition configuration-variant results. The top curve shows that 99% of time, the correct pose was achieved within ±15° (with a 180° front versus back direction ambiguity), whereas the correct object and pose were achieved 94% of the time with the directional ambiguity and 89% of the time with no ambiguity.

Table 5 Forced Recognition Confusion Matrix for Configuration Variants (36 Scatterers, ±9% Amplitude Tolerance)

| Test targets (serial number) | Identification results (configurations modeled) | | | |
	BMP2	(#c21)	T72	(#132)
BMP2 (#9563)	106	(98.1%)	2	
BMP2 (#9566)	107	(97.2%)	3	
T72 (#812)	11		92	(89.3%)
T72 (#S7)	6		88	(93.6%)

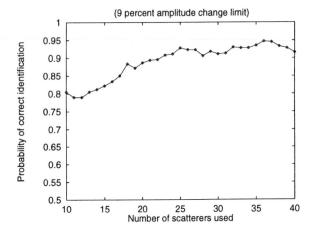

Figure 11 Effect of number of scattering centers used on recognition of config-uration differences.

Thus, the differences between the top and middle curves are the misidentifications; between the middle and the bottom are the cases where the direction is wrong by 180°.

Figures 14a–14d show scatterplot recognition results in BMP2–T72 vote space for configuration variants of the tracked BMP2 APC and tracked

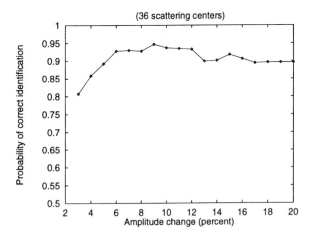

Figure 12 Effect of amplitude change tolerance on recognition of configuration differences.

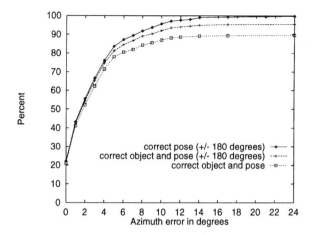

Figure 13 Forced recognition pose accuracy for configuration variants.

T72 tank and for various confusers: the wheeled BTR70 APC #c71, the tracked ZSU 23/4 antiaircraft gun #d08, and the wheeled BRDM2 APC #e71. The 45° line in Fig. 14 represents the decision boundary of the simplest decision rule: "the object with the most votes wins." In this forced recognition case, Fig. 14a, the overall recognition rate is 94.7%, where 2.3% of the BMP2s and 9.4% of the T72s are on the "wrong" side of the boundary and are misidentified. To handle unknown objects, in these cases fairly difficult "targetlike" confuser vehicles, we establish a minimum vote threshold for positive identifications; otherwise, the test object is labeled "unknown." In Figs. 14b–14d, the BTR70 APC is the most difficult confuser, the BRDM2 APC is somewhat less difficult, and the ZSU 23/4 antiaircraft gun is easy. For example, Fig. 14b illustrates that 99.6% of the BTR70 confuser false alarms could be eliminated with a 3000 vote threshold, but Fig. 14a shows that a 3000 vote threshold would eliminate more than half of the BMP2 and T72 identifications. In contrast, Fig. 14d shows that almost all of the ZSU confuser false alarms could be eliminated with a 1000 vote threshold without any reduction in the BMP2 and T72 identifications.

3.5.2 Articulation Experiments

In the articulation experiments the models are nonarticulated versions of T72 tank #a64 and ZSU23/4 antiaircraft gun #d08 and the test data are the articulated versions of these same serial number objects and BRDM2 APC #e71 as a confuser vehicle (all at 30° depression angle). The articulated

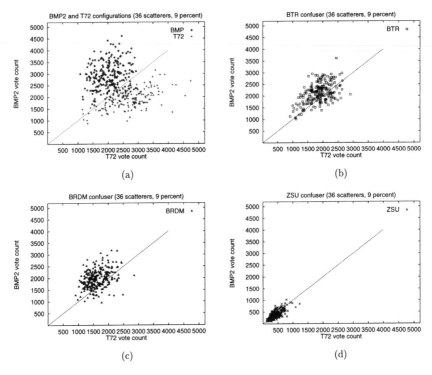

Figure 14   Scatterplots for recognition system results with configuration variants. (a) BMP vs. T72; (b) BTR confuser; (c) BRDM confuser; and (d) ZSU confuser.

object recognition results are shown in Table 6 using a 2100 common vote decision criterion for an overall 0.927 PCI at a 0.039 Pfa. (The overall forced recognition rate is 100% over a range from 14 to 40 scattering centers.) Figures 15a and 15b show recognition system scatterplot results in ZSU–T72 vote space for articulation of the ZSU 23/4 and T72 tank and for the BRDM2 confuser. Here, the results for the ZSU 23/4 and T72 are widely separated, giving 100% forced recognition results. Figure 15b shows that whereas the BRDM2 is always classified as a T72, a unique threshold of 2000–2500 T72 votes will eliminate most, if not all, of the false alarms at the cost of only a few T72s moved to the "unknown" classification. A common threshold applied to votes for either the T72 or the ZSU has a higher cost because many ZSUs are moved to "unknown," as shown in Table 6.

Figure 16 shows how the PCI varies with the percent articulation invariance (for a "within 1 pixel" location match) for the 6D recognition engine. The sets of curves are shown with different vote thresholds from 1700 to 2700 to generate failures that illustrate the effect of location invar-

Table 6 Example Articulated Object Confusion Matrix (36 Scatterers, ±9% Amplitude Tolerance, 2100 Vote Threshold)

Articulated test targets	Identification results		
	T72	ZSU	Unknown
T72 315° turret	94	0	4
ZSU 315° turret	0	84	10
BRDM2 (confuser)	10	0	248

iance on recognition rate. As expected, recognition failures generally occur for instances with relatively low invariance with articulation. The behavior shown in Fig. 16 for this 6D system with real MSTAR data and "within 1 pixel" location match is consistent with the previous results [42] for a 2D system with XPATCH simulated radar signature data and "exact" location match. Detailed comparisons between the articulated object recognition results for the 6D system and the earlier 2D system are presented in Ref. 43.

With our approach, in these experiments the effects of articulation generate noise and occlusion (scatterers are removed and others are added in region effected by articulation). Thus, we are only concerned with the major part (e.g., the tank or APC hull) and treat the smaller articulated turret as a 'don't care" region. An approach that explicitly models the small articulated part and recovers the pose of that part is given in Ref. 44. (This is a difficult task because of the relatively limited number of scatterers usually available on the smaller turret part.)

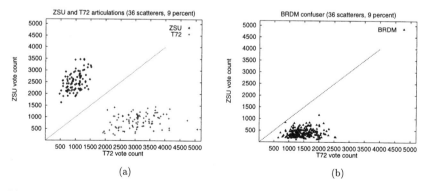

(a)   (b)

Figure 15 Scatterplots for recognition system results with articulation. (a) ZSU vs. T72; (b) BTDM confuser.

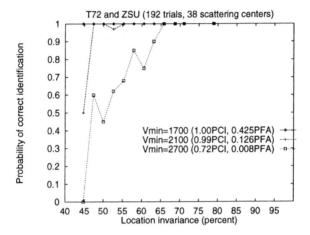

Figure 16   Recognition rate and articulation invariance.

3.5.3 Depression Angle Experiments

In the depression angle experiments, the models are T72 #132 and BMP2 #c21 at a 15° depression angle and the test data are the same serial number objects and the BTR70 #c71 confuser at 17°. The confusion matrix, shown in Table 7, for the depression angle results has an overall 0.822 PCI at 0.10 Pfa, obtained with a 2800 vote threshold. Figures 17a and 17b show scatterplot recognition results in BMP2–T72 vote space for the depression angle experiments. These results show better separation than the configuration variant results of Fig. 14a with a 99.3% forced recognition rate. The vote counts for depression angle change in Fig. 17a are typically higher (more away from the origin and away from the decision boundary) than in Fig. 14a for configuration variants, whereas the BTR confuser plots are generally similar. Thus, a common vote threshold for the depression angle cases eliminates false alarms at a lower cost than for the configuration-variant cases.

Table 7   Example Confusion Matrix for Depression Angle Changes (36 Scatterers, ±9% Amplitude Tolerance, 2800 Vote Threshold)

	Identification results [15° moldels]		
Depression angle 17° test targets	BMP2	T72	Unknown
BMP2 (#c21)	110	0	28
T72 (#132)	0	117	21
BTR70 (confuser)	15	8	207

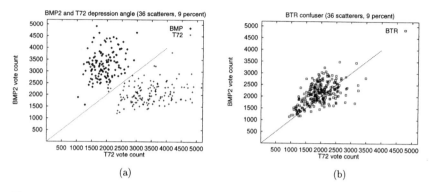

(a)   (b)

Figure 17   Scatterplots for recognition system results with depression angle change: (a) BMP versus T72; (b) BTR confuser.

3.5.4   ROC Curve Results

Receiver operating characteristic (ROC) curves can be generated from the scatterplot data in Figs. 14–17 by varying the vote threshold (typically from 1000 to 4000 in 50 vote increments). Figure 18 shows the significant effect on the configuration variant recognition ROC curves of using the different ZSU, BRDM, and BTR confusers whose scatterplot results were given in Fig. 14. Excellent results are obtained with the ZSU 23/4 confuser, whereas the BTR70 is a difficult case. Figure 19 shows the ROC curve

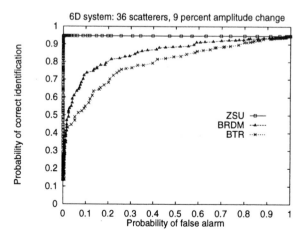

Figure 18   Effect of confusers on configuration variant ROC curve.

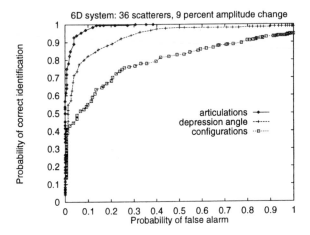

Figure 19 ROC curves for articulation, depression angle, and configuration variants.

recognition results for the articulation, depression angle change, and configuration-variants cases, all with the recognition system using the same operating parameters. The ROC curves in Fig. 19 show that the differences in configuration of an object type are a more difficult challenge for the recognition system than small depression angle changes, as both were generated using the BTR confuser. The excellent results for the articulation case are basically due to the dissimilarity of the ZSU, T72, and BRDM as seen in Fig. 15.

Figure 20 shows ROC curves for the same MSTAR T72 and BMP2 configuration variants with the BTR confuser using three different versions (2D, 6D, and 8D) of the recognition system. The earlier simpler 2D system (described in Ref. 41) used only relative locations of pairs of peaks; the 6D system is described in Section 3.4 and the 8D version (described in Ref. 37) uses an additional feature the shape factor (a measure of the relative sharpness) of the two peaks. Each of the systems was optimized for the forced recognition configuration-variant case: the 2D system at 20 scatterers; the 6D system at 36 scatterers; the 8D system at 50 scatterers (with a $\pm30\%$ shape factor change limit). Both the 6D and 8D system results are a substantial improvement over the earlier 2D system results. Although Fig. 20 shows that the 8D system gave worse results than the 6D system in the region below 0.1 Pfa, reoptimizing the operating parameters (e.g., using 45 scatterers) gives the 8D system somewhat better results in the region below 0.1 Pfa at the cost of a slightly reduced forced recognition rate.

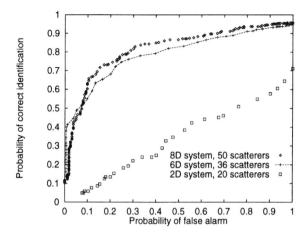

Figure 20   ROC curves for configuration variants with 2D, 6D, and 8D systems.

3.5.5   Integration of Multiple Recognizers

Instead of tuning the parameters of a single recognizer to achieve the optimum forced recognition performance, the results of multiple recognition systems with different parameters can be integrated to produce a better overall result. Figure 21 shows the results of integrating 9 recognizers (operating in a 3×3 neighborhood about the optimum parameters of 36 scatterers and a 9% maximum percent magnitude change) for the most difficult

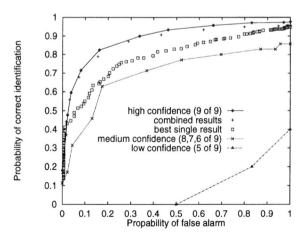

Figure 21   Effect of integrating multiple recognizers on ROC curves for configuration variants.

case of configuration variants. The upper ROC curve, labeled "high confidence," represents the combined 90.3% of the T72 tank and BMP2 APC
target cases and 70.1% of the BTR70 APC confuser cases where all nine of
the recognizers agreed. The second curve, labeled "combined results," represents the overall average of all the multiple recognizer results (five or more
of nine recognizers agree). It is a significant improvement over the best
single system result, previously shown as the configurations result in Fig.
19 as well as the BTR result in Fig. 18 and now replotted and labeled as
"best single result" in Fig. 21. For example, at a 15% false-alarm rate, the
combined result of multiple recognizers is a 79% recognition rate compared
to a 65% rate for the best single recognizer. The "medium confidence" curve
represents the 8.4% of the targets and 23.7% of the BTR70 APC confusers
where six to eight of the nine recognizers agreed. The "low confidence"
curve represents 1.3% of the targets and 6.2% of the BTR70 APC confusers
where only five of nine recognizers agreed.

3.5.6 Occluded Objects

Figure 22 shows the forced recognition performance of the system with
occluded objects in terms of PCI as a function of percent occlusion with
the "number of scattering centers used" as a parameter. Each point for a
specific number of scattering centers and percent occlusion is the average
PCI for all four occlusion directions, the four objects (BMP, BTR, T72,
and ZSU), and the number of available test azimuths. We defined the
available test azimuths as azimuths that had at least the "number of
scattering centers used" present in the data; thus, we avoid introducing

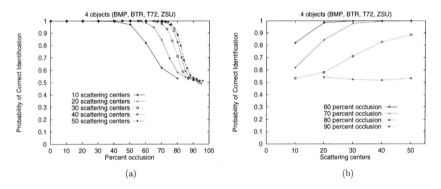

Figure 22   Effect of occlusion and number of scatterers on recognition rate:
(a) effect of occlusion; (b) effect of number of scatterers.

an uncontrolled variable: the number of scattering centers actually available for some instance of an object at a specific azimuth orientation. (In practice, if some target aspect did not have the appropriate number of scattering centers, the performance would degrade as if the "missing" scatterers were occluded.) The forced recognition results in Fig. 22 show a "breakpoint" at 60–75% occlusion for 20 scatterers or more. At very high occlusion levels, one would expect that the recognition results with four objects would approach 25%, due to chance. With the MSTAR data, we achieve 50% recognition because the ZSU is almost never confused with the other vehicles, so the three remaining vehicles at a little over 33% and the ZSU at over 90% yields an overall rate of about 50% recognition. The results shown in Fig. 22 (which use both scatterer location and magnitude features) are better than the predicted performance given in Ref. 45, which considers only location features.

Typical forced recognition results for 40 scattering centers and 70% occlusion are shown as a confusion matrix in the left half of Table 8. With 3410 correct identifications in 3420 trials, the overall PCI is 0.9971. The right half of Table 8 shows the pose accuracy results, where 99.18% of the time the pose is correct within $\pm 5°$, and in 98.68% of the cases the pose is exactly correct.

In the occluded data, for the true case (the actual object, azimuth used in the test instance), there are n valid scatterers of M scatterers used (where $n \leq M$ due to occlusion). Using an unweighted voting method and neglecting any random contribution of nonmatching points, the number of votes predicted for the true case, V, is given by $V = n(n - 1)/2$. Figure 23 shows that the actual number of votes received lies just slightly above the prediction curve and that the random contributions of nonmatching points are negligible. These results for the 6D algorithm are a significant improvement over the results for the earlier 2D algorithm [42], where there was a relatively large random contribution from the nonmatching points. Detailed comparisons between the occluded results for the 6D system, using real vehicle data from the MSTAR dataset and the earlier 2D system, using XPATCH-generated simulated SAR signature data, are presented in Ref. 46.

Using a vote threshold decision rule (i.e., the votes for the potential winning object exceed some threshold, v_{min}), recognition results were obtained for the occluded versions of the BMP, BTR, T72, and ZSU test vehicles as well as a similarly occluded BRDM2 confuser vehicle. Figures 24a, 24b, and 24c give the probability of correct identification, probability of false alarm, and probability of miss, respectively, as a function of v_{min} for 20 and 40 scatterers and 70% occlusion. The resulting ROC curves for 20 and 40 scatterers and 70% occlusion are shown in Fig. 24d. Figure 25 gives the ROC curves for 40 scatterers with 65–80% occlusion. An illustrative con-

Table 8 Forced Recognition Confusion Matrix for 70% Occluded Objects (40 Scatterers, e =Exact Pose, c =Pose Within ±5°)

70% occluded test targets	Identification results				Pose accuracy			
	BMP	BTR	T72	ZSU	BMP	BTR	T72	ZSU
BMP	769	1	2	0	768c, 768e			
BTR	1	774	1	0		773c, 770e		
T72	3	0	773	0			767c, 765e	
ZSU	1	1	0	1094				1084c, 1072e

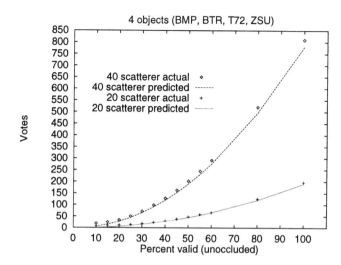

Figure 23 Occluded performance results versus prediction.

fusion matrix for 70% occlusion and $v_{min} = 65$ (40 scatterers) is shown in Table 9. The overall PCI is 0.997 and the P_f is 0.025.

Figure 26 compares the mean and standard deviation of the votes generated by the four test objects with the votes generated by the BRDM2 confuser vehicle for 40 scatterers as a function of the percentage of valid (unoccluded) scatterers. This shows that, with 40 scatterers, for above 30% valid data (or less than 70% occlusion), the occluded BRDM2 is not in competition with the actual object. However, although the target may be occluded, the confuser vehicle may not necessarily be occluded in the practical case. Hence, to cope with *unoccluded confusers*, one would need to set a threshold of about 200 votes for a valid identification (labeled by a in Fig. 26), which would then limit the ability to recognize targets to about 50% target occlusion (b in Fig. 26).

To test the combined effects of occlusion and positional noise, Gaussian noise with zero mean and standard deviation sigma (in units of 1-ft resolution pixels) is added to the range and cross-range locations of the scattering centers in the occluded test data. The overall recognition performance for 4 objects, using 40 scatterers with varying amounts of occlusion, is shown in Fig. 27 as a function of positional noise. Figure 27 confirms that the recognition system accommodates a ±1 pixel uncertainty in scattering center location for up to 70% occlusion.

All of the occlusion experiments described here involve recognizing four objects (BMP, BTR, T72, and ZSU). Because there are only two articu-

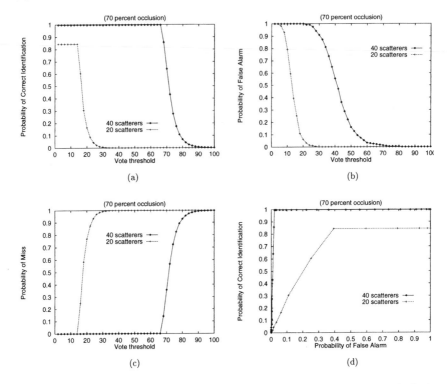

Figure 24 Vote threshold and receiver operating characteristic (70% occlusion): (a) probability of correct identification; (b) probability of false alarm; (c) probability of miss; (d) receiver operating characteristic.

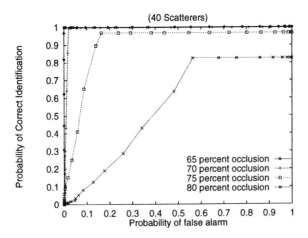

Figure 25 Effect of occlusion on receiver operating characteristics (40 scatterers).

Table 9   Typical Confusion Matrix for 70% Occluded Objects (40 Scatterers, $v_{min} = 65$)

| | \multicolumn{5}{c}{Identification results} | | | | |
70% occluded test targets	BMP	BTR	T72	ZSU	Unknown
BMP	769	1	2	0	0
BTR	1	774	1	0	0
T72	3	0	773	0	0
ZSU	1	1	0	1094	0
BRDM2	2	3	0	1	237

lated objects available (T72 and ZSU), it is useful to establish the effect of scaling the forced recognition problem from four occluded objects to two occluded objects prior to investigating the effect of occlusion on articulated objects. Figure 28 illustrates the effect of scaling on occluded object recognition. The 20- and 40-scatterer curves for 4 objects are the same data as previously shown in Fig. 22a. As one would expect, the results for the two-object case are better than the four-object case: The "break" point is less pronounced and the PCI is higher at very high occlusion levels.

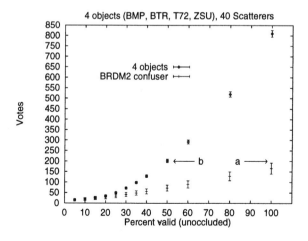

Figure 26   Occluded performance with targets versus unknown object (40 scatterers), where a is the unoccluded confuser threshold and b is the resulting operating point.

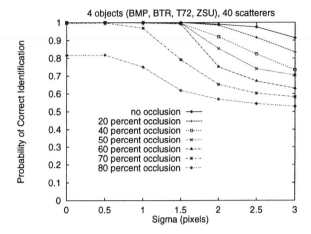

Figure 27   Effect of positional noise on occluded object recognition (40 scatterers).

3.5.7 Occluded Articulated Objects

In the articulated experiments, the models are nonarticulated versions of T72 #a64 and ZSU23/4 #d08 and the test data are the articulated versions of these same serial number objects (with the turret rotated to 315°) that are occluded in the same manner as earlier. The MSTAR articulated data is at 30° depression angle. Figure 29 shows the effect of occlusion on

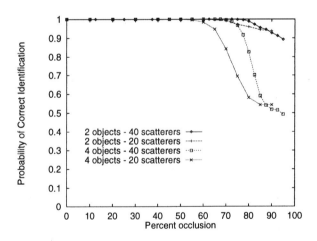

Figure 28   Effect of scaling on occluded object recognition.

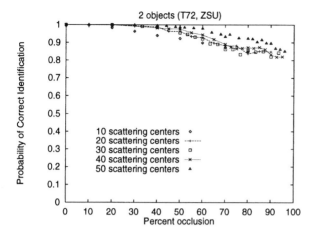

Figure 29   Effect of occlusion on MSTAR articulated object recognition.

recognition of these articulated objects for various numbers of scattering centers used.

3.6   CONCLUSIONS

The large variance in SAR scattering center locations with object pose (azimuth) can be successfully captured by modeling objects at small (e.g., 1°) azimuth increments. Although this requires a significant modeling effort, about 8 megabytes total of computer memory for each modeled object in our implementation, the resulting vote-based recognition process can be accomplished quickly (2–3 s) with efficient table look-ups. The locations and magnitudes of many scatterers are quasi-invariant with object configuration variations, articulations, and small changes in radar depression angle. Typically, about 50% of the scatterers are invariant within a 10% magnitude and one pixel location bound for configuration variants and articulations. A model-based recognition system, using an inexact match of the local features scatterer location and magnitude, can successfully handle difficult conditions with object configuration variants, articulation, and occlusion with significant forced recognition rates and excellent pose accuracy results. Using local features is essential for recognizing articulated and occluded objects, where global features like the object shape and major axis can be significantly altered. The use of additional features, scatterer magnitudes, and the ability to handle inexact matches of these quasi-invariant local features are the keys to the success of this 6D recognition system compared to the earlier 2D recognition system which used only relative scatterer locations and required an exact location match. Some of the confuser vehicles are sufficiently similar to the objects of interest (e.g., BTR70 APC versus T72 tank and BMP2 APC) that they pose a much more severe challenge than other confusers (such as the ZSU23/4 gun). The physical differences in the configuration-variant cases are a more difficult challenge than the articulation and depression angle cases, which involve the same physical object. Combining the results of multiple recognizers can give significantly improved performance over the single best recognizer. The possibility of an *unoccluded confuser* vehicle is an important practical limiting factor on the performance that can be achieved in recognizing highly occluded vehicle targets (e.g., 50% occlusion versus over 70% with occluded confusers). These algorithms achieve excellent occluded object recognition results for real SAR data with simulated occlusion, even with the combined effects of articulation and occlusion.

A significant limitation of this work is that we, like all other researchers in the field using the MSTAR data, have assumed flat terrain so that the major pose issue is the 360° unknown target azimuth (vehicle yaw). Because the radar signature changes significantly with small vehicle pose changes, the effects of vehicle pitch and roll due to real terrain (which were not included in the MSTAR data) need to be considered for an eventual product implementation. Fortunately, vehicle operational limitations (e.g., tracked vehicles cannot handle more than about 10° of roll) reduce the amount of target pitch and roll; however, 10° in both axes would add about two orders of magnitude more model data requirements to the problem. Scaling the results obtained for 2–4 objects to a more realistic 20–40 objects adds another order of magnitude.

The current work, which implicitly relies on the dissimilarity between different objects and on the invariance of the same object to changing conditions, can be extended to explicitly determine and utilize these measures for increased recognition performance. This becomes especially important as the problem is scaled to larger numbers of objects and larger numbers of models per object to handle target pitch and roll conditions. One approach is to explicitly discount model similarity between objects, as measured by feature space collisions. The intuition is that "ambiguous" features should be discounted. Another approach is to explicitly promote features that are invariant with changing conditions (e.g., configurations, articulations, or small pose changes). Here, the intuition is that the invariant features are more "reliable" and should be promoted. Higher-resolution data, instead of 1-ft resolution now (commonly) available, would greatly increase the discriminating power of the current scatterer location and magnitude features. Additional features such as valleys or ridges could also be used, especially with higher-resolution data. As more features and constraints are added, the number of system tuning parameters increases and optimizing recognition performance becomes difficult. An approach combining performance prediction [45] and adaptive learning of the recognition algorithm tuning parameters [40] needs to be developed to facilitate optimizing recognition system performance.

ACKNOWLEDGMENT

This work was supported in part by DARPA/AFOSR grant F49620-97-1-0184 and AFRL grant F33615-99-C-1440. The contents and information do not reflect positions or policies of the U.S. Government.

REFERENCES

1. D Carlson, B Kumar, R Mitchell, M Hoffelder. Optical trade-off distance classifier correlation filters (OTDCCFs) for synthetic aperture radar automatic target recognition. SPIE Proceedings: Algorithms for Synthetic Aperture Radar Imagery IV (E Zelnio, ed.). SPIE, Bellingham, WA, 1997, Vol 3070, pp 110–120.
2. D Casasent, S Ashizawa, SAR detection and recognition filters. SPIE Proceedings: Algorithms for Synthetic Aperture Radar Imagery IV. SPIE, Bellingham, WA, 1997, Vol. 3070, pp 125–136.
3. R Meth, R Chellappa. Automatic classification of targets in synthetic aperture radar imagery using topographic features. SPIE Proceedings: Algorithms for SAR Imagery III. (E Zelnio, R Douglas, eds.) SPIE, Bellingham, WA, 1996 Vol. 2757, pp 186–193.
4. T Ryan, B Egaas. SAR target indexing with hierarchical distance transforms. SPIE Proceedings: Algorithms for Synthetic Aperture Radar Imagery III. (E Zelnio, R Douglas, eds.) SPIE Bellingham, WA, 1996, Vol. 2757, pp 243–252.
5. J Verly, R Delanoy, C Lazott. Principles and evaluation of an automatic target recognition system for synthetic aperture radar imagery based on the use of functional templates. SPIE Proceedings: Automatic Target Recognition III. SPIE, Bellingham, WA, 1993, Vol. 1960, pp 57–71.
6. D Casasent, R Shenoy. Feature space trajectory for distorted-object classification and pose estimation in SAR. Opt Eng 36:2719–2728, 1997.
7. A Waxman, M Seibert, A Bernardon, D Fay. Neural systems for automatic target learning and recognition. Lincoln Lab J6(1):77–116.
8. JH Yi, B Bhanu, M Li. Target indexing in SAR images using scattering centers and the Hausdorff distance. Pattern Recogn Lett 17:1191–1198, 1996.
9. T Ross, S Worrell, V Velten, J Mossing, M Bryant. Standard SAR ATR evaluation experiments using the MSTAR public release data set. SPIE

Proceedings: Algorithms for Synthetic Aperture Radar Imagery V. (E Zelnio, ed.) SPIE, Bellingham, WA, 1998, Vol. 3370, pp 566–573.

10. B Bhanu. Automatic target recognition: state-of-the-art survey. IEEE Trans Aerospace Electron Syst 22:364–379, July 1986.

11. B Bhanu, D Dudgeon, E Zelnio, A Rosenfeld, D Casasent, I Reed. Introduction to the special issue on automatic target detection and recognition. IEEE Trans Image Process 6(1):1–6, 1997.

12. B Bhanu, T Jones. Image understanding research for automatic target recognition. Proceedings ARPA Image Understanding Workshop, 1992, pp 249–254.

13. D Dudgeon, R Lacoss. An overview of automatic target recognition. Lincoln Lab J 6(1):3–9, 1993.

14. L Novak, G Owirka, C Netishen. Performance of a high-resolution polarimetric SAR automatic target recognition system. Lincoln Lab J 6(1):11–24, 1993.

15. D Kreithen, S Halverson, G Owirka. Discriminating targets from clutter. Lincoln Lab J6(1):25–51, 1993.

16. D Casasent, R Shenoy. Synthetic aperture radar detection and clutter rejection MINACE filters. Pattern Recogn 30(1):151–162, 1997.

17. W Irving, A Willsky, L Novak. A multiresolution approach to discriminating targets from clutter in SAR imagery. SPIE Proceedings: Algorithms for Synthetic Aperture Radar Imagery II 84. (D Giglio, ed.). SPIE, Bellingham, WA 1995, Vol. 2487, pp 272–299.

18. J Starch, R Sharma, S Shaw. A unified approach to feature extraction for model-based ATR. SPIE Proceedings: Algorithms for Synthetic Aperture Radar Imagery III. (E Zelnio, R Douglas, eds.) SPIE, Bellingham, WA 1996, Vol. 2757, pp 294–305.

19. J Wissinger, R Washburn, D Morgan, C Chong, N Friedland, A Nowicki, R Fung. Search algorithms for model-based SAR ATR. SPIE Proceedings: Algorithms for Synthetic Aperture Radar Imagery III. (E Zelnio, R Douglas, eds.) SPIE, Bellingham, WA 1996, Vol. 2757, pp 279–293.

20. E Keydel, S Lee. Signature prediction for model-based automatic target recognition. SPIE Proceedings: Algorithms for Synthetic Aperture Radar Imagery III. (E Zelnio, R Douglas, eds.) SPIE, Bellingham, WA 1996, Vol. 2757, pp 306–317.

21. G Ettinger, G Klanderman, W Wells, W Grimson. A probabilistic optimization approach to SAR feature matching. SPIE Proceedings: Algorithms for Synthetic Aperture Radar Imagery III. (E Zelnio, R Douglas, eds.) SPIE Bellingham, WA 1996, Vol. 2757, pp 318–329.

22. W Jachimczyk, D Cyganski. Enhancements of pose-tagged partial evidence fusion SAR ATR. SPIE Proceedings: Algorithms for Synthetic Aperture Radar Imagery IV. (E Zelnio, ed.) SPIE, Bellingham, WA 1997, Vol. 3070, pp 334–345.

23. J Mossing, T Ross. An evaluation of SAR ATR algorithm performance sensitivity to MSTAR extended operating conditions. SPIE Proceedings:

Algorithms for SAR Imagery V. (E Zelnio, ed.) SPIE, Bellingham, WA 1998, Vol. 3370, pp 554–565.

24. K Ikeuchi, T Shakunga, M Wheeler, T Yamazaki. Invariant histograms and deformable template matching for SAR target recognition. Proceedings IEEE Conference on Computer Vision and Pattern Recognition, 1996, pp 100–105.

25. J Wissinger, R Ristroph, J Diemunsch. W Severson, E Freudenthal. MSTAR's extensible search engine and inferencing toolkit. SPIE Proceedings: Algorithms for SAR Imagery VI. (E Zelnio, ed.) SPIE, Bellingham, WA 1999, Vol. 3721, pp 554–570.

26. I Biederman. Recognition-by-components: A theory of human image understanding. Psychol Rev 94:115–147, 1987.

27. Y Hel-Or, M Werman. Recognition and localization of articulated objects. Proceedings IEEE Workshop on Motion of Non-Rigid and Articulated Objects, 1994, pp 116–123.

28. A Beinglass, H Wolfson. Articulated object recognition, or: How to generalize the generalized Hough transform. Proceedings of the IEEE Conference on Computer Vision and Pattern Recognition, 1991, pp 461–466.

29. S Stanhope, E Keydel, W Williams, V Rajlich, R Sieron. The use of the mean squared error matching metric in a model-based automatic target recognition system. SPIE Proceedings: Algorithms for Synthetic Aperture Radar Imagery V. (E. Zelnio, ed.) SPIE, Bellingham, WA 1998, Vol. 3370, pp 360–368.

30. L Kaplan, R Murenzi, E Asika, K Namuduri. Effect of signal-to-clutter ratio on template-Based ATR. SPIE Proceedings: Algorithms for Synthetic Aperture Radar Imagery V (E. Zelnio, ed.) SPIE, Bellingham, WA 1998, Vol. 3370, pp 408–419.

31. R Bhatnagar, R Dilsavor, M Minardi, D Pitts. Intra-class variability in ATR systems. SPIE Proceedings: Algorithms for Synthetic Aperture Radar Imagery V. (E. Zelnio, ed.) SPIE, Bellingham, WA 1998, Vol. 3370, pp 383–395.

32. Q Pham, T Brosnan, M Smith, R Mersereau. An efficient end-to-end feature based system for SAR ATR. SPIE Proceedings: Algorithms for Synthetic Aperture Radar Imagery V. (E, Zelnio, ed.) SPIE, Bellingham, WA 1998, Vol. 3370, pp 519–529.

33. D Kottke, P Fiore, K Brown, J Fwu. A design for HMM-based SAR ATR. SPIE Proceedings: Algorithms for Synthetic Aperture Radar Imager V. (E. Zelnio, ed.) SPIE, Bellingham, WA 1998, Vol. 3370, pp 541–551.

34. Khoros Pro v2.2 User's guide. Reading, MA: Addison-Wesley Longman, 1998.

35. D Dudgeon, R Lacoss, C Lazott, J Verly. Use of persistent scatterers for model-based recognition. SPIE Proceedings: Algorithms for Synthetic Aperture Radar Imagery (D. Giglio, ed.) SPIE, Bellingham, WA 1994, Vol. 2230, pp 356–368.

36. L Novak, S Halversen, G Owirka, M Hiett. Effects of polarization and resolution on SAR ATR. IEEE Trans Aerospace Electron Syst 33(1):102–115, 1997.

37. B Bhanu, G Jones III. Recognizing target variants and articulations in synthetic aperture radar images. Opt Eng 39(3):712–723, 2000.

38. Y Lamden, H Wolfson. Geometric hashing: A general and efficient model-based recognition scheme. Proceedings International Conference on Computer Vision, 1988, pp 238–249.
39. WEL Grimson. Object Recognition by Computer: The Role of Geometric Constraints. Cambridge, MA: MIT Press, 1990.
40. B Bhanu, Y Lin, G Jones, J Peng. Adaptive target recognition. Machine Vision Applic 11:289–299, 2000.
41. B Bhanu, G Jones III, J Ahn. Recognizing articulated objects and object articulation in SAR images. SPIE Proceedings: Algorithms for Synthetic Aperture Radar Imagery V. (E. Zelnio, ed.) SPIE, Bellingham, WA 1998, Vol. 3370, pp 493–505.
42. G Jones III, B Bhanu. Recognition of articulated and occluded objects. IEEE Trans Pattern Anal Machine Intell 21(7):603–613, 1999.
43. G Jones III, B Bhanu. Recognizing articulated objects in SAR images. Pattern Recogn 34(2), 469-485, 2001.
44. B Bhanu, J Ahn. A system for model-based recognition of articulated objects. Proceedings, International Conference on Pattern Recognition, 1998, pp 1812–1815.
45. M Boshra, B Bhanu. Predicting performance of object recognition. IEEE Trans Pattern Anal Machine Intell 22(9):956–969, 2000.
46. G Jones III, B Bhanu. Recognizing occluded objects in SAR images. IEEE Trans Aerospace Electron Syst 37(1): 316–328, 2001.
47. B Bhanu, Y Lin. Stochastic models for recognition of occluded objects. In: Advances in Pattern Recognition. FJ Ferri, JM Inesta, A Amin, P Pudil, eds. New York: Springer-Verlag 2000, pp 560–570.

4

Edge Detection and Location in SAR Images: Contribution of Statistical Deformable Models

Olivier Germain and Philippe Réfrégier
Ecole Nationale Supérieure de Physique de Marseille,
Domaine Universitaire de Saint-Jérôme, Marseille, France

4.1 INTRODUCTION

Remote sensing plays a key role in many domains devoted to observation of the Earth, such as oceanography, cartography, or agriculture monitoring. Among the different acquisition systems, Synthetic Aperture Radar (SAR) imagery has broadly opened the field of applications in the past 20 years. This active sensor emits a microwave illumination (1–10 GHz) and measures the backscattered component (see Refs. 1 and 2 for a detailed description of the system). It offers the advantage of acquiring high-resolution images of the Earth's surface, in any weather conditions, both day and night. However, the main drawback of SAR is the well-known speckle corruption, inherent to any active and coherent imaging technique, which limits the analysis of the image.

Segmentation aims at decomposing the image into a tesselation of homogeneous regions. It is a low-level processing which helps further steps like classification or pattern recognition [3]. Actually, classical segmentation algorithms do not perform well on SAR images and a class of new methods, dedicated to speckled images, has arisen. Many techniques have been proposed, including edge detection [4–7], region growing [8,9], and random Markov Fields (RMF) [10–12]. Operational segmentation chains generally consider a combination of these techniques. For instance, the method developed by Fjørtøft et al. [13,14] includes four steps: edge detection, edge extraction by watershed, region fusion, and contour refinement with RMF.

In this chapter, we will address two points:

- Speckle-dedicated edge detectors are of special interest in SAR image segmentation because they are used in a large number of algorithms. There have been many studies about their performance in edge detection, but the accuracy of edge location was only recently carefully addressed, to our knowledge [15–17]. We first provide a characterization of the spatial accuracy of these detectors.
- Second, we describe an improvement of the segmentation accuracy while keeping a low computing time. For this purpose, we consider a statistical approach based on active contours. On one hand, we evaluate the improvement brought by the original technique; on the other hand, we adapt it to specific problems of SAR images.

Section 4.2 is devoted to the characterization of speckle-dedicated filters. We first explain how the speckle corruption motivated the development of new edge detectors. We then address the spatial accuracy of these detectors, focusing on an important property; these detectors give a biased location in some situations.

In Section 4.3, we present the method of statistical active contour, which aims at segmenting a unique object in the scene. The improvement in spatial accuracy brought by this method is evaluated.

Multicomponent images are often available in SAR (i.e., multispectral, polarimetric, multidate images). We therefore propose in Section 4.4 an extension of statistical active contour to multicomponent images. Two different approaches are discussed and compared.

To deal with the case of multiregion segmentation, a generalization of the statistical active contour is discussed in Section 4.5. Here we consider a polygonal deformable partition called an active grid. This method is used as a postprocessing step and allows a fast refinement of a presegmentation.

4.2 CHARACTERIZATION OF EDGE DETECTORS IN SAR

4.2.1 Edge Detection in Speckled Image

Edge detectors aim at segmenting the image by finding out the transitions between homogeneous regions, rather than directly identifying them. They compute an *edge strength map* of the scene, in which the pixel intensity represents the "likelihood" of the presence of an edge at this position.

This is achieved by scanning the image with an analyzing window split in two half-windows and evaluating for each position the similarity between the pixels within the two half-windows. Hence, the edge detector response implicitly requires two choices: the attribute on which homogeneity is defined (intensity, texture, ...) and the measure of variability between two homogeneous regions.

Differential Edge Detectors

Classical edge detectors are based on the *difference* of average intensity computed in each half-window; for that reason, they are said to be *differential*. The most basic one, comparable to the Prewitt filter [18], has the following response:

$$d = |m_2 - m_1| \tag{1}$$

where m_1 and m_2 are the average intensities computed in the two half-windows of the analyzing window, (i.e., a simple rectangle symetrically split in two (Fig. 2b). More complex edge detectors like those of Shen and Castan [19], Canny and Deriche [20,21], or Paillou [22] also belong to this family. They just differ by the shapes of their analyzing window, which are determined to optimize given heuristical criteria.

It has been shown that differential edge detectors are not adapted to speckled images. This is due to the fact that speckle can be modeled as a random, exponentially distributed, multiplicative noise (see Refs. 23 and 24 for details on speckle models, in particular, the *fully developed* model), whereas these detectors are designed for additive Gaussian noise. As a result, their false-alarm rate is nonconstant but depends on the mean intensity of the region [4]. In other words, differential detectors detect more spurious edges in brighter zones than in darker ones (Fig. 1b).

A simple way of overcoming this problem is to apply the logarithmic transformation $I \to \log I$ to the original image. In particular, one can define a detector whose response is

$$d_{\log} = |\tilde{m}_2 - \tilde{m}_1| \tag{2}$$

where $\tilde{m}_1$ and $\tilde{m}_2$ represent the average intensities in the half-windows after the logarithmic transformation. Because of this processing, the speckle becomes additive and differential edge detectors have a constant false-alarm rate (CFAR) (see Fig. 1c).

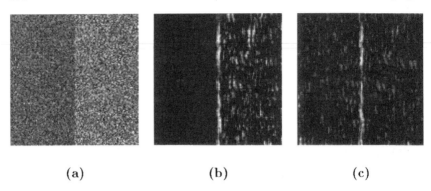

(a)   (b)   (c)

Figure 1 (a) Speckled image with an edge; (b) edge detection with d. The false-alarm rate depends on the mean intensity. (c) Edge detection with $d_{\log}$. The false alarm rate is constant.

Ratio Edge Detectors

To design CFAR edge detectors, Bovik [25] and Touzi [6] have proposed in the late 1980s using the ratio, rather than the difference, of the average intensities:

$$r = \frac{m_2}{m_1} \qquad (3)$$

After normalization, the response of the corresponding edge detector is

$$r_0 = \max\left(r, \frac{1}{r}\right) \qquad (4)$$

As for differential detectors, it is possible to define ratio edge detectors whose window shape is more elaborate. Thus, Fjørtoft et al. have followed Shen and Castan [19,26] to determine a ratio detector, optimal in the least square sense, for a certain stochastic image model where there is more than one edge present [13]. The corresponding window is symetrical and is exponentially shaped.

It was not until 1995 that Oliver et al. brought a clear theoretical justification for the use of the ratio of averages in SAR edge detection [7,27]. In the framework of statistical decision theory, they have used an optimal method to cope with this problem: the hypothesis testing by likelihood ratio [28]. Let $\mathbf{I}_1$ and $\mathbf{I}_2$ be the two samples of pixel intensities within the two half-windows, $\mathbf{I}$ the union of these two samples, and m the average of **I**. For a given position of the window, the following hypotheses are considered:

- $\mathcal{H}_0$: The window covers an homogeneous region. The reflectivity is therefore the same in the two half-windows and is equal to μ.
- $\mathcal{H}_1$: The window is on an edge. The reflectivities in the two half-windows are therefore different and equal to μ_1 and μ_2.

The two corresponding decisions are then as follows:

- $\hat{\mathcal{H}}_0$: There is no edge.
- $\hat{\mathcal{H}}_1$: An edge is present.

The Neyman–Pearson strategy [29] allows one to define the optimal decision rule, in the sense that it gives the highest probability of detection (PD) for a fixed probability of a false alarm (PFA). This rule consists of comparing the likelihood ratio

$$\Lambda = \frac{p(\mathbf{I}|\mathcal{H}_1)}{p(\mathbf{I}|\mathcal{H}_0)} \tag{5}$$

to a threshold t whose value depends on the desired PFA. Decision $\hat{\mathcal{H}}_1$ is taken if Λ is greater than t and decision $\hat{\mathcal{H}}_0$ in the other case:

$$\Lambda \underset{\hat{\mathcal{H}}_0}{\overset{\hat{\mathcal{H}}_1}{\gtrless}} t \tag{6}$$

One has $p(\mathbf{I}|\mathcal{H}_0) = p(\mathbf{I}|\mu)$ and $p(\mathbf{I}|\mathcal{H}_1) = p(\mathbf{I}_1|\mu_1)p(\mathbf{I}_2|\mu_2)$ because the two half-windows are assumed independent. If the fully developed speckle model is assumed [23,24], $p(\mathbf{I}|\mu)$, $p(\mathbf{I}_1|\mu_1)$, and $p(\mathbf{I}_2|\mu_2)$ are decreasing exponentials[*] and one obtains

$$\log \Lambda = N \log \mu - N_1 \log \mu_1 - N_2 \log \mu_2 + N\frac{m}{\mu} - N_1\frac{m_1}{\mu_1} - N_2\frac{m_2}{\mu_2} \tag{7}$$

where N_1, N_2, and N are respectively the number of pixels in $\mathbf{I}_1$, $\mathbf{I}_2$, and $\mathbf{I}$. Note that this approach is optimal (i.e., highest PD for a fixed PFA) only if the intensities are actually exponentially distributed and if the reflectivities μ, μ_1, and μ_2 are known. In practice, these parameters are, most of the time, a priori unknown and are called *nuisance* parameters: Their values are of no interest but must be dealt with [Eq. (7)]. A simple way of coping with this problem is to replace μ_1, μ_2, and μ by their maximum likelihood (ML)

---

[*] The same result is obtained if $p(\mathbf{I}|\mu)$, $p(\mathbf{I}_1|\mu_1)$, and $p(\mathbf{I}_2|\mu_2)$ are assumed to be gamma-distributed with the same order L. This case corresponds to multilook images [15].

estimates, which are m_1, m_2, and m. Thus, one obtains the *generalized* likelihood ratio:

$$\log \hat{\Lambda} = N \log m - N_1 \log m_1 - N_2 \log m_2 \tag{8}$$

This detector will be denoted the Generalized Likelihood Ratio Test (GLRT). Noting that $m = (N_1 m_1 + N_2 m_2)/(N_1 + N_2)$, it is possible to rewrite Eq. (8) as a function of r:

$$\log \hat{\Lambda} = -N_1 \log r + N \log \left(\frac{N_1 r + N_2}{N} \right) \tag{9}$$

Generally, a symmetrical window ($N_1 = N_2$) is chosen because it leads to better detection performance [15]. In this case, the previous expression becomes

$$\log \hat{\Lambda} = N \log \left[\frac{1}{2} \left(\frac{1}{\sqrt{r}} + \sqrt{r} \right) \right] \tag{10}$$

Consequently, there exists a bijective relation between $\hat{\Lambda}$ and r_0: These detectors are equivalent. In the following, we will denote by *ratio detector* each edge detector whose expression is a bijective function of Eq. (10).

4.2.2 Accuracy of Ratio Edge Detectors

Many studies have been performed to characterize the performance of the ratio detectors in *detecting* edges [6,14,27,30]. Here, we propose studying the performance of edge *location*. The purpose is to assess the influence of the analyzing window geometry (shape, size, orientation) on the location performance.

Models and Criterion

The edge that we consider is a straight, steep transition between two uniform regions with constant reflectivities μ_1 and μ_2 (Fig. 2). We assume that the intensities in the two regions are uncorrelated and exponentially distributed with means μ_1 and μ_2. The contrast of the edge is then defined as the ratio $\rho = \mu_2/\mu_1$. Furthermore, the edge can be tilted by an angle α.

The analyzing window (Fig. 2) is a $L_x \times L_y$ rectangle, vertically split into two symetrical regions. We define the shape parameter $\eta = L_y/L_x$ and the size $N = L_x L_y$.

The detection performance is classically assessed by the Receiver Operating Curve (ROC), which represents the Probability of False Alarm (PFA) versus the Probability of Detection (PD). For the location, we will look at the estimation of the edge position along the x axis. Let $\hat{\Lambda}(x, y)$ be

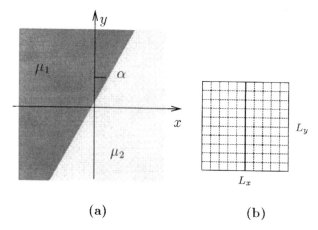

(a) **(b)**

Figure 2 (a) Edge model; (b) 10×10 analyzing window.

the response of the edge detector at the position (x, y). The edge position estimate is the one that maximizes the response along x:

$$x_e = \arg\max_x \hat{\Lambda}(x, 0) \tag{11}$$

The accuracy can therefore be assessed by the histogram of x_e computed over several realizations (1000 typically) of the speckled image. A sharp histogram centered on the true edge position is characteristic of a good location performance.

Detection Performance

It is clear that the location accuracy is partially related to the detection performance: a too-low contrast edge that it is hardly detectable will be all the more difficult to locate. It is therefore interesting to study some detection properties through the ROC.

Figure 3 illustrates the effects of the tilt between the window and the edge. When $\alpha = 0$, it is obvious that the detection performances do not depend on the window shape: Whatever η, when the window is centered on the boundary between the two regions, each region of the window contains homogeneous samples (i.e., all pixels belong to the sample probability density function (PDF). Things become different if the window orientation is no longer adapted to the edge. For high values of η (typically, $\eta > 1$), the detection efficiency drops rapidly when α is increasing. On the contrary, when η is lower ($\eta < 1$), the window is much more robust to orientation difference.

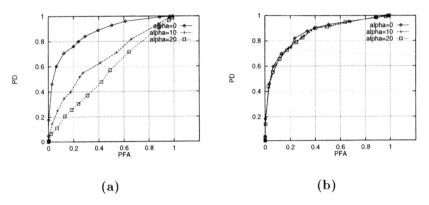

(a)   (b)

Figure 3   ROC for the detection of an edge of contrast $\rho = 1.6$ and variable tilt with two windows of the same size $N = 80$ but different shapes: (a) $\eta = 5$, (b) $\eta = 1/5$.

Location Performance

The shape of the window has a crucial influence in the edge location. Figure 4 shows the histograms of x_e, estimated with two differently shaped windows: a wide one (low η) and a narrow one (high η). When the orientation of the window is adapted to the edge ($\alpha = 0$), the narrow window gives a more accurate location of the edge than the wide one. When $\alpha \neq 0$, the location performance drops: The variance of the estimation increases and a bias

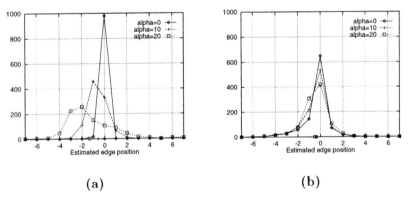

(a)   (b)

Figure 4   Histograms of the position estimate x_e of a α-tilted edge, with two windows of the same size $N = 80$ but different shapes: (a) $\eta = 5$, (b) $\eta = 1/5$. The contrast is $\rho = 4$.

appears. The increase of variance can be explained by the results of Fig. 3: As detection gets worse when α increases, the edge cannot be accurately located. Once again, the decrease of location performance is very pronounced for the narrow window, whereas the large one is more robust to the tilt of the edge.

To go further, we have computed histograms of x_e for different edges and windows. Only the average $\overline{x_e}$ and the standard deviation $\hat{\sigma}_{x_e}$ have been retained to characterize the location performance. These are represented in Fig. 5. On the diagram $(\overline{x_e}, \hat{\sigma}_{x_e})$, the closer to the origin a point is, the better the corresponding edge location. It is remarkable that the corruption of location when α increases depends on the contrast of the edge: We observe

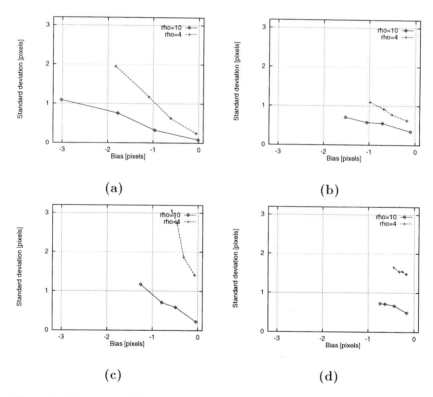

Figure 5 Variation of bias $\overline{x_e}$ and standard deviation $\hat{\sigma}_{x_e}$ in the estimation of the edge position x_e. Four windows are considered: (a) $N = 128$, $\eta = 2$; (b) $N = 128$, $\eta = 1/2$; (c) $N = 32$, $\eta = 2$; (d) $N = 32$, $\eta = 1/2$. One point in the diagram gives $\overline{x_e}$ and $\hat{\sigma}_{x_e}$ for an edge of contrast ρ and tilt α. For a given contrast, the points are linked and the values of α along the line are respectively from right to left: $0°$, $10°$, $20°$, $30°$.

a variance increase for low ρ and a bias increase for high ρ. The type of window has a strong influence on the performance. If narrow windows yield the best location when $\alpha = 0$ (Figs. 5b and 5d), wide windows are more robust to edge orientation (when $\alpha \neq 0$), whereas the corruption is striking for narrow windows (Figs 5a and 5c). In addition, the increase of variance with α is all the more important when N is small (Figs. 5c and 5d). On the contrary, the increase of bias with α is all the more important when N is large (Figs. 5a and 5b).

4.2.3 A Phenomenological Model for the Location Bias

In the previous subsection, we have observed that edge location with ratio edge detector is biased when the orientation of the window does not correspond to the edge [17]. Here, we show that the bias is also observed in some other situations and we give a general model to explain these results [16].

We first provide examples which reveal that the edge location with the ratio edge detector is biased in various situations. For this purpose, we synthetized speckled edge images and processed them with the GLRT edge detector, using a vertical 10×10 window (Fig. 6). Edges were extracted with a constrained watershed algorithm [31]. In the three first columns of Fig. 6, the speckle is monolook and uncorrelated. The intensity distributions of the two regions are exponential laws with different means. In the fourth column, the speckle is correlated and the image has been generated according to Ref. 32.

When the edge is an "ideal" step (Fig. 6, column 1), the GLRT edge detector yields a correct location. On the other hand, for different types of "nonideal" edges, we observe a bias toward the darker side. In the first example, the step edge was tilted so than the orientation of the window is no longer adapted (Fig. 6, column 2). This case was already observed in the previous subsection. In the second example, the edge is not a straight line but a sinuous frontier (Fig. 6, column 3). The third example (Fig. 6, column 4) shows the location bias when the speckle is correlated by the point spread function (PSF) of the acquisition system. One can note that the occurence of the bias depend on whether the edge is "ideal" (the edge is a step, vertically oriented) or not. We conjecture that when a "nonideal" edge is processed, the estimation of the means near the frontier is somehow degraded, yielding a false location of the edge. To explore this conjecture more precisely we propose a simple phenomenological model to interpret the results observed in Fig. 6 and give an approximated expression of the bias.

In a rigorous manner, the bias is defined as the expectation of the difference between the estimated and the real (say $x = 0$) edge positions:

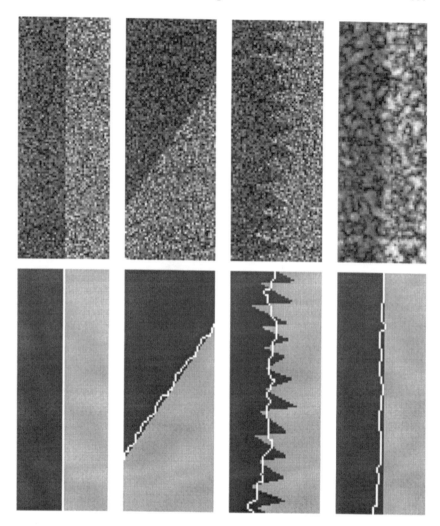

Figure 6 Edge location for different situations. In all cases, the contrast is $\rho = 4$. First row: Speckled edges; second row: edge location performed by GLRT filtering (10×10 vertical windows) followed by an edge extraction step (watershed algorithm). The result is shown on the speckle-free edge. First column: ideal, straight, step edge; second column: edge tilted by $35°$; third column: sinuous edge; fourth column: edge with correlated speckle (the point spread function is a circular kernel of diameter $d = 5$).

$$\langle x_e \rangle = \left\langle \underset{x}{\text{argmax }} \hat{\Lambda}(x, 0) \right\rangle \tag{12}$$

In our phenomenological model, we propose approximating the bias by applying the edge detector on the speckle-free image (i.e., an image made of two regions of constant intensities μ_1 and μ_2). Thus, one obtains a deterministic location δ, which is taken as an approximation of $\langle x_e \rangle$. This model amounts to applying the operator $\langle \cdot \rangle$ on the speckled image (to eliminate the speckle) instead of x_e. We also consider that the image is continuous i.e., the coordinates x and y take real values). Let $M_1(x, y)$ and $M_2(x, y)$ be the averages computed in the two half-windows on the speckle-free image at position (x, y). If $\rho > 1$, the phenomenological expression of the bias is

$$\delta = \underset{x}{\text{arg max}} \frac{M_2(x, 0)}{M_1(x, 0)} \tag{13}$$

When the window is in the neighborhood of the tilted edge, the half-windows contain a mixture of intensities μ_1 and μ_2 (Fig. 7). To stay as general as possible, we do not make any assumption on the spatial repartition of the intensity in the analyzing window. We simply note that the behavior of the plots $M_1(x, 0)$ and $M_2(x, 0)$ differ for ideal and nonideal edges (Figs. 7 and 8). In these figures, we have set $\mu_2 > \mu_1$ and we have introduced a phenomenological positive parameter a that gives the size of the "mixture zone," which is the zone near the edge where the two half-windows contain a mixture of intensities μ_1 and μ_2.

In the case of an "ideal" edge, $M_1(x, 0)$ is constant $[M_1(x, 0) = \mu_1]$ for any $x \leq 0$ and increases linearly with x (until it reaches μ_2). Similarly, $M_2(x, 0)$ is constant $[M_2(x, 0) = \mu_2]$ for any $x \geq 0$ and decreases linearly (until it reaches μ_1) when x decreases. Things are slightly different for a "nonideal" edge. $M_1(x, 0)$ remains constant $[M_1(x, 0) = \mu_1]$ as long as the subwindow $\mathcal{R}_1$ does not enter the zone containing a mixture of the two intensities μ_1 and μ_2 (i.e., for $x \leq -a$). Then, $M_1(x, 0)$ increases with x toward μ_2 but not in a linear manner. Similarly, $M_2(x, 0)$ remains constant $[M_2(x, 0) = \mu_2]$ as long as $x \geq a$ and nonlinearly decreases toward μ_1 when x decreases. We propose to model the behavior of $M_1(x, 0)$ and $M_2(x, 0)$ in the interval $x \in [-a, a]$. As mentioned earlier, this behavior is not necessarily linear and we therefore use quadratic expansions:

$$M_1(x, 0) \simeq \mu_1 + \beta(x + a)^2 H(x + a) \tag{14}$$

$$M_2(x, 0) \simeq \mu_2 - \beta(x - a)^2 H(-x + a) \tag{15}$$

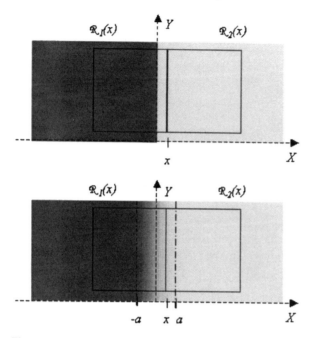

Figure 7 Top: "Ideal" edge; bottom: "nonideal" edge with a "mixture zone" of length 2*a*.

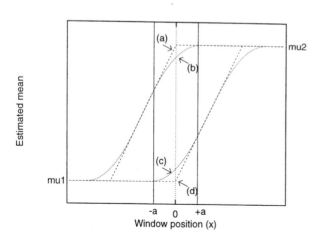

Figure 8 Variation of the estimated means as the window is shifted across the edge: (a) $M_2(x, 0)$ for an "ideal" edge; (b) $M_2(x, 0)$ for a "nonideal" edge; (c) $M_1(x, 0)$ for a "nonideal" edge; (d) $M_1(x, 0)$ for an "ideal" edge. The means are computed on the speckle-free image of Fig. 7.

where β is a positive constant and H is the Heaviside function. Combining Eqs. (13) and (15) and expanding the calculation to the second order, one obtains the following expression:

$$\delta = a \frac{1 - \rho}{1 + \rho} \quad \text{with} \quad \rho = \frac{\mu_2}{\mu_1} \tag{16}$$

which does not depend on β.

Although it is very simple, this model permits one to describe three main facts:

- The bias is toward the darker side (see the sign of δ).
- The absolute value of the bias increase with the edge contrast ρ [approximately as $|(1 - \rho)/(1 + \rho)|$].
- The asymptotic value of the bias (as $\rho \to \infty$) is given by a phenomenological parameter a which is related to the size of the mixture zone within the sliding window.

The relevance of this model also lies in its generality: It allows the description of several different cases for which biased location occurs. One only need to determine the size of the mixture zone. Thus, in the case of correlated speckle (Fig. 6, column 4), the mixture zone size is equal to the PSF width d: $2a = d$. If the edge is tilted by an angle α with respect to the window (Fig. 6, line 2), then it can easily be seen that the mixture zone has a size $2a = L_y \tan \alpha$, where L_y is the height of the window. For a sinuous edge (Fig. 6, column 3), a is related to the variance of the edge position.

We performed simulations to test the reliability of this model. For that purpose, we used edges corrupted by correlated speckle (like the one presented in Fig. 6, column 4). The PSF is a circular kernel of diameter d. For a given set (ρ, d), 1000 realizations of the edge image are generated to estimate the bias. This estimation is compared to the bias measured on the speckle-free image and to the value given by expression (16). The results are shown in Fig. 9. Please note that no scaling or free parameter have been adjusted to obtain these plots. There is rather good agreement between the estimated bias and the measure on the speckle-free image; this shows that the phenomenological model we considered is a good way of evaluating the bias. On the other hand, expression (16) does not fit very well with the estimation, especially for large d. This is due to the fact that this expression has been obtained with a second-order expansion. However, the general behavior of the curve remains well described.

It is worth noting that the bias is indeed inherent to the use of the ratio and that differential edge detectors are not biased. In Fig. 10, we have compared the location of a vertical edge with the ratio detector $\hat{\Lambda}$ and the CFAR differential detector $d_{\log}$. In both cases, the window has a size

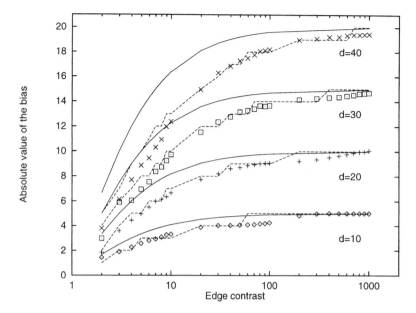

Figure 9 Bias of the edge position estimate as a function of the edge contrast ρ, for different sizes d (in pixels) of the PSF. For each d, three curves are plotted: bias estimated from 1000 realizations of the speckled image (points), bias measured on the speckle-free image (dashed curve), and approximated expression of the bias (solid curve). The sliding window is 128×128.

of 10×10 and is tilted by $30°$ with respect to the vertical. When the contrast is high, $d_{\log}$ leads to an unbiased location, unlike $\hat{\Lambda}$ (Fig. 10, top). However, $\hat{\Lambda}$ has better location accuracy than $d_{\log}$: When the contrast is lower, the variance of location is clearly higher for $d_{\log}$ than for $\hat{\Lambda}$ (Fig. 10, bottom).

4.2.4 Multiorientation Edge Detection

Up to now, we have addressed the location of an edge with a fixed orientation, analyzing the consequence of ignoring this orientation. In practice, the image is, of course, processed with several windows of different orientations which compute a response $\hat{\Lambda}$, assuming a given orientation of the edge.

Types of Windows

Different kinds of window are presented in Fig. 11. One can work with a fixed frame and let only the frontier rotate within the frame [6]. The draw-

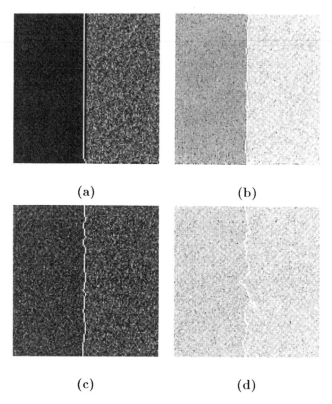

(a) (b)

(c) (d)

Figure 10 Comparison of edge detectors $\hat{\Lambda}$ (first column) and $d_{\log}$ (second column) for the location of a vertical edge with a 10×10 window tilted by $\alpha = 30°$. First row: high contrast ($\rho = 50$); second line: low contrast ($\rho = 2$).

back of this choice is that the window shape is not invariant. Another choice is to apply the rotation to the entire window (Fig. 11, second row). As shown in this figure, the discrete rotation may slightly modify the respective sizes of the half-windows (Fig. 11, third row).

Fusion of the Responses

Suppose that n_d windows with orientations $\alpha_k = [(k - 1)/n_d] \times 180°$ are considered. There are several ways of fusing their responses. For a position (x, y), the responses are arranged in a n_d-dimensional vector and the global response can be viewed as the norm of this vector:

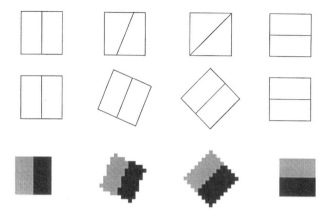

Figure 11 Directional windows for $\alpha = 0°$, $22.5°$, $45°$, and $90°$. First row: Rotation modifies the frontier and not the frame of the window; second row: rotation modifies the entire window; third row: discrete versions of the windows presented in the second row.

$$\left\| \hat{\Lambda} \right\|_{\gamma, n_d} = \left[\sum_{k=1}^{n_d} \left| \hat{\Lambda}(\alpha_k) \right|^{\gamma} \right]^{1/\gamma} \tag{17}$$

where γ is the order of the chosen norm. Three orders are often considered:

$$\left\| \hat{\Lambda} \right\|_{\infty, n_d} = \max_{k \in [1, n_d]} \left| \hat{\Lambda}(\alpha_k) \right| \tag{18}$$

$$\left\| \hat{\Lambda} \right\|_{1, n_d} = \sum_{k=1}^{n_d} \left| \hat{\Lambda}(\alpha_k) \right| \tag{19}$$

$$\left\| \hat{\Lambda} \right\|_{2, n_d} = \sqrt{\sum_{k=1}^{n_d} \left| \hat{\Lambda}(\alpha_k) \right|^{2}} \tag{20}$$

The norm of order 2 [Eq. (20)] is often used by analogy with the gradient modulus of the differential edge detectors [13]. The norms of orders ∞ [Eq. (18) and 1 [Eq. (19)] can be justified in the framework of decision theory. If the edge orientation is unknown, it becomes a nuisance parameter which prevent the evaluation of the numerator of Eq. (5). This one can therefore be written as $p(\mathbf{I}|\mathcal{H}_1, \alpha_k)$. Two solution are then possible:

- Replacing α_k with its ML estimate, that is the orientation which maximizes $p(\mathbf{I}|\mathcal{H}_1, \alpha_k)$. The numerator is then equal to $\max_{\alpha_k} p(\mathbf{I}|\mathcal{H}_1, \alpha_k)$.

- Adopting the marginal Bayesian approach [33], which consists of integrating the likelihood upon all of the possible values of orientations. In this case, the numerator is equal to $\sum_{\alpha_k} p(\mathbf{I}|\mathcal{H}_1, \alpha_k)$.

Note that norm ∞ is analogous to the solution when the nuisance parameter is estimated, and norm 1 is similar to the marginal Bayesian approach. One can note that, rigorously speaking, the Bayesian formalism cannot be applied to this problem. Indeed, for the kind of window we are using (Fig. 11, second row), the sample $\mathbf{I}$ depends on the nuisance parameter α_k. The Bayesian formalism would be rigorously valid only if one used the windows of the first row of Fig. 11.

Experimental Study

In the following, we show an experimental comparison to answer the following two questions:

- Which norm performs better?
- How many orientations should one consider?

The image in Fig. 12 was segmented with $\|\hat{\Lambda}\|_{\gamma, n_d}$, followed by watershed edge extraction. The size of the windows is 18×18 and the orientations are $\alpha_k = [(k - 1)/n_d] \times 180°$. Three values of n_d have been tested: $n_d = 2, 4, 8$. The quality of segmentation is assessed by computing the number of misclassified pixels $\mathcal{N}$. For fixed γ and n_d, the histogram of $\mathcal{N}$ is computed over 1000 realizations of the image model.

Results are presented in Fig. 13. Clearly, whatever the number of orientations, the three norms yield similar results. Regarding the number of orientations, if the performance is increased when passing from $n_d = 2$ to $n_d = 4$; the improvement is less obvious when passing from $n_d = 4$ to $n_d = 8$. This shows that using a great number of different orientations is not necessarily a good way of improving multidirectional edge detection. Indeed, when n_d increases, both the PD and the PFA increase. The question is, Which increases faster?

4.3 CONTRIBUTION OF THE STATISTICAL ACTIVE CONTOUR

4.3.1 The Statistical Active Contour

Recently, a new segmentation technique based on active contours and developed in the framework of statistical estimation theory has been proposed [34–36]. For a given image model, this method has properties of optimality

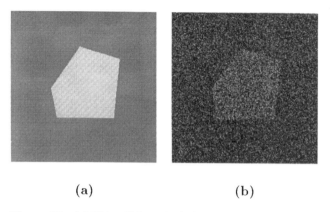

(a)   (b)

Figure 12   (a) 256×256 synthetic image with a polygonal object; (b) image with speckle. The contrast is $\rho = 1.7$.

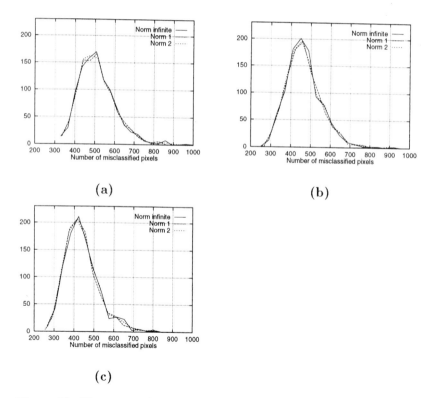

Figure 13   Histograms of number of misclassified pixels $\mathcal{N}$ obtained for the three norms $\gamma = \infty$, 1, 2 and for a variable number of orientations: (a) $n_d = 2$, (b) $n_d = 4$, (c) $n_d = 8$.

for the estimation of the frontier of an object. Here, we recall the principle of the Statistical Active Contour (SAC) for speckled images.

Image Model

We assume that the scene $\mathbf{s} = [s_i, i \in [1, N]]$ is only composed of two connex regions: the object and the background. The object region is identified by a function $\mathbf{w}$ such that

$$\begin{cases} w_i = 1 & \text{if the pixel } i \text{ belongs to the object} \\ w_i = 0 & \text{otherwise} \end{cases} \tag{21}$$

This function thus defines a partition of the image into two regions; namely the object $\Omega_a = \{i | w_i = 1\}$ and the background $\Omega_b = \{i | w_i = 0\}$, having respectively $N_a(\mathbf{w})$ and $N_b(\mathbf{w})$ pixels. We then assume that the pixel intensities in the object (respectively in the background) are realizations of an independent, homogeneous, spatially uncorrelated random field $\mathbf{a}$ (respectively $\mathbf{b}$). The intensity of the pixel i in the scene can therefore be written

$$s_i = a_i w_i + b_i(1 - w_i) \tag{22}$$

If one adopts the model of a fully developed speckle, the probability density function (PDF) of $\mathbf{a}$ and $\mathbf{b}$ are exponential laws* of means μ_a and μ_b (which are the reflectivities of the regions):

$$p\left(\frac{a_i}{\mu_a}\right) = \frac{1}{\mu_a} e^{-a_i/\mu_a} \tag{23}$$

$$p\left(\frac{b_i}{\mu_b}\right) = \frac{1}{\mu_b} e^{-b_i/\mu_b} \tag{24}$$

Statistical Estimation of the Shape

The goal of segmentation is to determine the shape $\mathbf{w}$ of the object in the scene $\mathbf{s}$. Adopting the latter image model, the problem can be transposed in the framework of statistical estimation. We have to estimate the parameter $\mathbf{w}$ knowing the data $\mathbf{s}$. The likelihood of $\mathbf{s}$ for a certain shape $\mathbf{w}$ can be written

$$P[\mathbf{s} | \mathbf{w}, \mu_a, \mu_b] = \prod_{i | w_i = 1} p\left(\frac{s_i}{\mu_a}\right) \prod_{i | w_i = 0} p\left(\frac{s_i}{\mu_b}\right) \tag{25}$$

---

* The same result is obtained if the PDFs of $\mathbf{a}$ and $\mathbf{b}$ are gamma laws with the same order L. This case corresponds to multilook images [35].

Note that in the latter equation, we do not know the nuisance parameters μ_a and μ_b. A simple way of handling this is to replace the parameters by their ML estimates, namely

$$\hat{\mu}_a = m_a(\mathbf{w}) = \frac{1}{N_a(\mathbf{w})} \sum_{i|w_i=1} s_i \tag{26}$$

$$\hat{\mu}_b = m_b(\mathbf{w}) = \frac{1}{N_b(\mathbf{w})} \sum_{i|w_i=0} s_i \tag{27}$$

After substituting Eqs. (26) and (27) into Eq. (25), one easily shows that maximizing the likelihood is equivalent to minimizing the following criterion:

$$J(\mathbf{w}, \mathbf{s}) = N_a(\mathbf{w}) \log m_a(\mathbf{w}) + N_b(\mathbf{w}) \log m_b(\mathbf{w}) \tag{28}$$

The shape $\mathbf{w}$ which minimizes this criterion then realizes the optimal segmentation in the ML sense.

Optimization Scheme

$\mathbf{w}$ is defined as the interior of a polygonal contour with N_n nodes. To perform the minimization of $J(\mathbf{w}, \mathbf{s})$, we start from an initial contour and use a simple iterative algorithm. For each iteration:

- Consider an elementary deformation by randomly moving a node of the polygon. More precisely, the node is randomly chosen and randomly moved by (δ_x, δ_y), where δ_x and δ_y are integers uniformly drawn in the set $[-A_{\max}, A_{\max}]$.
- Compute $J(\mathbf{w}, \mathbf{s})$ for the new $\mathbf{w}$.
- Accept the deformation if it has lowered J; cancel it otherwise.

This process is continued until $J(\mathbf{w}, \mathbf{s})$ does not decrease anymore.

To improve the convergence of the SAC, we have adopted a simple multiresolution scheme [37], which consists in progressively increasing the number of nodes of the contour. The idea is to apply successively the optimization algorithm with finer levels of resolution (i.e., larger N_n). In general, the initial contour has few nodes (typically four or five) and it is possible to obtain an approximate segmentation of the object, even if the initialization is far from the solution. This result is then used to initialize a second step in which N_n is increased to refine the segmentation. Generally, the multiresolution scheme includes three steps. To increase N_n, the level of resolution is specified at each step. This parameter is given by d_n, which represents the minimal distance between two successive nodes. Note that the previous nodes are not affected; we simply go along the contour and add a new node each time the distance to the last node is greater than d_n.

There can be an interest in using different amplitudes of deformation (parameter A_{max}) from one step to another. Generally, high amplitudes are used at low resolution (first step) and A_{max} is then decreased for steps at finer resolution.

Regularization of the Contour

If the number of nodes is large, the minimization of $J(\mathbf{w}, \mathbf{s})$ can lead to a very irregular contour. To avoid this, it is interesting to introduce regularization constraints to the contour. This approach is well justified in the framework of statistical estimation because the regularization constraints are viewed as an a priori knowledge of the parameter to be estimated. This knowledge is translated into a prior:

$$P[\mathbf{w}] = U_0 \exp\left(-\frac{1}{2\xi^2} U_{int}(\mathbf{w})\right) \tag{29}$$

with

$$U_{int}(\mathbf{w}) = \sum_{i=1}^{N_n} d^2(i) \tag{30}$$

where $d(i)$ is the distance between the node i and the center of the segment defined by its two neighbors (nodes $i - 1$ and $i + 1$). U_0 is a normalization constant guaranteeing the prior to be a PDF and ξ is a parameter. $\mathbf{w}$ can now be estimated in the Maximum a posteriori (MAP) sense by searching for the shape that maximizes $P[\mathbf{s}|\mathbf{w}]P[\mathbf{w}]$. This is equivalent to minimizing the following criterion:

$$J\lambda(\mathbf{w}, \mathbf{s}) = (1 - \lambda)J(\mathbf{w}, \mathbf{s}) + \lambda U_{int}(\mathbf{w}) \tag{31}$$

where λ is a parameter depending on ξ and can vary between 0 and 1. λ controls the smoothness degree of the contour by balancing the influences of the likelihood $J(\mathbf{w}, \mathbf{s})$ and of the prior $U_{int}(\mathbf{w})$.

Summary

To conclude, we recall the different parameters of the SAC method. Generally, the initial contour has few nodes (say four or five) and the convergence is made in three steps.

- Maximum amplitude of the deformation (A_{max}). We use a large value to permit a large deformation of the contour.

- Internode distance (d_n). This depends on the type of object to segment (complex or simple shape) and the desired resolution.
- Regularization parameter (λ). It is generally null for the two first steps and set to $\lambda = 0.1$ in the third one if N_n is large.

4.3.2 Segmentation Results

We illustrate the segmentation results on two genuine monolook, intensity SAR images (Figs. 14 and 15). The scenes correspond to agricultural regions and the goal is to segment a field appearing on a more or less homogeneous background.

The parameters of the SAC are the same for the two images. The convergence is made in three steps. $A_{max} = 7$ in the three steps; λ is null in the two first steps and is set to 0.1 in the third one. The internode distances are $d_n = 20$ after step 1 and $d_n = 10$ after step 2. It can be seen in Figs. 14 and 15 that the fields are correctly segmented with this protocol.

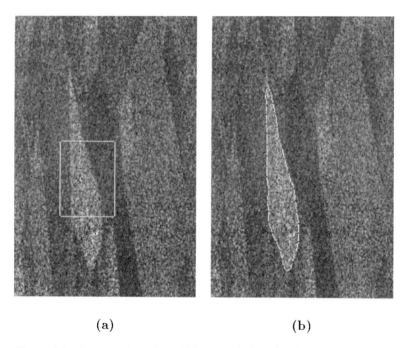

(a)   (b)

Figure 14 Segmentation of a field in an agricultural region near Bourges (France). The image size is 301×201. (a) Contour initialization ($N_n = 4$); (b) result. Computing time: 350 ms.

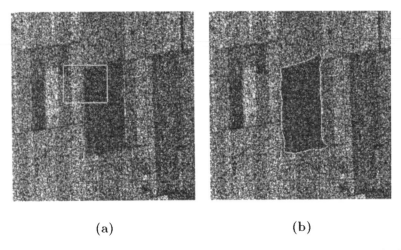

(a)   (b)

Figure 15 Segmentation of a field in an agricultural region of Ukraine. The image size is 250×250. (a) Contour initialization ($N_n = 4$); (b) result. Computing time: 280 ms.

4.3.3 Study of Segmentation Refinement

We have seen in Section 4.2 that the spatial accuracy of the GLRT edge detector is limited when the analyzing window is not adapted to the edge. Because of its deformable properties, the SAC is likely to fit the shape of the edge and, therefore, to accurately locate it. However, this method is supervised in the sense that the contour must be initialized. Thus, it is interesting to use both methods in cooperation. In the first step, the edges are roughly located with the GLRT $\hat{\Lambda}$ and the watershed algorithm. This result initializes the active contour, which, after convergence, refines and regularizes the segmentation. Here, we study the contribution of the SAC to segmentation refinement.

Protocol

The segmentation scheme (GLRT edge detection, watershed extraction, and refinement with SAC) is tested on a 128×128 synthetic image (Fig. 16). This image follows the model described in Section 4.3.1: It includes a single object and is corrupted with a fully developed speckle. The contrast of the object is $\rho = \mu_a/\mu_b$.

We perform edge detection with four differently oriented windows of size 8×8. The responses of the four detectors are fused by taking the norm ∞ [Eq. (18)]. A closed boundary can be extracted using the watershed

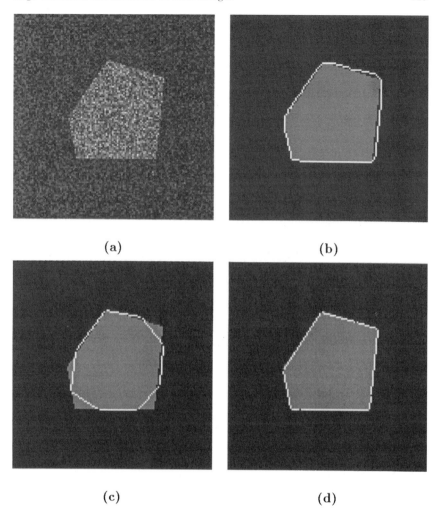

Figure 16 Segmentation of the synthetic image with a high contrast ($\rho = 10$). (a) Original speckled image; (b) GLRT edge detection performed on (a); (c) initial active contour obtained by sampling the contour in (b); (d) active contour after convergence.

algorithm constrained with markers [31]. Then, this boundary is vectorized; that is, we progress along the boundary and sample it with a given internode distance ($d_n = 25$) to define the initial active contour. As the initialization is close to the searched solution, only one step is used in the optimization. Because the contour has few nodes, no regularization is necessary ($\lambda = 0$).

To assess the accuracy of segmentation, we generate 1000 realizations of the image model and segment each of them. The accuracy is then evaluated by the following:

- The histogram of misclassified pixels after segmentation computed over the 1000 segmentations
- The image obtained by "stacking" the 1000 estimated contours

These data permit the comparison of the segmentation accuracy before and after the refinement of SAC.

Results

Figures 16 and 17 illustrate the different steps in the segmentation of the synthetic image. When the contrast is high (Fig. 16), the estimation of the object contour is biased after the first step and the SAC allows the correction of this bias. When the contrast is low (Fig. 17), the variance on the contour location is important and the SAC permits one to regularize the estimation.

These results are confirmed by the histograms of misclassified pixels and the stacked images. When $\rho = 10$, the segmentation does not really change from one realization to another. This is the reason why histograms are particularly sharp around their mean values (Fig. 18) and contour stacks are thin (Fig. 19). The segmentation refinement essentially consists in the bias correction. One can also observe that the corners of the object are located more accurately in Figs. 19. When $\rho = 2$ (see Figs. 20 and 21), there is more variability in the segmentations. As a consequence, the histograms are no sharper and the contour stacks are broader. However, the improvement of the SAC is still very clear. Indeed, the use of a polygonal contour intrinsically regularizes the segmentation and improves its accuracy.

4.4 STATISTICAL ACTIVE CONTOUR FOR MULTICHANNEL IMAGES

The availability of multichannel images can help to reduce the speckle effect and facilitates segmentation. Multichannel information can be obtained when the acquisition of the scene is performed with several sensors (polarimetric or multispectral image) or if the same scene is acquired at different dates (multidate image), which is often the case in SAR imagery. We therefore describe an extension of the SAC method to multichannel images and we compare the performance of two particular solutions.

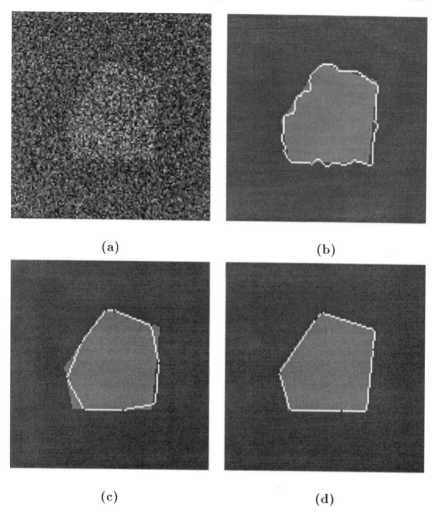

Figure 17 Segmentation of the synthetic image with a low contrast ($\rho = 2$). (a) Original speckled image; (b) GLRT edge detection performed on (a); (c) initial active contour obtained by sampling the contour in (b); (d) active contour after convergence.

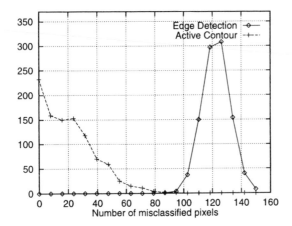

Figure 18 Histograms of the number of misclassified pixels after segmentation of the synthetic image by edge detection and after refinement with SAC. The contrast was $\rho = 10$.

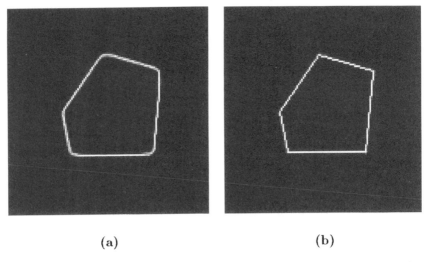

(a)   **(b)**

Figure 19 Stack of the 1000 estimated contours of the synthetic image by edge detection (a) and after refinement with SAC (b). The contrast was $\rho = 10$.

4.4.1 Extension of the SAC to multichannel images

Consider a set $\mathbf{S} = \{\mathbf{s}^{(\ell)} | \ell \in [1, P]\}$ of P images with N pixels. Thus, the intensity at pixel i for the channel ℓ is denoted $s_i^{(\ell)}$. We intend to segment a target appearing in each of the P channels which are assumed to be independent of each other. For each channel ℓ, the intensities of the target and the background are supposed to follow a white gamma distribution of mean $\mu_a^{(\ell)}$ and $\mu_b^{(\ell)}$, respectively. The orders of the gamma distributions are assumed to be the same, whatever the channel and the region (target or background). Considering this model, the shape $\mathbf{w}$ of the object is estimated in the ML sense. The likelihood is

$$
P[\mathbf{S}|\mathbf{w}, \boldsymbol{\mu}_a, \boldsymbol{\mu}_b, L] = \prod_{\ell=1}^{P} \left\{ \prod_{i|w_i=1} p\left(\frac{s_i^{(\ell)}}{\mu_a^{(\ell)}}, L\right) \prod_{i|w_i=0} p\left(\frac{s_i^{(\ell)}}{\mu_b^{(\ell)}}, L\right) \right\} \tag{32}
$$

where $\boldsymbol{\mu}_a = \{\mu_a^{(\ell)} | \ell \in [1, P]\}$ and $\boldsymbol{\mu}_b = \{\mu_b^{(\ell)} | \ell \in [1, P]\}$ are two vectors containing the reflectivities of the object and the background in each channel ℓ. To cope with the nuisance parameters $\boldsymbol{\mu}_a$ and $\boldsymbol{\mu}_b$, we now consider two different solutions derived from the maximum a posteriori estimation technique. This requires a prior $P(\boldsymbol{\mu}_a, \boldsymbol{\mu}_b)$ which represents an assumption made on the possible values of the nuisance parameters. We study two different priors that lead to different solutions.

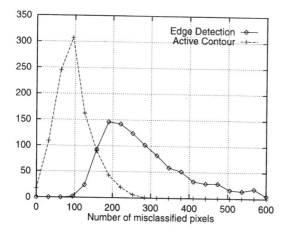

Figure 20 Histograms of the number of misclassified pixels after segmentation of the synthetic image by edge detection and after refinement with SAC. The contrast was $\rho = 2$.

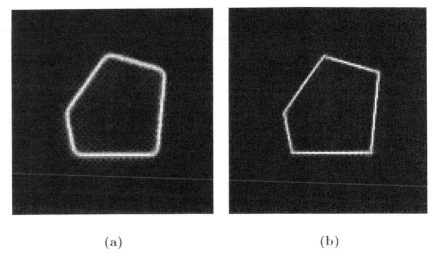

(a)   (b)

Figure 21   Stack of the 1000 estimated contours of the synthetic image by edge detection (a) and after refinement with SAC (b). The contrast was $\rho = 2$.

In the first solution, we make no prior assumption about the nuisance parameters so that $P(\boldsymbol{\mu}_a, \boldsymbol{\mu}_b)$ is constant. In that case, the nuisance parameters are estimated in the ML sense:

$$\hat{\mu}_a^{(\ell)} = m_a^{(\ell)}(\mathbf{w}) = \frac{1}{N_a(\mathbf{w})} \sum_{i|w_i=1} s_i^{(\ell)} \tag{33}$$

$$\hat{\mu}_b^{(\ell)} = m_b^{(\ell)}(\mathbf{w}) = \frac{1}{N_b(\mathbf{w})} \sum_{i|w_i=0} s_i^{(\ell)} \tag{34}$$

Maximizing the likelihood [Eq. (32)] is then equivalent to minimize the following criterion:

$$J^{\mathrm{vect}}(\mathbf{w}, \mathbf{S}) = N_a(\mathbf{w}) \sum_{\ell=1}^{P} \log\left[m_a^{(\ell)}(\mathbf{w}) \right] + N_b(\mathbf{w}) \sum_{\ell=1}^{P} \log\left[m_b^{(\ell)}(\mathbf{w}) \right] \tag{35}$$

This is equivalent to minimizing $\sum_{\ell=1}^{P} J(\mathbf{w}, \mathbf{s}^{(\ell)})$. In the following, it will be referred to as the *vectorial* algorithm.

The second solution consists in assuming that the nuisance parameters are the same in all channels. In other words, we introduce the following prior: $P(\boldsymbol{\mu}_a, \boldsymbol{\mu}_b) = \psi(\boldsymbol{\mu}_a)\psi(\boldsymbol{\mu}_b)$, where ψ is a function equal to 1 if all the components of its vectorial argument are identical, and equal to 0 otherwise. In practice, this assumption can be fulfilled in the case of a multidate image

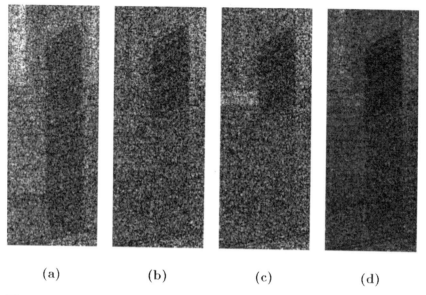

(a)   **(b)**   **(c)**   **(d)**

Figure 22 Three-date intensity monolook SAR image of an agricultural region: (a) date 1 (16/09/91); (b) date 2 (25/09/91); (c) date 3 (01/10/91); (d) sum of the three dates.

for which no major modification of the scene occurred between the acquisition of the different dates. Thus, the disposition of elementary scatterers may have varied enough so that the realization of speckle differ from one date to another but without modifying the mean reflectivity of the zones. The MAP estimation of the nuisance parameters leads to

$$\hat{\mu}_a^{(\ell)} = \frac{1}{PN_a(\mathbf{w})} \sum_{\ell=1}^{P} \sum_{i|w_i=1} s_i^{(\ell)}, \qquad \forall \ell \in [1, P] \tag{36}$$

$$\hat{\mu}_b^{(\ell)} = \frac{1}{PN_b(\mathbf{w})} \sum_{\ell=1}^{P} \sum_{i|w_i=0} s_i^{(\ell)}, \qquad \forall \ell \in [1, P] \tag{37}$$

In this case, one should minimize the criterion:

$$J^{\text{scal}}(\mathbf{w}, \mathbf{S}) = N_a(\mathbf{w}) \log\left[\sum_{\ell=1}^{P} m_a^{(\ell)}(\mathbf{w})\right] + N_b(\mathbf{w}) \log\left[\sum_{\ell=1}^{P} m_b^{(\ell)}(\mathbf{w})\right] \tag{38}$$

which is equivalent to minimizing $J(\mathbf{w}, \sum_{\ell=1}^{P} \mathbf{s}^{(\ell)})$. It consists in summing the P channels of $\mathbf{S}$ and applying the SAC on the unique resulting image. The solution will therefore be denoted the *scalar* algorithm.

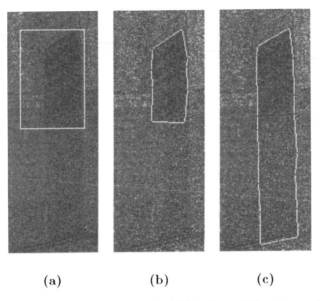

(a)   (b)   (c)

Figure 23   Segmentation of a field in the multidate image of Fig. 22: (a) contour
initialization; (b) result of the scalar algorithm; (c) result of the vectorial algorithm.
The results are shown on the sum of the three dates.

4.4.2 Comparison of the Two Solutions

We have seen that two particular solutions could be proposed to simply
extend the SAC method to multichannel images. For the *scalar* approach,
we have assumed that the reflectivities are the same for all of the channels,
whereas the *vectorial* approach is subject to no particular assumption. It is
thus interesting to compare the performance of these two solutions when the
reflectivities of the object and the background are the same and when they
are different.

A comparative study has therefore been carried out on synthetic
speckled images. It is detailed in the Appendix. The conclusions of this
study are the following:

- When the reflectivities are different, the vectorial algorithm always
  provides much better results than the scalar algorithm.
- When the reflectivities are the same, both algorithms have the same
  performance, even if the scalar algorithm is the optimal solution in
  that case.

This result highlights the good properties of the vectorial algorithm, which is
always at least as effective as the scalar algorithm.

4.4.3 Results on Real Data

The scalar and vectorial algorithms have been applied on a genuine mono-look, multidate SAR image (see Fig. 22). Three available dates are exploited to segment a field in an agricultural region. The field appears clearly in the first date with a weaker contrast than in the two others. The convergence is made in three steps. $d_n = 30$ after the first step and $d_n = 20$ after the second one. In the three steps, $\lambda = 0$ and $A_{\max} = 7$. The initialization is far from the object and does not lead to a correct segmentation with the scalar algorithm. The vectorial algorithm is therefore more efficient in that case because it yields a satisfying result with the same initialization as the scalar algorithm (see Fig. 23).

4.5 SEGMENTATION WITH STATISTICAL ACTIVE GRID

The SAC method presented in Section 4.3 permits the segmentation of a unique object in the scene. In this section, we address the more general problem of segmenting an image made of several different regions. We propose an extension of the SAC to a deformable partition with a fixed number of regions. This Statistical Active Grid (SAG) allows a semisupervised segmentation of the image.

4.5.1 The Statistical Active Grid

Segmentation Criterion

Let us consider that the scene[*] $\mathbf{s} = \{s(x, y)\}$ is a tessellation of R statistically independent and simply connected regions. The hypothesis of a fully developed speckle is used again, and in each region Ω_r ($r \in \{1, 2, \ldots, R\}$), we assume that the pixel intensities are realizations of independent and identically distributed random variables with an exponential PDF of parameter μ_r. Let $\mathbf{w} = \{w(x, y)\}$ be an R-valued function that denotes a partition of the image into R regions Ω_r, so that $w(x, y) = r$ if and only if (x, y) belongs to region Ω_r. We propose to estimate $\mathbf{w}$ in the maximum likelihood (ML) sense. For a given partition $\mathbf{w}$ of the scene, the log-likelihood can be written

$$\log P[\mathbf{s}|\mathbf{w}, \mu_1, \mu_2, \ldots, \mu_R] = \sum_{r=1}^{R} \sum_{(x,y) \in \Omega_r} \log P[s(x, y)|\mu_r] \tag{39}$$

---

[*] In this section, we need to differentiate the spatial coordinates of pixels and we will use $s(x, y)$ instead of s_i.

Note that we will not address the estimation of the number of regions R: This parameter will be assumed known a priori. To deal with the nuisance parameters μ_r, we estimate them in the ML sense:

$$\hat{\mu}_r = m_r(\mathbf{w}) = \frac{1}{N_r(\mathbf{w})} \sum_{(x,y) \in \Omega_r} s(x, y) \tag{40}$$

where $N_r(\mathbf{w})$ is the number of pixels in region Ω_r. After replacing the parameters μ_r by their ML estimates, the maximization of the likelihood is equivalent to minimizing the criterion

$$J(\mathbf{s}, \mathbf{w}) = \sum_{r=1}^{R} N_r(\mathbf{w}) \log m_r(\mathbf{w}) \tag{41}$$

Partition Model

The partition $\mathbf{w}$ is induced by a polygonal grid—that is to say, a set of nodes linked by segments to define the boundaries of polygonal regions. $\mathbf{w}$ is then entirely defined by the N_n nodes and the grid topology. However, there is still an ambiguity in properly defining the whole partition (i.e., to attribute each pixel to one and only one region). This ambiguity deals with the pixels located on the boundaries of the regions. These pixels can a priori belong to one or the other region and we describe in the following subsections the convention we have chosen to cope with this problem.

Optimization

The minimization of the criterion is done in the same manner as for the SAC (Section 4.3). The grid is simply deformed by moving the nodes and the minimization of Eq. (41) is then performed by accepting each move that has lowered the criterion. The multiresolution scheme is still used to improve the convergence: The number of nodes is progressively increased, according to a given internode distance d_n, in a three-step process.

Topology Representation

The active grid has N_n nodes and R polygonal regions. This grid is described by two structures:

- One structure contains the spatial coordinates of the N_n nodes. This structure changes during the convergence.
- The other one is relative to the grid topology (i.e., the relationship between nodes and regions). This structure remains invariant during the convergence.

The topology of the grid is represented by an oriented, valued graph (Fig. 24). To each node in the grid corresponds a vertex in the graph. When two nodes M and M' are linked by a segment, the valuation of the corresponding are $\overrightarrow{MM'}$ is the label of the region on the left side of the edge $\overrightarrow{MM'}$. This graph gives access to two types of information. Given a region, one can determine the list of nodes that define its polygonal boundary. Given a node, one can determine the different regions to which it belongs.

Projection of the Partition on the Discrete Lattice

As we pointed out previously, an ambiguity arises when partitioning the discrete lattice from a given state of the grid: The pixels located on the frontier between two regions might be assigned to both regions. To overcome it, we have to modify the convention that defines the interior of a discrete contour. In the SAC approach, the interior of a contour is composed of the pixels strictly inside the contour plus the pixels of the contour. This convention can no longer be considered for the active grid because, otherwise, the pixels on segments separating two regions would belong to both regions. We have adopted a new convention that is actually analogous to considering that frontiers between regions are located on the dual lattice of the pixels sites. The region Ω_r corresponding to a contour C_r is defined as the strict interior of the contour translated by $(1/2, 1/4)$ (see Fig. 25). With this convention, the different regions of the grid do not overlap. The fast algorithm of Ref. 36 can still be implemented with the main conventions recalled in Figs. 26 and 27.

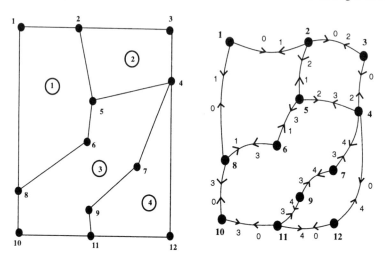

Figure 24 Representation of the grid topology with an oriented valued graph. Example of grid with 12 nodes and 4 regions (left) and its representation (right).

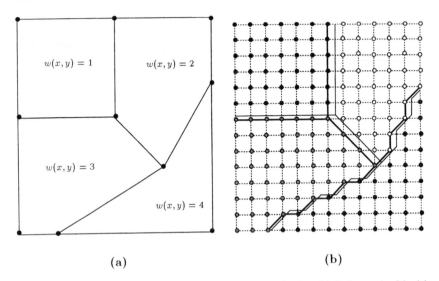

(a)   (b)

Figure 25   Projection of the partition on the discrete lattice: (a) Polygonal grid with 10 nodes and 4 regions; (b) corresponding partition of the discrete lattice. The pixels marking the boundaries between regions are linked together (thick lines). After translating these lines by (1/2, 1/4) (thin lines), the four regions are defined without ambiguity (pixels in black, white, dark gray, and bright gray).

4.5.2   Results

Segmentation Refinement

The SAG is a more supervised technique than the SAC. Indeed, in addition to the initialization problem, it requires the knowledge of the number of regions in the image. Here, we intend to use it in cooperation with the GLRT edge detector, as a postprocessing stage to refine edge location. A more quantitative study of segmentation accuracy improvement with SAG can be found in Ref. 38. The improvement of accuracy is here illustrated on two extracts of genuine SAR images (Figs. 28 and 29). Edge detection is performed with the four 10×10 windows of Fig. 11. The active grid is initialized with the result of edge detection and converges in three steps ($d_n = 30$ and $d_n = 20$ after the first and second steps). The active grid clearly improves the location of the field boundaries.

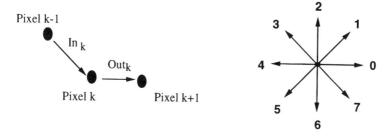

Figure 26 Notations used for the coding presented in Fig. 27. If one goes anticlockwise around the contour, In_k and Out_k are the vectors formed by a pixel k with its previous $(k - 1)$ and next $(k + 1)$ neighbors (left). These vector directions are represented with the Freeman code (right).

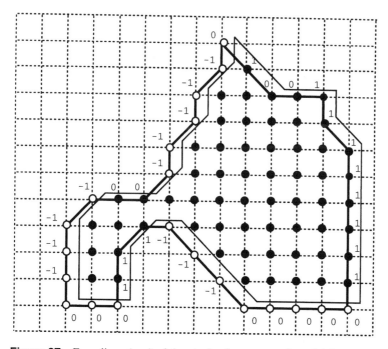

Figure 27 Encoding $c(x, y)$ of the pixels of a contour C_r useful for a fast algorithm implementation (see Ref. 36). Pixels belonging to the interior Ω_r are represented in black.

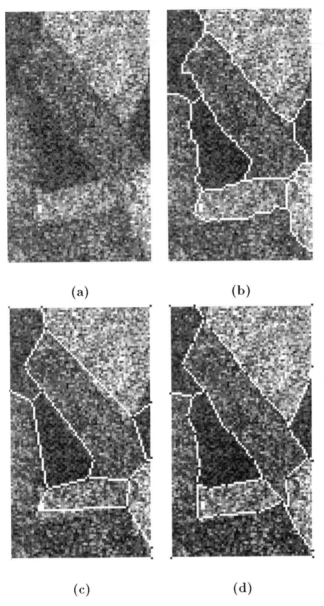

(a)   (b)

(c)   (d)

Figure 28 Segmentation of an agricultural region with eight fields. (a) original three-look SAR image (size: 70 × 119 pixels); (b) segmentation obtained by edge detection and watershed edge extraction; (c) initialization of the active grid obtained by sampling the presegmentation presented in (b); (d) active grid after convergence (computing time on Pentium III-700: 240 ms).

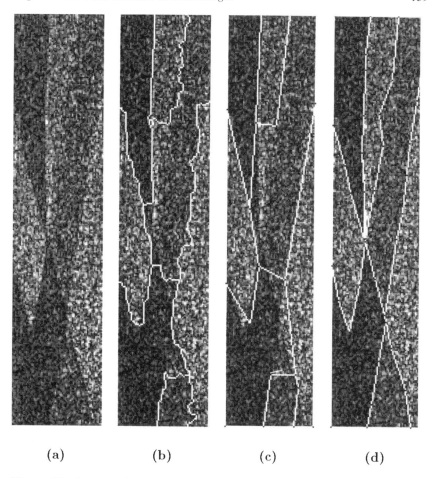

(a)   (b)   (c)   (d)

Figure 29 Segmentation of an agricultural region with seven fields: (a) original monolook SAR image (size: 65×288 pixels); (b) segmentation obtained by edge detection and watershed edge extraction; (c) initialization of the active grid obtained by sampling the presegmentation presented in (b); (d) active grid after convergence (computing time on Pentium III-700: 380 ms).

Computing Time

Finally, we evaluate the performance of the fast algorithm of Ref. 36 on some large images with a great number of regions. In this subsection, all of the presented computing times correspond to user time (not CPU time) and were measured on a PC under Linux (Mandrake 7.0) with a 700 MHZ Pentium III processor and a 256Mo RAM.

To characterize the computing time, we considered some "checkboard images," corrupted by speckle where the reflectivities of the square regions are alternatively 1 and 3 (Fig. 30). The grid was initialized as shown in Fig. 30. The convergence is done in two steps and $d_n = 32$ after the first step. This means that we always work with the same final resolution. Figure 30 illustrates an example of segmentation obtained for 512×512 image with 256 regions. Table 1 gives the corresponding computing time. These are relatively small, regarding the important sizes of the images as well as their number of regions.

4.6 SUMMARY

Edge detection in SAR images is a difficult problem due to the speckle corruption of the image. Classical differential edge detector do not perform well on these images and operators based on the ratio of averages have been introduced to improve edge detection. In this chapter, we have characterized the spatial accuracy of such edge detectors. We have shown that it was limited each time there is inadequacy between the edge and the analyzing window. In particular, these detectors yield a biased location of the edge in this condition. This bias is toward the darker side of the edge, increases with the contrast, and is bounded by a parameter related to the degree of inadequacy between edge and window.

We have then described a method to improve segmentation accuracy. The idea is to use edge detection and statistical active contours in cooperation. The active contour is used as a postprocessing step: It is initialized by the results of edge detection and, after convergence, permits the refinement of edge location. The technique has also been extended to efficiently cope with multichannel images. Finally, we have described a method that generalizes the concept of active contour to the case of a deformable partition called "active grid." Thanks to a fast algorithm (see Ref. 36), the SAG allows efficient and quick segmentation refinement in images with several regions.

APPENDIX: COMPARATIVE STUDY OF *VECTORIAL* AND *SCALAR* ALGORITHMS FOR MULTICHANNEL IMAGE SEGMENTATION

In this appendix, we compare the scalar and the vectorial algorithms in two different situations: when the reflectivities of the object and the background are the same and when they are not.

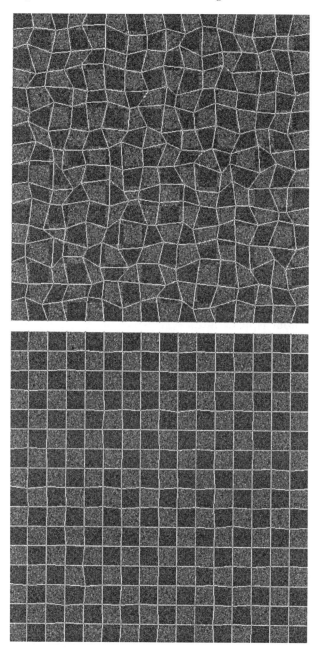

Figure 30 Segmentation of a 512 × 512 synthetic image with 256 regions: Top: initialization; bottom: result. Computing time on Pentium III-700: 1.7 s.

Table 1 Computing Time (Measured on Pentium III-700) of the SAG for Images of Variable Sizes and Number of Regions

Image size	No. of regions	No. of nodes	Computing time
1024 × 1024	256	290 and then 560	3.3 s
1024 × 1024	64	80 and then 366	1.6 s
512 × 512	256	290 and then 290	1.7 s
512 × 512	64	80 and then 148	1.0 s

Note: The convergence is made in two steps (the number of nodes increases in the second step) and the final internode distance is fixed to $d_n = 32$.

Protocol

The algorithms are applied on a two-channel synthetic image (Fig. 31) in which the speckle is fully developed. Their performance is assessed by computing the average and the standard deviation of the number of misclassified pixels after the segmentation of 1000 realizations of the image model. The convergence of the contour is made in three steps in which $A_{max} = 3$ and $\lambda = 0$. The internode distances are $d_n = 10$ and $d_n = 5$.

The two-channel image is characterized by four parameters, which are the mean intensities of the object and background in each channel, namely $\mu_a^{(1)}$, $\mu_b^{(1)}$, $\mu_a^{(2)}$, and $\mu_b^{(2)}$. Actually, it is possible to reduce the number of effective parameters. As the vectorial algorithm works simultaneously on the two channels, it is obvious that the two relevant parameters in this case are the ratios of the object and background reflectivities in each channel:

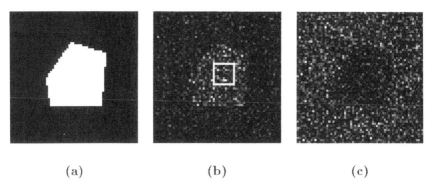

(a)   (b)   (c)

Figure 31 (a) 64 × 64 image. The size of the object is 737 pixels. (b, c) Realization of the image model for the parameters $\mu_a^{(1)} = 3$, $\mu_b^{(1)} = 1$, $\mu_a^{(2)} = 1$, and $\mu_b^{(2)} = 1/2$.

$$\rho^{(1)} = \frac{\mu_a^{(1)}}{\mu_b^{(1)}}, \qquad \rho^{(2)} = \frac{\mu_a^{(2)}}{\mu_b^{(2)}} \tag{42}$$

As a matter of fact, if the reflectivities $\mu_a^{(\ell)}$ and $\mu_b^{(\ell)}$ are multiplied by the same factor $\xi^{(\ell)}$, the criterion $J^{\mathrm{vect}}(\mathbf{w}, \mathbf{S})$ remains the same to within an additive constant. However, these two parameters do not suffice to describe the problem when the scalar algorithm is considered. Here, only one parameter can be eliminated. Indeed, if the four reflectivities $\mu_a^{(\ell)}$ and $\mu_b^{(\ell)}$ are multiplied by the same constant ξ, the criterion $J^{\mathrm{scal}}(\mathbf{w}, \mathbf{S})$ remains the same to within an additive constant. In addition to $\rho^{(1)}$ and $\rho^{(2)}$, we consider a third parameter, $\rho^{(s)}$, which is a measure of the contrast in the sum image. The three parameters necessary to entirely describe the problem are then

$$\rho^{(1)} = \frac{\mu_a^{(1)}}{\mu_b^{(1)}}, \qquad \rho^{(2)} = \frac{\mu_a^{(2)}}{\mu_b^{(2)}}, \qquad \rho^{(s)} = \frac{\mu_a^{(1)} + \mu_a^{(2)}}{\mu_b^{(1)} + \mu_b^{(2)}} \tag{43}$$

Note that, from these, it is possible to come back to the initial parameters if one of them is normalized:

$$\mu_b^{(2)} = 1, \qquad \mu_a^{(2)} = \rho^{(2)}, \qquad \mu_b^{(1)} = \frac{\rho^{(2)} - \rho^{(s)}}{\rho^{(s)} - \rho^{(1)}}, \qquad \mu_a^{(1)} = \frac{\rho^{(2)} - \rho^{(s)}}{\rho^{(s)} - \rho^{(1)}} \rho^{(1)} \tag{44}$$

First Case: Reflectivities Differ

The two algorithms are first compared in the case where the reflectivities in the channels differ. For a given set $(\rho^{(1)}, \rho^{(2)})$, the performance of the vectorial algorithm is fixed. This performance should then be compared with the scalar one, which depends on the third parameter $\rho^{(s)}$. For a given value of $(\rho^{(1)}, \rho^{(2)})$, it is easy to check that

$$\min\left(\rho^{(1)}, \rho^{(2)}\right) \leq \rho^{(s)} \leq \max\left(\rho^{(1)}, \rho^{(2)}\right) \tag{45}$$

We have therefore presented the results in the following way. The comparison is made for a fixed $(\rho^{(1)}, \rho^{(2)})$ and the performance of the scalar algorithm is evaluated as a function of $\rho^{(s)}$. This can be compared to the performance of the vectorial algorithm (which does not vary with $\rho^{(s)}$). Some representative values of $(\rho^{(1)}, \rho^{(2)})$ have been picked for the comparison (Fig. 32).

We consider points A and B in Fig. 32 which illustrate the situation where $\log \rho^{(1)}$ and $\log \rho^{(2)}$ have the same sign. Figure 33 shows that whatever $\rho^{(s)}$, the vectorial algorithm leads to better or equivalent performance than the scalar algorithm.

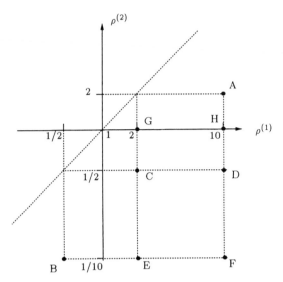

Figure 32   The contour initialization is shown in Fig. 31b. Representation of the set of parameters $(\rho^{(1)}, \rho^{(2)})$ for which the comparison was made. Each point A, B,..., H) represents a set $(\rho^{(1)}, \rho^{(2)})$.

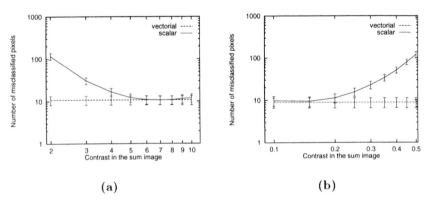

Figure 33   Comparison of the scalar and vectorial algorithms performance as a function of $\rho^{(s)}$, in the case where the two channels have logarithmic contrast of the same sign: (a) $\rho^{(1)} = 10$, $\rho^{(2)} = 2$ (point A); (b) $\rho^{(1)} = 1/2$, $\rho^{(2)} = 1/10$ (point B).

Then, we consider points C, D, E, and F in Fig. 32 ($\log \rho^{(1)}$ and $\log \rho^{(2)}$ have opposite signs). Here, again, the performance of the scalar algorithm is always worse than the vectorial one (Fig. 34). The worse value is of course obtained when $\rho^{(s)} = 1$.

Finally, points G and H in Fig. 32 are considered. In this case, one of the two channels does not contain any information. The vectorial algorithm still gives the best performance (Fig. 35).

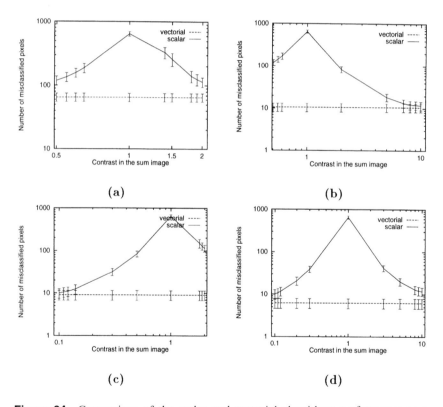

(a)   (b)

(c)   (d)

Figure 34 Comparison of the scalar and vectorial algorithms performance as a function of $\rho^{(s)}$, in the case where the two channels have logarithmic contrast of opposite sign: (a) $\rho^{(1)} = 2$, $\rho^{(2)} = 1/2$ (point C); (b) $\rho^{(1)} = 10$, $\rho^{(2)} = 1/2$ (point D); (c) $\rho^{(1)} = 2$, $\rho^{(2)} = 1/10$ (point E); (d) $\rho^{(1)} = 10$, $\rho^{(2)} = 1/10$ (point F).

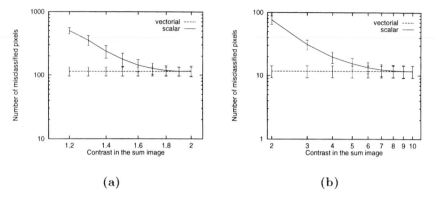

(a)   **(b)**

Figure 35   Comparison of the scalar and vectorial algorithms performance as a function of $\rho^{(s)}$, in the case where one of the two channels has no information (null logarithmic contrast): (a) $\rho^{(1)} = 2$, $\rho^{(2)} = 1$ (point G); (b) $\rho^{(1)} = 10$, $\rho^{(2)} = 1$ (point H).

Second Case: Reflectivities Are the Same

Let us switch to the case for which the scalar algorithm is optimal, (i.e., when $\mu_a^{(1)} = \mu_a^{(2)}$ and $\mu_b^{(1)} = \mu_b^{(2)}$). Here, the relevant parameter is $\rho = \mu_a^{(1)}/\mu_b^{(1)} = \mu_a^{(2)}/\mu_b^{(2)}$. The performance of both algorithms are therefore tested as a function of ρ (Fig. 36). It turns out that the performance of the vectorial algorithm is similar to the scalar one.

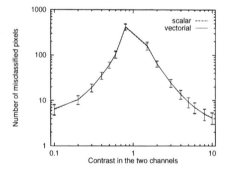

Figure 36   Comparison of the scalar and vectorial algorithms performance as a function of the contrast ρ, in the case where reflectivities are the same in the two channels.

ACKNOWLEDGMENTS

This work was supported by the French Space Agency (CNES) which supplied the SAR data. These data were acquired with satellite ERS-1 and are delivered by the European Space Agency (ESA) in the SLC format. The authors are grateful to all the members of the Physics and Image Processing Group for fruitful discussions.

REFERENCES

1. BC Barber. Theory of digital imaging from orbital synthetic aperture radar. Int J Remote Sensing 6:1009–1057, 1985.
2. CJ Oliver, S Quegan. Principles of SAR image formation. In: Understanding SAR Images. Norwood: Artech House, 1998, pp 11–42.
3. R Fjørtoft, F Séry, D Ducrot, A Lopès, C Lemaréchal, C Fortier, P Marthon, E Cubero-Castan. Segmentation, filtering and classification of SAR images. Proceedings VIII Latin American Symposium on Remote Sensing, 1997.
4. VS Frost, KS Shanmugan, JC Holtzman. Edge detection for SAR and other noisy images. Proceedings International Geoscience and Remote Sensing Symposium, 1982, Vol. FA2, pp 4.1–4.9.
5. AC Bovik, DC Munson. Optimal detection of object boundaries in uncorrelated speckle. Opt Eng 25(11):1246–1252, 1986.
6. R Touzi, A Lopès, P Bousquet. A statistical and geometrical edge detector for SAR images. IEEE Trans Geosci Remote Sensing 26(6):764–773, 1988.
7. CJ Oliver, I McConnell, D Blacknell, RG White. Optimum edge detection in SAR. In: SPIE Proceedings of the Conference on Satellite Remote Sensing, 1995, Vol. 2584, pp 152–163.
8. RG White. Change detection in SAR imagery. Int J Remote Sensing 12(2):339–360, 1991.
9. R Cook, I McConnell. Mum (merge using moment) segmentation for SAR images. In: SPIE Proceedings European Symposium on Remote Sensing, SAR Data Processing for Remote Sensing. SPIE, 1994, Vol. 2316, pp 92–103.
10. P Kelly, H Derin, K Hartt. Adaptative segmentation of speckled images using a hierarchical random field model. IEEE Trans Acoustics Speech Signal Process 36(10):1628–1641, 1988.
11. E Rignot, R Chellappa. Segmentation of synthetic-aperture-radar complex data. J Opt Soc Am A, 8(9):1499–1509, 1991.
12. R Cook, I McConnell, D Stewart, C Oliver. Segmentation and simulated annealing. In: SPIE Proceedings European Symposium on Remote Sensing, SAR Image Analysis, Modelling and Techniques. SPIE, 1996, Vol. 2955, pp 30–37.

13. R Fjørtoft, A Lopès, P Marthon, E Cubero-Castan. An optimum multiedge detector for SAR image segmentation. IEEE Trans Geosci Remote Sensing 36(3):793–802, 1998.

14. R Fjørtoft. Segmentation d'images radar par détection de contours. PhD thesis, Institut National Polytechnique de Toulouse (France), 1999.

15. O Germain, Ph Réfrégier. Edge detection and localisation in SAR images: A comparative study of global filtering and active contour approaches. In: SPIE Proceedings of the Conference on Image and Signal Processing for Remote Sensing. SPIE, 1998, Vol. 3500, pp 111–121.

16. O Germain, Ph Réfrégier. On the bias of the likelihood ratio edge detector for SAR images. IEEE Trans Geosci Remote Sensing 38(3):1455–1458, 2000.

17. O Germain, Ph Réfrégier. Edge location in SAR images: Performance of the likelihood ratio filter and accuracy improvement with an active contour approach. IEEE Trans Image Process 9(12), 72–78, January 2001.

18. WK Pratt. Digital Image Processing. New York: Wiley, 1978.

19. J Shen, S Castan. An optimal linear operator for edge detection. Proceedings IEEE Computer Society Conference on Computer Vision and Pattern Recognition (CVPR'86), 1986, pp 109–114.

20. JF Canny. A computational approach to edge detection. In: Readings in Computer Vision: Issues, Problems, Principles and Paradigms. San Matêo, CA: Morgan Kaufmann, Publishers, 1986, pp 184–203.

21. R Deriche. Using canny's criteria to derive a recursive implemented optimal edge detector. Int J Computer Vision 1(2):167–187, 1987.

22. Ph Paillou. Detecting step edges in noisy SAR images: a new linear operator. IEEE Trans Geosci Remote Sensing 35(1):191–196, 1997.

23. JW Goodman. Statistical properties of laser speckle patterns. In: Laser Speckle and Related Phenomena. Heidelberg: Springer-Verlag, 1975, pp 9–75.

24. CJ Oliver, S Quegan. Fundamental Properties of SAR Images. In: Understanding SAR Images. Norwood: Artech House, 1998, pp 75–122.

25. AC Bovik. On detecting edges in speckle imagery. IEEE Trans Acoustics Speech Signal Process 36(10):1618–1627, 1988.

26. J Shen, S Castan. An optimal linear operator for step edge detection. CVGIP, Graphics Models Image Process 54(2):112–133, 1992.

27. CJ Oliver, D Blacknell, RG White. Optimum edge detection in SAR. IEE Proc Radar Sonar Navigat 143(1):31–40, 1996.

28. Azzalini. Statistical Inference—Based on the Likelihood. New York: Chapman & Hall, 1996.

29. J Neyman, ES Pearson. The problem of the most efficient tests of statistical hypothesis. Phil Trans R Soc A231:289–333, 1933.

30. A Lopès, R Touzi, E Nezry, H Laur. Structure detection and statistical adaptative speckle filtering in SAR images. Int J Remote Sensing 14(9):1735–1758, 1993.

31. L Vincent, P Soille. Watersheds in digital spaces: An efficient algorithm based on immersion simulations. IEEE Trans Pattern Anal Machine Intell 13(6):583–598, 1991.

32. M Tur, KC Chin, JW Goodman. When is speckle noise multiplicative? Appl Opt 21(7):1157–1159, 1982.
33. CP Robert. The Bayesian Choice—A Decision-Theoretic Motivation. New York: Springer-Verlag, 1996.
34. O Germain, Ph Réfrégier. Optimal snake-based segmentation of a random luminance target on a spatially disjoint background. Opt Lett 21(22):1845–1847, 1996.
35. Ph Réfrégier, O Germain, T Gaidon. Optimal snake segmentation of target and background with independent gamma density probabilities, application to speckled and preprocessed images. Opt Commun 137:382–388, 1997.
36. C Chesnaud, Ph Réfrégier, V Boulet. Statistical region snake-based segmentation adapted to different physical noise models. IEEE Trans Pattern Anal Machine Intell 21(11):1145–1157, 1999.
37. C Chesnaud, V Pagé, Ph Réfrégier. Robustness improvement of the statistically independent region snake-based segmentation method. Opt Lett 23(7):488–490, 1998.
38. O Germain, Ph Réfrégier. Statistical multiregion snake-based segmentation. In SPIE's 45th Annual Meeting, Mathematical Modeling, Estimation and Imaging, 2000, Vol. 4121.

5

View-Based Recognition of Military Vehicles in Ladar Imagery Using CAD Model Matching

Sandor Z. Der
U.S. Army Research Laboratory, Adelphi, Maryland

Qinfen Zheng and Rama Chellappa
University of Maryland, College Park, Maryland

Brian Redman
Arete Associates, Tucson, Arizona

Hesham Mahmoud
Wireless Facilities, Inc., San Diego, California

5.1 INTRODUCTION

Laser radars (ladars) typically output both range and intensity images. The range images provide explicit three-dimensional (3D) information about objects. Intensity returns measure the reflectivity of objects. There are two kinds of ladar systems: direct-detection ladar and coherent ladar. A direct-detection ladars ends out a simple pulse for each pixel, receives the returning signal energy, and estimates the range to that pixel by detecting the pulse in the return signal. A coherent ladar system sends out a signal on a carrier frequency, just as a standard AM or FM radio does. An advantage of direct-detection over coherent systems is that direct detection gives improved range accuracy for a given set of lasers parameters, because direct-detection ladars are less susceptible to speckle and turbulence distortion of the signal. An advantage of coherent systems is that they can measure target motion using Doppler processing, whereas direct detection systems can measure motion only by sending repeated pulses and measuring the difference in range. Ladars are day/night sensors that operate well in a wide variety of weather conditions. They complement many traditional sensors (passive infrared, visible) and are useful in many applications. Figure 1 gives a simplified block diagram of ladar operation.

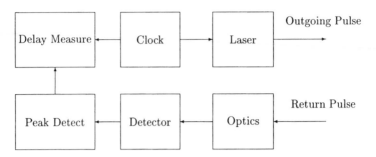

Figure 1 Ladar architecture.

Both coherent and direct-detection ladars typically return a range image, consisting of the estimated range to each pixel in the image, and an intensity image, consisting of the measured energy received for that pixel. The range image contains a number of sensor effects. Range error is simply the error in the measured time delay between the sending and receiving of the signal. When insufficient energy returns to the receiver (e.g., at long ranges) to distinguish the returning signal from background and detector noise, either a dropout or an anomaly will result. A dropout occurs when the maximum receiver energy fails to exceed a threshold, so the ladar reports no range estimate for that pixel. An anomaly occurs when a background or detector noise spike exceeds the dropout threshold and exceeds the genuine return signal, resulting in a wildly incorrect range value. A good introduction to laser sensing can be found in Ref. 1.

Ladars have been used extensively in manufacturing and outdoor environments. For indoor applications, low-power, eye-safe ladars can give accurate range measurements for quality control and robot positioning. For outdoor applications, ladars provides a convenient and accurate ranging capability, but they are limited by atmospheric attenuation and some obscurants, so that accuracy suffers as range increases. The algorithm described in this chapter is geared toward a direct-detection ladar with high range resolution but relatively low cross-range resolution (10 × 20 pixels in a target-sized region). The application is identification of military vehicles in outdoor images.

5.2 SYNTHETIC LADAR IMAGERY

The advantages and disadvantages of synthetic data are well known. Synthetic data requires an upfront investment of effort to develop the data generation algorithms, which may or may not be more than the effort required to obtain corresponding real data. Once the algorithms are generated, the effort to produce additional data is usually much smaller than for real data, so synthetic datasets tend to be larger and, thus, more densely sample the probability space of the data. Synthetic data do not require waiting for the right environmental conditions, so that appropriate data can be obtained at will. For this reason, it is often used to fill in gaps in a real dataset. The chief disadvantage is that the data are only as good as the model used to develop the simulation, so that the utility of the synthetic data varies greatly depending on application. Blind use of synthetic data can lead to optimistic estimates of system performance, or worse, to a system that has been tuned to perform well on synthetic data, but deteriorates on real data. Because the estimation of the tails of distributions of real data is difficult and the extreme values of random phenomena tend to be the ones that cause system failure, the estimation of failure rates using synthetic data is problematic.

Computer vision algorithm development efforts frequently suffer from a lack of adequate data for training and testing of algorithms. This is particularly true for algorithms that have a large number of free parameters, which require large datasets to ensure that the parameters are not overfitted to the data. Data acquisition is often extremely expensive, forcing algorithm designers to develop algorithms that are too simple for the problem or to overfit the data, or to use synthetic imagery.

Many computer vision researchers have used synthetic data to assist in the design and evaluation of their algorithms. Computer-generated synthetic data are becoming more common as the processing capability of computers increases. Jones and Bhanu [2] have used computer-generated synthetic aperture radar (SAR) imagery from the XPATCH [3] radar signature prediction code to identify features that are invariant to target articulation. These features are then used to identify targets in various articulations. Miller et al. [4] used computer-generated multisensor data to evaluate an algorithm for simultaneous detection, tracking, and recognition of objects.

Synthetic imagery generated from model boards has also been common. A number of authors [5–7] have used the model-board-derived TRIM2 [8] synthetic forward-looking infrared (FLIR) database. These algorithms are based primarily on the geometrical shape of targets and, therefore, are not highly sensitive to the inadequacies of synthetic data. Liu et al. [9] use both RADIUS [10] model board imagery and computer-generated synthetic imagery to evaluate algorithms which simultaneously locate buildings and estimate their size and shape parameters. They also address the issue of validation of computer recognition software using the Kolmogorov–Smirnov comparison of empirical and theoretical probability distributions.

5.2.1 Synthetic Image Generation Algorithm

Laser radars and other active sensors tend to be easier to characterize than most passive sensors, because of reduced sensitivity to environmental conditions. For laser radars in the near-infrared, the background noise is caused by stray solar radiation in the narrow passband of the optical receiver. This radiation tends to be small compared to the reflected laser energy and is easily characterized. Similarly, the effects of atmospheric scintillation, speckle, and receiver noise have been extensively studied, and appropriate models exist. The rest of this section describes the simulation for laser radar returns.

Component Disassembly

In order to simulate the effects of target shape, speckle, and scintillation, we decomposed the laser pulse into components along the range and the two cross-range dimensions. Thus, the pulse can be expressed as $U(x, y, z)$, where x and y are the horizontal and vertical cross-range dimensions and z is the range dimension (the direction that the pulse is traveling). The beam's irradiance profile is assumed to be Gaussian, and the pulse shape in the z dimension is an input to the model and can be written as $V(z)$ or, alternatively, as $V(ct)$ where c is the speed of light. Thus, the outgoing pulse is decomposed as

$$U(x, y, z) = P_s G(x, y) V(z) \tag{1}$$

where P_s is the total pulse power, x, y, and z take on discrete values, $G(x, y)$ is the proportion of energy contained within a component located at (x, y) under the two-dimensional Gaussian curve at location (x, y), and $V(z)$ is the discrete pulse shape in the range dimension shown in Fig. 2. The integral of $G(x, y)$ and $V(z)$ are both 1 because they are normalized to unity.

Each cross-range component (x, y) corresponds to the energy contained in a 5-cm × 5-cm square area (this size can be set arbitrarily, of course) at target range; thus, the number of cross-range components used by the simulation is dependent on the range and the beam divergence of the sensor. Table 1 shows the portion of energy in each cross-range component for a pulse that spreads to an area of 25 cm × 25 cm at target range. The distance that each component travels is determined by a geometric model, which is an input to the simulation, and has resolution of 5 cm × 5 cm, matching the cross-range decomposition of the pulse. At the target, each component is treated as if it encounters a resolved planar surface perpendicular to the line of sight to the sensor. However, a pulse may see nonperpendicular surfaces and unresolved surfaces as a set of components at different ranges. The pulses are not split by frequency because we assume

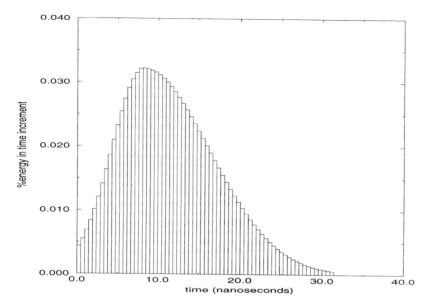

Figure 2   Simulated laser pulse shape.

a quasimonochromatic source. For a broadband source, such as a semicon-
ductor laser, accurate modeling of speckle would require either breaking up
the pulses by frequency or using a different distribution than the one we use
to model speckle and turbulence modulation.

The standard radar equation governs the return power received for
each component sent [11]. Therefore, he received component power is given
by

$$P_R = P_c e^{2\alpha R} \frac{d^2}{4R^2} \epsilon \rho \tag{2}$$

Table 1   Portion of Total Pulse Energy in Cross-Range
Components with a Gaussian Irradiance Profile $[G(x,y)]$

	$x = -0.1$	$x = -0.05$	$x = 0.0$	$x = 0.05$	$x = 0.1$
$y = 0.1$	0.003	0.014	0.022	0.014	0.003
$y = 0.05$	0.014	0.059	0.095	0.059	0.014
$y = 0.0$	0.022	0.095	0.154	0.095	0.022
$y = -0.05$	0.014	0.059	0.095	0.059	0.014
$y = -0.1$	0.003	0.014	0.022	0.014	0.003

where P_R is the received component power, P_c is the transmitted component power, α is the atmospheric extinction coefficient, R is the range to the target, d is the effective diameter of the receiver's clear aperture, ϵ is the receiver's optical efficiency, and ρ is the diffuse reflectivity of the resolved target. The resolved target assumption is applied only at the level of the component; the pulse itself may be unresolved.

Speckle and Turbulence Modulation

The power received from each component is modulated by speckle and turbulence. Speckle is applied to each component, using the exponential distribution to modulate the received power because we assume a quasimonochromatic source. For broadband sources, the method described by Parry [12] could be used to compute the probability density function (PDF). Thus, the power received from a component in the presence of speckle is

$$P_{\text{speckle}} = S(P_R) \tag{3}$$

where S is an exponentially distributed random variable with parameter P_R;

$$\text{Prob}[S(\lambda) = s] = \frac{e^{-s/\lambda}}{\lambda} \tag{4}$$

where λ is the mean of the exponential random variable [13]. Similarly, turbulence is applied by modulating the power of each component with a lognormal random variable. Some authors have argued that other distributions for turbulence are more appropriate, especially in high turbulence, where the K distribution matches the data well [14], although it tends to underestimate probabilities of high irradiances [15]. Experimental evidence suggests that in the presence of significant aperture averaging, the statistics of the irradiance are lognormal even in the high-fluctuation regime [16]. Many other PDFs have been suggested, but none have been shown to fit the data under all conditions [15]. The simulation has been written using modular code, making it easy to change the PDF if consensus is reached. The mean μ of the variable is the power prior to the application of the turbulence. The mean normalized variance σ_I^2 is determined by [17]

$$\sigma_I^2 = \gamma_{\text{target}} \sigma_{I\text{point}}^2 \tag{5}$$

where the term γ_{target} accounts for the target averaging of turbulence, $\sigma_{I\text{point}}^2$ is the intensity fluctuation for a point target without aperture averaging:

$$\sigma_{I\text{point}}^2 = 1.23 c_n^2 (k)^{7/6} (2R)^{11/6} \tag{6}$$

where R is the range from the target to the sensor, k is the wave number, and c_n^2 is the turbulence refractive index structure constant, which is dependent on atmospheric conditions. The value of c_n^2 used in the simulation is a user-controlled input parameter. Typical values are obtained from Ref. 11. The target-averaging term γ_{target} is determined using [17]

$$\gamma_{target} = \left(\frac{\rho_l}{r_{eff}}\right)^{7/3} \tag{7}$$

where the value

$$r_{eff} = \min\left[r_{tgt}, r_{beam}(R), r_{fov}(r)\right] \tag{8}$$

is a measure of the effective averaging area and ρ_l is the long-term turbulence cell size. It is calculated using

$$\rho_l = \frac{\sqrt{R/k}}{\sqrt{1 + R/k\rho_0^2}} \tag{9}$$

where [17]

$$\rho_0 = \rho_0^{(p)}\left\{\left[\left(1 - \frac{R}{f_{xmt}}\right)^2 + \left(\frac{R}{z_B}\right)^2\left(1 + \frac{\delta^2}{3}\right)\frac{1}{1 + \delta^2}\right]\right.$$

$$\left.\left[1 - \frac{13}{3}\left(\frac{R}{f_{xmt}}\right)^2 + \frac{11}{3}\left(\frac{R}{f_{xmt}}\right)^2 + \frac{1}{3}\left(\frac{R}{z_B}\right)^2 2\frac{(1 + \delta^2/4)}{(1 + \delta^2)}\right]\right\} \tag{10}$$

$$\rho_0^{(s)} = \left(0.5k^2 c_n^2 R\right)^{-3/5} \tag{11}$$

$$\rho_0^{(p)} = \left(1.46k^2 c_n^2 R\right)^{-3/5} \tag{12}$$

$$z_B = \left(\frac{kr_{b0}}{2}\right)\left[\frac{1}{(\rho_0^{(s)})^2} + \frac{1}{4r_{b0}^2}\right]^{-1/2} \tag{13}$$

$$F_{xmt} = \frac{-r_{b0}}{\tan(\theta_{b1/2}) - \tan(\lambda/2\pi r_{b0})} \tag{14}$$

$$\delta = \frac{2r_{b0}}{\rho_0^{(p)}} \tag{15}$$

The above Rytov solution predicts that the variance of the intensity fluctuation increases indefinitely as the range or the structure construct c_n^2 increases. Empirical measurements show that the normalized variance of intensity fluctuations saturate at approximately 1, whereas the normalized irradiance variance is unbounded in the Rytov solution [18]. The empirical curve of this saturation is shown in Fig. 3. This curve is stored in the form of a look-up table in the simulation.

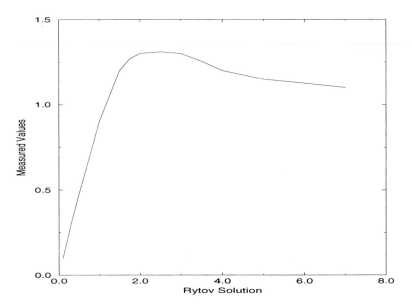

Figure 3   Saturation of intensity fluctuation.

The received power in a component after turbulence is then

$$P_{\text{turbulence}} = e^{G(a,b)} P_{\text{speckle}} \tag{16}$$

where G is a Gaussian random variable with mean a and variance b. The parameters a and b are determined from the mean and variance of the lognormal distribution using

$$b = \sqrt{\log\left(\frac{\sigma^2}{\mu e^2 + 1}\right)} \tag{17}$$

$$a = \log(\mu) - \frac{b^2}{2} \tag{18}$$

The speckle and turbulence modulation of the component intensities is not independent from component to component. The correlation length of the speckle and turbulence in the cross-range dimension are calculated, and components are grouped together so that the combined components are the size of a turbulence or speckle cell. The turbulence cell size is equivalent to the value ρ_l calculated earlier. The speckle cell size in the target plane is

$$l_{\text{speckle}} = \frac{\lambda R}{\sqrt{A_O}} \tag{19}$$

where A_O is the area of the receiver optics. The correlation time of speckle and turbulence for a single path is assumed to be greater than the pulse width, so that turbulence and speckle for a given pulse is a function only of cross-range dimensions (x, y), not of the range dimension z.

Once speckle and turbulence have been applied, the components are summed in the cross-range dimension to form the return signal at the detector. Each component is shifted in range corresponding to the range to the target for that component. The returning power at the detector then has the form

$$U_r(z) = \sum_{x,y} P_{\text{turbulence}}(x, y) V(z - 2T(x, y)) \tag{20}$$

where the cross-range dependence of $P_{\text{turbulence}}(x, y)$ is made explicit and $T(x, y)$ represents the range to the target at location (x, y).

Detector and Amplifier Noise

The return signal power is multiplied by ξ_{ann}, which is a function of the geometric radius of the laser spot image, turbulence blur circle, and diffraction-limited blur radius. The factor is calculated as [1]

$$\xi_{\text{ann}} = \frac{2\pi F_{\text{gc}}(R)}{\pi r_b^2(R)} \int_0^{r_b(R)} \xi_a(R, r) r \, dr \tag{21}$$

$$\xi_a(R, r) = \Lambda\left(r_D, \frac{r_{\text{aper}} f_{\text{rcvr}}}{R}, \frac{r f_{\text{rcvr}}}{R}\right) - \Lambda\left(r_D, \frac{r_{\text{obs}} f_{\text{rcvr}}}{R}, \frac{r f}{R}\right)$$

$$\left[\pi\left(\frac{r_{\text{aper}} f_{\text{rcvr}}}{R}\right)^2\right] \tag{22}$$

$$\Lambda(a, b, c) \doteq b^2 \Psi(a, b, c) + a^2 \Psi(b, a, c) - ac \sin(\Psi(b, a, c)) \tag{23}$$

$$\Psi(b, a, c) \doteq \arccos\left(\frac{c^2 + b^2 - a^2}{2bc}\right) \tag{24}$$

$$F_{\text{gc}}(R) = \min\left(\frac{A_d}{A_{\text{blur}}(R)}, 1\right) \tag{25}$$

where A_d is the area of the detector, r_{aper} is the radius of the receiver aperture, r_{obs} is the radius of the obscuration, and $F_{\text{gc}}(R)$ is the detector geometric compression factor, and

$$A_{\text{blur}}(R) = \pi r_{\text{blur}}(R)^2 \tag{26}$$

$$r_{\text{blur}}(R) = r_{\text{geom}}(R) + r_{\text{turb+diff}}(R) \tag{27}$$

The diffraction-limited blur radius and turbulence blur circle radius for the short exposure case can be taken as

$$r_{trb+diff} = r_{diff}\left[1 + \frac{r_{diff}^2}{r_{trb}}\right]^{-1/2} \tag{28}$$

$$r_{diff} = \frac{1.22\lambda f_{rcvr}Q}{d_{rcv}} \tag{29}$$

$$r_{trb} = \frac{2}{\pi\nu_{trb}} \tag{30}$$

where ν_{trb} is the turbulence cutoff spatial frequency given by the solution ν of the following equation [19]:

$$\left(\frac{1}{3.44}\right)^{3/5}\left(\frac{r_0}{\lambda f_{rcvr}}\right) = \nu\left[1 - \alpha\left(\frac{\lambda f_{rcvr}\nu}{d_{rcv}}\right)^{1/3}\right]^{3/5} \tag{31}$$

and

$$r_0 = 2.1\left(1.46k^2 c_n^2 R\right)^{-3/5} \tag{32}$$

is Fried's coherent aperture diameter due to turbulence. The geometric radius of the laser spot image is

$$r_{geom} = \frac{f_{rcvr}}{R}r_b(R) \tag{33}$$

where the laser spot radius, $r_b(R)$, is calculated as [20]

$$r_b(R) = \left[\frac{R^2}{k^2 r_{b0}^2} + r_{b0}^2\left(1 - \frac{R}{f_{xmt}}\right)^2 + \frac{4R^2}{k^2\rho_l^2}\right]^{1/2} \tag{34}$$

if $R > k\min(\rho_l^2, 4r_{b0}^2)$ or $\rho_l > 2r_{b0}$, and

$$r_b(R) = \left\{\frac{R^2}{k^2 r_{b0}^2} + r_{b0}^2\left(1 - \frac{R}{f_{xmt}}\right)^2 + \frac{4R^2}{k^2\rho_l^2}\left[1 - 0.62\left(\frac{\rho_l}{2r_{b0}}\right)^{1/3}\right]^{6/5}\right\}^{1/2} \tag{35}$$

$$f_{xmt} = \frac{-r_{b0}}{\tan(\theta_{b1/2}) - \tan(\lambda/2\pi r_{b0})} \tag{36}$$

otherwise. In the above equation, r_{b0} is the radius of the beam waist, f_{rcvr} is the focal length of the receiver, f_{xmt} is the effective focal length of the transmitter, $\theta_{b1/2}$ is the half-angle transmitted beam divergence, Q is the quality factor of the optics, d_{rcv} is the diameter of the effective clear aperture, λ is the wavelength of the laser, k is the wave number, and R is the range to the target.

The return signal is subject to background noise, shot noise, amplifier noise, and dark current noise, all of which are independent identically distributed (iid) Gaussian. The detector noise, consisting of dark, shot, and background noise, is calculated as [21]

$$\text{NEP}_{\text{Detector}} = \frac{\sqrt{2e_c B[I_{\text{ds}} + (_{\text{db}} + I_b + I_s)M^2 F]}}{\mathcal{R}} \tag{37}$$

where $\mathcal{R}$ is the responsivity of the detector, B is the electrical bandwidth, I_{ds} is the surface dark current, I_{db} is the bulk dark current at unity gain, $I_b = \mathcal{R}P_b$ is the current due to the background illumination, $I_s = \mathcal{R}P_{\text{turbulence}}$ is the current due to the received signal, M is the detector gain, and F is the excess noise factor due to the detector gain. P_B is the background power calculated from

$$P_B = \rho h_{\text{sun}} T_r A_r \sin\left(\frac{\theta_{\text{fov}}}{2}\right)^2 \Delta\lambda \tag{38}$$

where A_r is the area of the receiver, T_r is the transmission of the receiver, ρ is the background reflectance, θ_{fov} is the field of view, h_{sun} is the background solar irradiance, and $\Delta\lambda$ is the optical bandwidth.

Amplifier noise is calculated assuming a basic resistor-capicitor (RC) filtered amp. It follows that

$$\text{NEP}_{\text{Amp}} = \sqrt{\frac{4kTBN}{R_L \mathcal{R}^2}} \tag{39}$$

where k is Boltzmann's constant, T is the temperature in degrees Kelvin, B is the electrical bandwidth, N is the noise factor for the electronics, and R_L is the load resistor calculated from

$$R_L = \frac{1}{2\pi BC} \tag{40}$$

where C is the capacitance of the detector.

The noises are added in quadrature to determine the overall noise figure for the detectors

$$\text{NEP}_{\text{Total}} = \sqrt{\text{NEP}_{\text{Amp}}^2 + \text{NEP}_{\text{Detector}}^2} \tag{41}$$

The NEP is the standard deviation of the Gaussian distribution of the additive noise in the optical power domain. The simulation generates a Gaussian random variable with zero means and a standard deviation equal to NEP. This Gaussian noise is added to the signal and is then filtered according to the electrical bandwidth of the sensor. The noisy signal is then passed to the pulse detector.

The pulse detection portion of the simulation is modified according to the sensor being simulated. Commonly used pulse detection algorithms include matched filtering followed by peak detection, threshold detection of the rising edge of the pulse, or averaging the detection of the rising and falling edge of the pulse. Quantization error of the system clock is simulated by shifting the transmitted and received pulse by a random variable that is uniformly distributed across ± one-half of a clock cycle.

Figure 4 shows the effect of the noise on the simulated received pulse. Figure 5 shows a simulated ladar range image exhibiting the effects of the signal fluctuations and noise included in the simulation.

5.2.2 Verification and Validation of Synthetic Imagery

Verification is used to ensure that an algorithm code performs as the designer intended. It does not serve to test the assumptions upon which the algorithm was designed; hence, real data are not required to verify an algorithm. Validation is used to determine if a simulation accurately reflects the real world and, hence, requires real data. Validation can serve to both verify that a simulation accurately encodes the assumptions of the designer

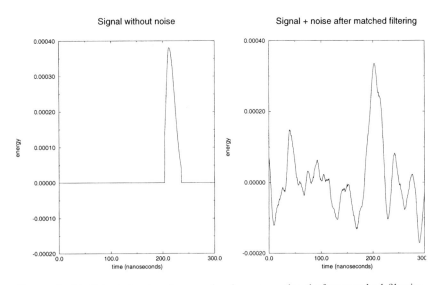

Figure 4   Undistorted pulse shape and noisy return signal after matched filtering.

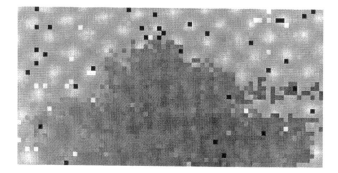

Figure 5   Input range image and resulting ladar image. (Black pixels indicate dropouts and white pixels indicate anomalies.)

and that the design assumptions are reasonable. A separate verification step serves to eliminate the possibility of error due to mistakes in coding, so that remaining error between simulated and real data can safely be assumed to be due to the inadequacies of the assumptions used in the simulation. This maximizes the use of the limited real data.

This section discusses verification and validation of the laser radar simulator. The simulator was used to help design a laser radar to be used with an automatic target recognizer (ATR) algorithm [22]. The simulator provided performance predictions of each potential laser radar design. The simulator was also used to create a set of synthetic images for the development and evaluation of a recognition algorithm. The performance of the algorithm on synthetic and real images is reported.

Verification

The purpose of verification is to ensure that the simulation code performs as the designer intended. Verification does not attempt to test the assumptions of the design. Hence, verification does not require real data. It does require that the simulation be run with a wide range of inputs and that the outputs be compared to ensure that the input–output relationships perform as expected.

It was especially necessary to do this in our case because part of the purpose of the simulation was to aid in the design of a laser radar. Parameter studies were performed by running the simulation on a number of proposed preliminary designs. The synthetic images were input to an ATR to estimate recognition performance, to aid analysis of performance/ cost trade-offs. Figure 6 shows the procedure. Figure 7 gives an example of probability of recognition versus range curves for a fictitious laser radar. The only difference in the parameters is that the first figure represents a laser radar with a pulse energy of 5 mJ, and the second a pulse energy of 8 mJ. It can be seen that performance is improved with the increased pulse energy, as expected. The curves give recognition probability for a two-class problem; thus, a probability of 0.5 corresponds to a completely random decision. Each point of the curve was generated by producing 216 synthetic images of each of the two targets, corresponding to 3 images of the target at each of the 72 aspect angles (sampled every 5°), and using the 432 images as input to an ATR, which compares each synthetic range image to noise free range images obtained directly from computer-aided design (CAD) models. The probability of recognition was calculated as the number of times the best match corresponded to the correct target, but most necessarily the correct aspect angle. The method should be interpreted as a predictor of which laser radar will produce superior ATR performance rather than an accurate esti- mator of ATR performance. Accurate estimation of performance would require a more complex background, which would allow the ATR to take target silhouette discontinuities into account, and would require more tar- gets. As the method is implemented, estimated performance is highly depen- dent on the choice of two targets to be discriminated.

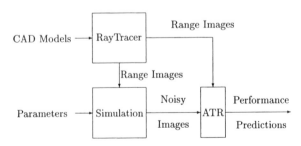

Figure 6 ATR performance predictions.

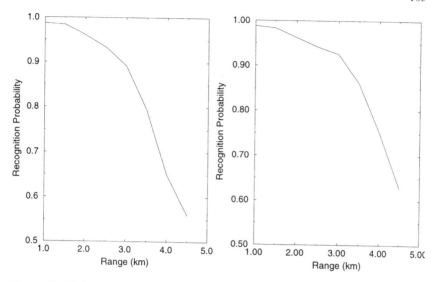

Figure 7   Estimated recognition performance with pulse energy of 5 and 8 mJ.

There were far too many simulation parameters to densely sample each parameter and run the simulation for every combination of parameter values. Instead, the parameter–performance relationship were verified systematically by varying each parameter in turn from a few canonical designs. Table 2 shows a partial list of parameters and the parameter values for a fictitious laser radar. The range of values of parameter could possibly take was determined by the designer, and a number of values in that range was used as inputs to the simulation. The resulting range error versus parameter curve would be compared to expected *qualitative* performance. For example, the dark noise current density can take on values from zero to infinity. For the LADAR of Table 2, the primary noise source is amplifier noise, so the expected behavior as the dark noise is increased is that the range error would be constant for small values of dark noise, increase steadily once the dark noise values were of the same order of magnitude as the amplifier noise, and then become unbounded when the dark noise drowns the signal. Figure 8 is a flow diagram of the procedure.

Knowledge of the physics behind the simulation suggests other parameter tests for verification. For example, increasing the optics size serves to increase the signal-to-noise ratio and also to reduce speckle effects by integrating over the speckle random field. These two effects can be isolated by simultaneously increasing the optics diameter and reducing the laser pulse energy so that the signal-to-noise ratio is kept constant. Thus, the depen-

Table 2 Partial List of Simulation Parameters

Wavelength	1.5 μm
Beam divergence	125 μrad
Pulse width	15 ns
Pulse energy	250 μJ
Optics diameter	0.19 m
Pointing σ	0.000042 rad
Optics quality factor	1.1
Receiver field of view	0.00019 rad
Dark noise current	5E-8 A
Electrical bandwidth	40 MHz
Optical bandwidth	12E-9 m
Receiver transmission	0.75
Noise factor	4.
Load resistor	50 Ω
Range to object	2.0 km
Atmospheric extinction	0.0002731 km^{-1}
Target reflectivity	0.38
Turbulence correlation constant	10E-13

dence of speckle distortion on the optics size can be isolated. The expected result is that the range error will decrease sharply when the optics is approximately the same size as the speckle correlation length, but then become constant once the aperture is much larger than the speckle correlation length. Figure 9 shows the resulting simulated performance. The correlation length of the speckle in this case is 0.015 m, which corresponds to the knee of the curve, as expected.

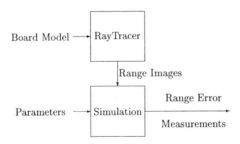

Figure 8 Range error predictions.

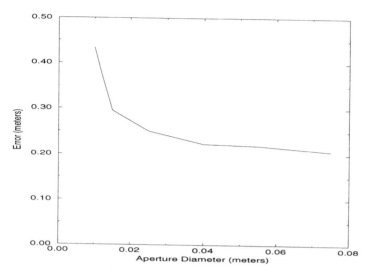

Figure 9 Simulated range error versus aperture size. (Laser power was varied to keep average return power constant. The speckle correlation cell diameter was 0.015 m.)

Validation

As a check on the quality of the simulation, we gathered a small amount of real data. We painted a 4 × 8-ft sheet of plywood carc green and positioned it perpendicular to the line of sight to the sensor. We captured laser radar images of the plywood and extracted an empirical probability density function from the images. Figure 10 shows simulated range error as a function of range. The two ×'s mark the range's standard deviation for the real laser radar. Additional simulation runs showed that the ladar is limited by quantization error of the clock at short ranges and by the signal-to-noise ratio at longer ranges. Many laser radar systems, including the Lockheed Martin Vought system used for this test, avoid the sharp increase in error at longer ranges by returning a no range value for return pulses that do not exceed several times the ambient root-mean-square (rms) noise level. Such a pixel is called a dropout. The percentage of dropout pixels would then increase as the range increases, whereas the range error on nondropout pixels would increase much more slowly. Figure 11 is a laser radar range image of two plywood boards in the bottom right corner and an M1 tank in the lower left. The range image is in ambiguous gray tone so that range differences are

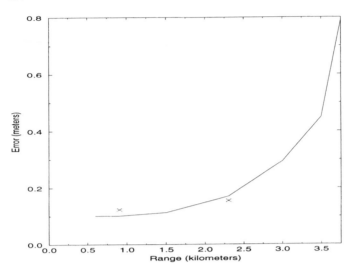

Figure 10   Simulated range error versus range. (The ×'s mark the real data points.)

visible but brighter pixels do not necessarily correspond to greater ranges. The dynamic range of laser radar range images is too great for human interpretation using a standard gray tone.

The plywood target data that we collected have characteristics that limit its use for validation purposes. The Lockheed Martin Vought laser radar has a relatively large aperture, so aperture averaging makes the effects of speckle and scintillation almost negligible. The detector noises are dominated by amplifier noise, so background and shot noise effects are not adequately tested. Although the comparison of real and simulated data was useful in validating the simulation, it is far from thorough. A thorough empirical validation would require laser systems having at least several different values of each sensor parameter, to ensure that each sensor parameter is accurately modeled. Of course, this would be extremely expensive.

The plywood reflection boards only served to validate the simulation for the simplest possible geometry, in which the entire beam falls on a planar object perpendicular to the line of sight of the sensor. It would be quite useful to validate the simulation on more complex geometries, which would test the simulation's ability to model the distortion of the pulse due to the different path lengths of the subpulses.

This experiment would require accurate measurement of the location of the beam footprint on the object and a detailed geometric description of

Figure 11 Ambiguous gray-tone range image of plywood boards at right and M1 at left.

the object. Suppose that the object is two planar boards, perpendicular to the line of the sight of the sensor, offset so that one board is 0.5 m closer than the other. The laser radar is positioned so that a portion of the beam falls on each of the boards, and the empirical PDF of the range returns is obtained by repeatedly measuring the range to the object. In order to calculate the estimated PDF using the simulation, the portion of the beam that falls on each board would need to be measured. One might be tempted to survey the location of the boards and of the laser radar and then to use a gimbal to measure the pointing angle of the laser radar. This is not practical because it would require a gimbal or theodylite with an accuracy an order of magnitude smaller than the beam width of the laser. It would also require that the laser in the laser radar be aligned to similar accuracy, which, generally, is not a good assumption.

A more realistic option would be to use an infrared imager to view the boards and see the area of laser illumination. The infrared imager would not need to be colocated with the laser radar. It could be placed close to the boards to get a better image. Background noise could be reduced by placing a notch filter in front of the optics that matches the wavelength of the laser. Background noise could be further reduced by using an imager with short sample times. Unfortunately, we were unable to perform these experiments due to time and equipment constraints.

Comparison of Simulated Data to Real Data

As a check on the quality of the simulation, we gathered a small amount of real data. We painted a 4 × 8-ft sheet of plywood carc green and positioned it perpendicular to the line of sight to the sensor and followed the procedures described under "Validation" on p. 167. Many laser radar systems, including the Lockheed Martin Vought system, avoid the sharp increase in error at longer ranges by returning no range value for return pulses that do not exceed several times the ambient rms noise level. Such a pixel is called a dropout. The percentage of dropout pixels would then increase as the range increases, whereas the range error on nondropout pixels would increase much more slowly. Table 3 gives a partial list of sensor parameters.

To test the speckle and scintillation portion of the simulation, it was necessary for us to reduce the aperture size until aperture averaging was insufficient to make speckle and scintillation effects negligible. The aperture size was reduced while the output power of the laser was increased so that the average return power remained a constant; thus, the primary cause of variation in performance is the reduction in aperture averaging that occurs at smaller aperture sizes. Figure 9 shows the resulting simulated performance.

Realistic Portrayals of Background

The simulation as currently configured is not suitable for designing and evaluating detection algorithms because of its simplistic handling of the background. Realistic backgrounds are challenging to laser radar detection algorithms because of the presence of target competitive shapes in the form of natural objects such as foliage and rocks. Tree lines, in particular, will frequently contain challenging shapes. Creating a realistic synthetic background requires accurate geometric modeling of terrain and foliage and a knowledge of the reflectance properties of natural objects.

A great deal of effort has gone into the creation of synthetic imagery in natural backgrounds (e.g., Refs. 24 and 25). Much of the work relates to synthetic infrared imagery and consists of geometric models of the terrain and natural objects, and associated thermal models applied to the geometric models. The laser radar simulation requires only a highly detailed geometric model, so the infrared work is readily applicable to

Table 3 Sensor Description

Laser type	NdYAG
Wavelength	1.574 μm
Beam divergence	125 μrad
Pulse width	10 ns
Pulse energy	250 μJ

laser radar. In particular, high detailed geometric tree models are available in Refs. 26 and 27. Detailed synthetic terrain geometry models are available in Ref. 28. Random tree placement models that locate trees in biologically plausible configurations exist in Refs. 24, 25, and 29. Computer-aided-design target models are already being used in our simulation. Combining these models requires mostly software integration work: converting the geometric models into one CAD format, applying the tree placement models and using the results to insert trees into the CAD model of the terrain, ray tracing the CAD model, and then applying the simulation to the resulting range image.

The complex geometry of the background model would make it much more difficult to validate using the method, described earlier, using known geometries. Validation of the background imagery could be accomplished in a number of alternative ways. The false-alarm rate of a detection algorithm on synthetic and real images could be compared. This would preferably be done with synthetic imagery that was designed specifically to match the geographic region where the real imagery was collected, as tree density and type and terrain roughness will clearly influence the false-alarm rate of a detection algorithm. An alternative validation scheme might be to match a texture model to tree lines, foliage canopies, and grass fields of both synthetic and real imagery and compare the associated texture parameters.

5.3 THE ATR ALGORITHM

The problem of object recognition in ladar range imagery has been an active research area for a number of years. The surface-fitting approach to object recognition has been used extensively. These methods segment the image into regions that match one of a set of predefined surface primitives. Some authors have used only planes as surface primitives [30–32]; others have used planar, cylindrical, and conic surfaces [33–35], and some have used more complex surfaces such as superquadrics [36]. One advantage to these surface-fitting methods is that reducing the image to a relatively small number of surface primitives greatly reduces the computational complexity associated with object recognition. Unfortunately, the method is sensitive to range error, especially for higher-order surfaces. As ladar range images are prone to noise at great ranges, object recognition by surface primitive extraction is useful primarily for indoor or short-range applications. Also, this approach is not practical when the cross-range resolution of the ladar is low, as in our application. Other authors [37,38] have used extracted edges which are then compared to a target model library. Another approach [39,40] has been to use detection theory, assuming a planar background and a known target shape, to calculate a likelihood ratio.

Koksal et al. [38] extracted from the ladar range image and then used a maximum a posteriori (MAP) estimate of the pose given the observed edges. By reducing the range image to edges and then using the expectation maximization algorithm to solve for the pose, the computational complexity of the matching algorithm is small. Edge matching algorithms tend to work best when the laser radar has good cross-range resolution so that a distinguishable silhouette is obtainable but does not require good range resolution. In the application for which we developed our algorithm, the range resolution is good, but the cross-range resolution is poor, requiring us to make maximum use of the available pixels.

Target recognition in pulsed ladar imager is challenging because of sensor noise, occlusion, background clutter, and, in our application, low spatial (cross-range) resolution. Range error induced by target surface speckle and atmospheric turbulence has led sensor designers to use pulse ladar systems that mitigate these sources of error, but ladar imagery for this scenario remains noisy and has relatively low resolution. The recognition task becomes more challenging when the number of candidate templates is large (because of many potential target types, position uncertainty, target orientation, and target articulation) and when the targets resemble each other.

The approach we take is to fit/verify the sensed range images to the range templates extracted from CAD target models. Each range template consists of the entire target, resulting in less sensitivity to range noise at the cost of increased computational complexity. In the following sections, we briefly discuss the laser simulator used for predicting target signatures. We then present details of the ladar ATR system. Next, experimental results on simulated ladar imagery as well as on real images are presented, followed by a conclusion and discussion of potential future improvements.

There are three major components of the ATR system: prescreening, surface fitting, and boundary fitting. Given a ladar image, several 1D projections of the range data are extracted and compared to corresponding projections previously extracted from target templates. Templates with significantly different projections are ruled out from further processing. Boundary checking and shape fitting are performed on each of the remaining templates. A coarse-to-fine matching scheme is used to generate the best match for each template. The target template with the highest combined matching score from shape and boundary fitting is declared to be the recognized target.

5.3.1 Prescreening

The idea behind our prescreening algorithm is to extract a set of reliable (e.g., shift invariant) features from the ladar image and compare them to the corresponding feature sets predicted from CAD target models. If the extracted features are similar to the predicted features from a template, then that template remains in the candidate list. The features we use are 1D projections of the ladar imagery along the vertical, horizontal, and range axes. Figure 12 shows projections of a target template (M1 at 45° and 0.2-m/ pixel resolution) on the vertical axis. Figure 13 shows average range curves for seven targets at 45° (projections are along the horizontal axis). As shown in Fig. 13, the 1D projections of different targets are quite different, providing good discrimination for prescreening. Any 3D translation of a target will result in a shift of the 1D curves that will not affect our matching scores. In addition, these 1D projections are not sensitive to small target rotations. Figure 14 shows average range curves for a target (M1) at 5° increments. The curves for the same target at similar aspect angles are very similar. Similar plots and conclusions can be obtained for projections along the depth axis (relative cumulative histogram).

We obtain the match score by correlating the projections from the data with the template projections, using a weighted correlation measure to improve the robustness of the matching technique. The similarity between average range curves is measured using

$$m_c = \frac{N_{f\cap g}}{N_{f\cup g}} \times \frac{\sum_i\left[\left(n_{fi} + n_{gi}\right)\left(\overline{f_i} - \mu_f\right)\left(\overline{g_i} - \mu_g\right)\right]}{\sigma_{\bar{f}}\,\sigma_{\bar{g}}\sum_i\left(n_{fi} + n_{gi}\right)} \qquad (42)$$

where g stands for a curve extracted from a target template and f stands for the corresponding curve extracted from the ladar image, $\overline{f_i}$ is the value of average range at the ith projection slot, μ_f is the mean of the average range for curve f, n_{fi} is the number of pixels for curve f at the ith projection slot, $\sigma_{\bar{f}}$ is the standard deviation of curve $\bar{f}$, $N_{f\cap g}$ is the total number of pixels common to curves f and g, and $N_{f\cup g}$ is the total number of pixels in curves f and g.

Because the size of the feature set is much smaller than the original ladar image, the prescreener can significantly improve the speed of the ATR system. In our implementation, the prescreening step filters out more than 80% of the candidate templates in the training set, without rejecting any of the targets in the training set.

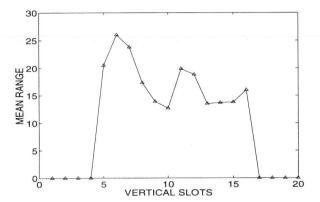

Figure 12   Examples of 1D projection used in prescreener (M1 at 45° azimuth, 1D projections onto the vertical axis).

5.3.2   Surface Fitting

Surface fitting is used to compare the 3D surface of the target templates to the ladar imagery. The PDF of the range return of a laser radar is a unimodal distribution centered on the correct range value with a flat tail that does not taper to zero as range error increases. The width of the peak of the distribution is determined by the width of the laser pulse and the distortion caused by the target geometry, whereas the height of the tail is determined by the probability that noise in the system will cause a spurious peak that is greater than the peak caused by the reflected laser pulse.

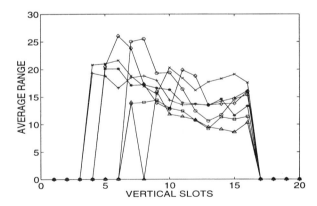

Figure 13   Plots of average range for seven targets at 45° azimuth.

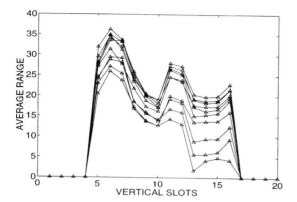

Figure 14 Plots of average range for Ml from $0°$ to $45°$ at $5°$ increments.

The expected width of the peak of the range error density function, determined by simulation or experiment, is used to set a threshold, τ_{srf} on range error. If the range difference between the template and data is smaller than the threshold, the pixel is counted as a fit. The ratio of the total number of the fitted pixels (M) and total number of valid ladar pixel over the target (N) is used to score the shape fitness. The M of N metric is a natural choice for ladar range imagery, because the large tails of the PDF causes standard metrics like mean square error to be dominated by outliers. We make a distinction between a nonfit pixel whose range is closer (the distance between sensor and returned energy peak is smaller) than the range predicted by the target template versus a nonfit pixel whose range is farther than predicted by the target template (the penetration case). Pixels that are closer may be due to partial occlusion, which is a relatively common occurrence, whereas pixels that are farther can only occur if the ladar has returned an anomaly (a range estimate that is based entirely on receiver and background noise, rather than a physical object reflecting energy), which is rarer. For this reason, we apply a penalty to the matching score for anomalous pixels.

The surface match score is computed as

$$M_j^{\mathrm{srf}} \doteq \sum_i I\big(|R(i) - T_j(i)| \le \tau_{\mathrm{srf}}\big) \tag{43}$$

where $I()$ is the indicator function that takes the value 1 if the inequality in the argument is true and 0 otherwise, and i is the index of the pixel values that fall on a particular target pose j. $R(i)$ is the input range image and $T_j(i)$

is the range template for the *j*th target pose, scaled to the size of the range image.

The penalty term for internal pixels that show a range farther than the template, thus indicating an anomaly, is given by

$$K_j \doteq \sum_i I\big(R(i) - T_j(i) \geq \tau_{\text{pen}}\big) \tag{44}$$

where τ_{pen} is a threshold for determining whether a range discrepancy between image and template is counted as a normal range error or an anomaly.

5.3.3 Boundary Fitting

A discontinuity in range values is likely to exist along the boundary of the target. Boundary fitting is used to determine if a consistent discontinuity exists for each hypothesized template. The bottom of the targets are excluded from the boundary fitting because there should be no range discontinuity between the ground and targets, and partial occlusion of the bottom of the target occurs frequently. The *M* of *N* metric is again used to measure boundary consistency, as

$$M_j^{\text{sil}} \doteq \sum_k I\left(\left|R\big(S_j^{\text{off}}(k)\big) - R\big(S_j^{\text{on}}(k)\big)\right| \geq \tau_{\text{sil}}\right) \tag{45}$$

where the boundary pixel pairs $S_j^{\text{off}}()$ and $S_j^{\text{on}}()$ are determined from the CAD model, and τ_{sil} is a parameter that is determined as described in the next subsection.

5.3.4 Match Evaluation

As mentioned earlier, incorrect small templates tend to have internal match scores but not be supported by a boundary consistency check. On the other hand, boundary fitting alone tends to have good scores on cluttered background areas. Correctly matched templates tend to achieve good scores on both shape and boundary fitting. The internal and boundary match scores are combined as

$$M_j \doteq \omega_{\text{srf}} \frac{M_j^{\text{srf}} - \omega_{\text{pen}} K_j}{N_j^{\text{srf}}} + (1 - \omega_{\text{srf}}) \frac{M_j^{\text{sil}}}{N_j^{\text{sil}}} \tag{46}$$

where $\omega_{\text{srf}} \in [0, 1]$ determines the weighting of the internal match versus the silhouette match.

The only free parameters of the algorithm are ω_{srf}, τ_{srf}, τ_{sil}, and ω_{pen}. The parameter τ_{srf} probably should not be considered a free parameter, because it should be set to approximately the distance between the center and knee of the PDF of the range error. The PDF of the range error can be roughly measured without any target data. The variance of this unimodal PDF will increase as the range increases, but this can be measured for a given sensor and recorded. Intuition can aid in setting several of the other parameters as well. For example, there is no point in setting τ_{sil} greater than the largest dimension of the largest target, because this will merely eliminate silhouette "hits" on genuine edges without eliminating any internal target discontinuities that could be confused with a silhouette edge. Likewise, τ_{sil} should be set higher than the standard deviation of the range error so that nonanomalous pixels illuminating a continuous surface are not counted as silhouette edges. The parameter τ_{pen} also needs to be set larger than τ_{srf} and also larger than the largest dimension of the largest target, so that a slightly misaligned but correct template is not eliminated because a pixel falls on a different portion of the target than the template would indicate. These settings can be summarized as

$$\tau_{srf} \approx \sigma_R \tag{47}$$

$$\sigma_R \leq \tau_{sil} \leq D_L \tag{48}$$

$$\max(\tau_{srf}, D_L) < \tau_{pen} \tag{49}$$

where σ_R is the standard deviation of the range error and D_L is the largest dimension of the largest target in the target set. For the parameters τ_{sil} and τ_{pen}, one might instead set them individually for each pixel of each target hypothesis by looking for maximum differences in template range values over a local area and replacing D_L in the above formulas.

The penalty $\omega_{penetrate}$ for pixels that give a range far greater than the hypothesized template (thus indicating an anomaly) should be set with knowledge of the probability of anomaly, which will typically be range dependent and rare on most sensors. The parameter ω_{srf} is more difficult, as the ideal value will be scenario dependent. If the targets are embedded in a tree line but unoccluded, the internal match is the most important. If the target is not near clutter and the range error is high, then the silhouette match becomes more important. Likewise, the ideal setting is dependent on sensor parameters. For higher cross-range resolutions, the boundary match is relatively more significant, whereas high range resolution makes the internal match more significant. One reasonable way to set ω_{srf} would be (1) to have a simple algorithm attempt to determine if the target is near clutter (is there a target-size blob isolated from its background?), (2) to look up the expected range error knowing the range to the hypothesized target, and then (3) to set ω_{srf} accordingly. As we did not have access to a large dataset with targets embedded in clutter, we could not test this notion and decided to set ω_{srf} by experiment on the synthetic training set. The value we currently use is 0.75.

5.3.5 Active Vision Approach to ATR

We investigate the situation where we have the opportunity to capture another ladar chip of the object after the initial recognition process. The operation of the ATR system is as follows: (1) The ladar sensor captures a ladar chip at low resolution that covers the whole object; (2) ATR is performed on the low-resolution ladar chip and a list of potential object types is reported along with their matching scores; (3) based on the top two competing object types, the second ladar chip is captured at twice the cross-range resolution but half the field of view; and (4) the ATR system analyzes the second ladar chip to confirm the recognition of the first ATR or select a different hypothesis. A valid question is how we can properly use this second look to confirm or improve the recognition results obtained using the first look. We investigate the optimum aim point for the second look.

We divide the template into eight overlapping regions, as shown in Fig. 15. These regions cover the whole template except for the bottom part, which is excluded from boundary-fitting evaluation in the ATR algorithm because it may be occluded by ground clutter. We measure the similarity between any two templates on each of the regions shown in Fig. 15. The region in which the top two competing templates differ the most is chosen as the region for the second view.

The similarity between two templates is measured based on surface shape and object silhouettes. For surface shape similarity, we first compute the range variance for each region. For a testing region i, we assume that there are R_{1i} valid object points in the first template and R_{2i} valid object points in the second template and that the range variances for the first and second templates are σ_{1i}^2 and σ_{2i}^2, respectively. The surface variance measure (SVM) for region i is defined as

$$\text{SVM}_i = \frac{\max(R_{1i}, R_{2i})}{\max[\min(R_{1i}, R_{2i}), 1]} |\sigma_{1i}^2 - \sigma_{2i}^2| \tag{50}$$

where the weighting factor before the variances difference is introduced to give more weight to regions with different numbers of valid template pixels. We ignore regions with insufficient numbers of template pixels in either of the competing templates. For silhouette similarity, we computed the boundary variance measure (BVM) defined as

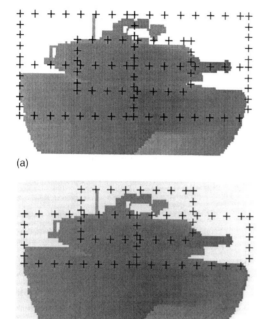

(a)

(b)

Figure 15 The eight overlapping regions illustrated on M1 template at a 30°
azimuth angle: (a) five overlapping regions; (b) the remaining three regions.

$$\text{BVM}_i = \Delta_{xi} + \Delta_{yi} \tag{51}$$

where i is the index for a region and

$$\Delta_{xi} = \frac{\max(L_{1i}, L_{2i})}{\max[\min(L_{1i}, L_{2i}), 1]} \left| \sigma^2_{1x_i} - \sigma^2_{2x_i} \right| \tag{52}$$

$$\Delta_{yi} = \frac{\max(L_{1i}L_{2i})}{\max[\min(L_{1i}, L_{2i}), 1]} \left| \sigma^2_{1y_i} - \sigma^2_{2y_i} \right| \tag{53}$$

L_{1i} and L_{2i} are the number of border pixels in region i for each template. $\sigma^2_{1x_i}$
and $\sigma^2_{2x_i}$ are the variances of border pixels along the horizontal coordinate.
Similarly, $\sigma^2_{1y_i}$, $\sigma^2_{2y_i}$ are the variances of the border pixels along the vertical
coordinate.

The combined difference measure (DM) for testing region i is defined
as

$$DM_i = \frac{SVM_i}{\max_{1 \le j \le 8}(SVM_j)} + \frac{\omega_3(BVM_i)}{\max_{1 \le j \le 8}(BVM_j)} \qquad (54)$$

where ω_3 is a weighting factor. In our current implementation, we use $\omega_3 = 0.5$. The region with the maximum DM is selected as the region for the second look.

5.4   ATR EVALUATION

The algorithm was trained on synthetic imagery and tested on real imagery. This section describes the experiments on simulated imagery, including testing using square and nonsquare pixels, the test using multilook ATR, and, finally, the test on real imagery. The ATR parameters were set using the square pile data described first. The other tests on simulated imagery were performed to determine if performance would differ for alternative hardware designs. The test on real data served as a form of validation that the algorithm would work in the real world and that the laser simulation could adequately predict good ATR parameter settings. For each of the experiments, the ATR-matched input ladar images with CAD models of seven targets, as shown in Table 4. The M1 and M60 are tanks, the M113 and Bmp2 are tracked armored personnel carriers, the Btr60 is a wheeled armored personnel carrier, the Zil131 is a tracked air-defense vehicle, and the M35 is a truck. Templates were generated for horizontal viewing angles and $5°$ increments of aspect angle, for each target, resulting in 72 templates per target for a total of 504 templates.

5.4.1   Test on Simulated Images

Simulated ladar images corresponding to all target types at a $0°$ elevation angle and $5°$ increments on aspect angles were generated and tested. The targets tested are listed in Table 4. Thus, there are 7 targets, each with 72 views, giving 504 images. Table 5 is the confusion matrix for the tests on simulated images. Both targets T6 and T7 are trucks. Figure 16 shows one of the errors reported in Table 5. The ladar image simulated from the template

Table 4   Target Lexicon

T1	T2	T3	T4	T5	T6	T7
M1	M113	Btr60	Bmp2	M60	Zil131	M35

of M35 at 225° (shown in Fig. 16) is wrongly recognized at Zil131 at 230° (whose template is shown in Fig. 16b).

5.4.2 Test on Multilook ladar ATR

We generated ladar chips of size 10×20 and 20×40 with the same footprint $2.5 \times 10 \, m^2$. A 10×20 ladar chip was first fed into our ATR system to generate the top 10 potential target/pose combinations. An optimum 10×20 subregion is then chopped from the corresponding high-resolution 20×40 ladar chip based on the identities of the top two competing candidates. The 10×20 high-resolution ladar chip is then fed into our system. For the second-look chip, we only match among the top 10 competing templates reported from the previous ATR. The matching scores from the first and the second looks are combined to generate the final recognition decision. Table 6 is the confusion matrix for the testing of simulated ladar images using our ATR algorithm with second-look capability. A comparison of Tables 5 and 6 shows that the ATR with the second look has improved performance.

Table 7 shows detailed scores for an ATR of a simulated ladar chip (M113A2 at 180°). The ATR on the first look is incorrect and the match to the correct template is ranked sixth. The ATR on the second look ranks the correct template at the top. The second highest match is given to a template of the same object at 5° azimuth rotation. Table 8 shows the detailed scores of another ATR example. The ladar chip tested is BMP2 at 0°. The ATR on the first look had a 5° azimuth angle error. The ATR on the second look moves the correct template to the top rank. In this example, the matches for the correct object at nearby azimuth angles remain high, whereas matches for incorrect objects drop below threshold.

(a)　　(b)

Figure 16 An example of misrecognition for a simulated image: (a) template of M35 at 225°; (b) template of Zil131 at 230°.

Table 5 Confusion Matrix of Classification on Synthesized
10×20 Ladar Imagery

	T1	T2	T3	T4	T5	T6	T7	Accuracy
T1	72	0	0	0	0	0	0	100%
T2	0	72	0	0	0	0	0	100%
T3	0	0	72	0	0	0	0	100%
T4	0	0	0	72	0	0	0	100%
T5	0	0	0	0	72	0	0	100%
T6	0	0	0	0	0	71	1	98.6%
T7	0	0	0	0	0	2	70	97.2%
Total								99.4%

5.4.3 Test on Real Images

The system has been blind tested on a dataset containing 276 real ladar images of M1A2 tanks and M113 armored personnel carries at various orientations. The inputs to the system were 10 × 20 ladar images cued by a human operator. The system matched each input image to stored templates of seven target types. The system achieved above 90% accuracy in recognition. Table 9 lists the confusion matrix for the test on real ladar images.

There are a number of reasons for the deterioration of performance between the synthetic and real data. The synthetic imagery was generated at the cardinal angles (every 5° in azimuth, starting at 0°), and the templates used in the recognition algorithm were also at the same pose. This should be

Table 6 Confusion Matrix for ATR with Second-Look Capability

	M1	M113	BTR60	BMP2	M60	ZIL131	M35	%
M1	72	0	0	0	0	0	0	100
M113A2	0	72	0	0	0	0	0	100
BTR60	0	0	72	0	0	0	0	100
BMP2	0	0	0	72	0	0	0	100
M60	0	0	0	0	72	0	0	100
ZIL131	0	0	0	0	0	72	0	100
M35	0	0	0	0	0	0	72	100
Total								100

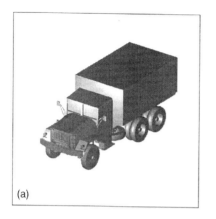

Figure 17 Illustration of ladar image simulation: (a) a rendering of M35 CAD model; (b) range template of M35 at $(315°, 0°)$ and 0.1 m/pixel; (c) synthesized ladar image at resolution of 0.4 m/pixel.

Table 7 An example of Second Look Correcting the First-Look ATR Error (Ground Truth is M113A2 at 180°)

Template	First-look score	First-look rank	Second-look score	Second-look rank
ZIL131-185	0.9764	1	0.7928	3
ZIL131-175	0.9744	2	0.5525	7
M113A2-175	0.9454	3	0.6165	5
M113A2-185	0.9153	4	0.7962	2
M35-175	0.8857	5	0.6760	4
M113A2-180	0.8580	6	**0.9681**	1
ZIL131-170	0.8111	7	0.4258	10
ZIL131-190	0.8103	8	0.4905	9
M35-185	0.7995	9	0.5045	8
M35-180	0.7979	10	0.6154	6

Table 8 An example of Second Look Confirming the First-Look
ATR Output (Ground Truth is BMP2 at $0°$

Template	First-look score	First-look rank	Second-look score	Second-look rank
BMP2-5	0.9770	1	0.9460	2
BMP2-0	0.8071	2	**0.9766**	1
BMP2-355	0.6931	3	0.7175	3
M1-180	0.5983	4	0	—
M1-175	0.5970	5	0	—
M1-190	0.5664	6	0	—
M1-185	0.5664	7	0	—
M1-165	0.5447	8	0	—
M1-195	0.5356	9	0	—
M1-200	0.5120	10	0	—

corrected so that the synthetic images are generated at random azimuth
angles. We believe this is the primary reason for performance deterioration
and that the inevitable imperfection in modeling the real world probabilis-
tically is a secondary factor, but this should be determined by further simulation. We expect that the synthetic performance would decrease so that it is
more similar to the real data if the pose was random.

5.5 CONCLUSIONS

We have reported a methodology for the development of algorithms for the
recognition of objects of known geometry in laser radar imagery. We have
made extensive use of synthetic imagery because of the lack of available real

Table 9 Confusion Matrix of Classification on Real
10×20 Ladar Imagery

	T1	T2	T3	T4	T5	T6	T7	Accuracy
T1	110	4	0	2	6	1	0	89.4%
T2	6	142	1	1	0	0	3	92.8%
Total								91.3%

data. Use of synthetic imagery for the training of algorithms require great care; for this reason, we have taken care to verify and validate the synthetic image generation code.

We have reported a ladar ATR system based on fitting range images to 3D templates extracted from CAD target models. A projection-based pre-screener was designed to filter out more than 80% of candidate templates. An M of N pixel matching scheme was used for internal shape matching, whereas a silhouette-fitting scheme is used for boundary matching. This model-based system was trained on synthetic data and tested on a set of 276 real ladar images of military vehicles at various orientations and ranges. The system achieves above 90% accuracy in recognition of 0.4-m cross-range resolution ladar images. We also described a preliminary design of an active vision ladar ATR algorithm that used the results of processing the first image to determine a good aim point for a second ladar image of the target.

REFERENCES

1. R Measures. Laser Remote Sensing, Fundamentals and Applications. New York: Wiley, 1984.
2. G Jones, B Bhanu. Invariants for the recognition of articulated and occluded objects in SAR images. Proceedings of the Image Understanding Workshop, 1996, pp 1135–1144.
3. D Andersch, S Lee, H Ling, C Yu. XPATCH: A high frequency electromagnetic scattering prediction code using shooting and bouncing rays. Proceedings of Ground Target Modeling and Validation Conference, 1994, pp 498–507.
4. M Miller, U Grenander, J O'Sullivan, D Snyder. Automatic target recognition organized via jump-diffusion algorithms. IEEE Trans Image Process 6(1):157–174, 1997.
5. J Antoine, K Bouyoucef, P Vandergheyst, R Murenzi. Target detection and recognition using two dimensional continuous isotropic and anisotropic wavelets. In Automatic Object Recognition V. SPIE Aerosense, Orlando, FL, 1995, Vol. 2485, pp 20–31.
6. S Der, R Chellappa. Probe-based automatic target recognition in infrared imagery. IEEE Trans Image Process 6(1):92–102, 1997.
7. A Kramer, D Perschbacer, R Johnston, T Kipp. Relational template matching algorithm for FLIR automatic target recognition. Proceedings SPIE Automatic Target Recognition, 1993.
8. R Harr. Training Collection at the Center for Night Vision and Electro-Optics. In Proceedings of SPIE Signal and Image Processing Systems Performance Evaluation. SPIE Aerosense, Orlando, FL, 1991.
9. X Liu, T Kanungo, R Haralick. Statistical validation of computer vision software. Proceedings of the Image Understanding Workshop, 1996, pp 1533–1540.

10. V Ramesh, K Thornton, D Nadadur, A Bedekar, X Liu, W Hudson, X Zhang, R Haralick. Groundtruthing the RADIUS model-based imagery. In RADIUS: Image Understanding for Imagery Intelligence. San Francisco: Morgan Kaufmann, 1997, pp 469–480.

11. J Shapiro, B Capron, R Harney. Imaging and target detection with a heterodyne-reception optical radar. Appl Opt 20:3292–3313, 1981.

12. G Parry. Speckle patterns in partially coherent light. In J Dainty, ed. Laser Speckle and Related Phenomena. New York: Springer-Verlag, 1985.

13. J Goodman. Statistical properties of laser speckle patterns. In J Dainty, ed. Laser Speckle and Related Phenomena. New York; Springer-Verlag, 1985.

14. G Parry. Measurement of atmospheric turbulence induced intensity fluctuations in a laser beam. Opt Acta 28(5):715–728, 1981.

15. J Churnside, S Clifford. Log-normal Rician probability-density function of optical scintillations in the turbulent atmosphere. J Opt Soc Am A 4(10):1923–1930, 1987.

16. R DeWitt. The distribution of irradiance fluctuations that result from atmospheric turbulence. PSR Note N406. Santa Monica, CA: Pacific Sierra Research Corp, 1981.

17. R Lutomirski et al. Degradation of laser systems by atmospheric turbulence. Defense Advanced Research Projects Agency Report R-9171-ARPA/RC, 1973.

18. A Ishimaru. Wave Propagation and Scattering in Radom Media. New York: Academic Press, 1978, Vol. 2.

19. J Goodman. Statistical Optics. New York; Wiley, 1985.

20. R Fante. Electromagnetic beam propagation in turbulent media. Proc IEEE 63(12):1669–1693, 1975.

21. P Webb, R McIntyre, J Conradi. Properties of avalanche photodiodes. RCA Rev 35:234–278, 1974.

22. Q Zheng, S Der. Model-based target recognition in LADAR imagery. IEEE CVPR, 1998.

23. S Der, B Redman, R Chellappa. Simulation of error in optical radar range measurements. Appl Opt 36(27):6869–6874, 1997.

24. S Der, G Dome, J Horger, M Lorenzo, R Moulton. A multi-sensor digital terrain board for processor test and evaluation. IEEE MILCOM, 1992.

25. J Penn, H Nguyen, T Kipp, C Kohler, G Huynh, M Sola. The CREATION scene modeling package applied to multispectral missile seekers and sensors. Proceedings of the Multispectral Missile Seekers Conference, 1995.

26. M Aono, T Kunii. Botanical tree image generation. IEEE Computer Graphics Applic 4(5):10–34, 1984.

27. J Weber, J Penn. Creation and rendering of realistic trees. Proceedings of SIGGRAPH, 1995.

28. H Peitgen, D Saupe, eds. The Science of Fractal Image. New York: Springer-Verlag, 1988.

29. L Balick. A forest canopy height surface model for scene simulation. Simulation 49:5–12, 1987.

30. M Hebert, J Ponce. A new method for segmenting 3-D scenes into primitives. Proc. IJCPR, 836–838, 1982.
31. D Milgram, C Bjorklund. Range image processing: planar surface extraction. Proc. IJCPR-5, 912–919, 1980.
32. C Chu, J Aggarwal. Image interpretation using multiple sensing modalities. IEEE Trans PAMI 14(8):840–846, 1992.
33. R. Bolle, D Cooper. On optimally combining pieces of information, with application to estimating 3-D complex-object position from range data. IEEE Trans PAMI 8(5):619–638, 1986.
34. R Hoffman, A Jain. Segmentation and classification of range images. IEEE Trans PAMI 9(5):608–620, 1987.
35. R Rimey, F Cohen. A maximum likelihood approach to segmenting range data. IEEE J Robotics Autom 4(3):277–286, 1988.
36. F Solina, R Bajcsy. Recovery of parametric models from range images: the case for superquadrics with global deformations. IEEE Trans PAMI 12(2):131–147, 1990.
37. J Verly, R Delanoy, D Dudgeon. Model-based system for automatic target recognition from forward-looking laser-radar imagery. Opt Eng 31(12):2540–2552, 1992.
38. A Koksal, J Shapiro, W Wells. Model-based object recognition using laser radar range imagery. Proc SPIE 3718, 1999.
39. M Mark, J Shapiro. Multipixel, multidimensional laser radar system performance. Laser Radar II. SPIE, 1987, Vol. 783, pp 109–122.
40. T Green, J Shapiro. Detecting objects in three-dimensional laser radar range images. Opt Eng 33:865, 1994.

6
Distortion-Invariant Minimum Mean Squared Error Filtering Algorithm for Pattern Recognition

Francis Chan
Naval Undersea Warfare Center, Newport, Rhode Island

Bahram Javidi
University of Connecticut, Storrs, Connecticut

6.1 INTRODUCTION

A distortion-invariant filtering algorithm permits the use of a smaller set of filters to detect particular targets when the exact size, orientation, or illumination is unknown. Although many filtering algorithms have been proposed to detect targets in noise [1–16] this technique incorporates distortion invariance [17–20], the effects of environmental degradation, and the additive overlapping and nonoverlapping background noises in the design of the optimal filter using the minimum mean-squared-error criterion [11,21–23]. The distortion-invariant filter has the advantage of detecting a target with a predefined set of distortions. It saves processing time because only one filter has to be used to detect a distorted target instead of going through the entire training set. However, its performance is lower than that of an optimal filter synthesized for one particular reference target aspect [4,9]. Environmental compensation is included in the filter design because of the adverse effects associated with atmospheric propagation [22,24,25].

This chapter describes a filtering algorithm that takes into the account the modulation transfer function (MTF) [25] of the propagation environment due to aerosols and turbulence in order to optimize the detection of targets in clutter and system noise. The receiver operating characteristics (ROCs) peak-to-output energy ratio, peak-to-correlation intensity [21,22], and a number of other metrics are used as the measure of effectiveness of this algorithm.

6.2 MATHEMATICAL MODEL OF THE OBSERVED SCENE

The observed scene is typically presented to the processor as a two-dimensional sampled array of pixel values. In this case, the target is immersed in nonoverlapping background clutter, blurred by the aerosols and turbulence between the target plane and the sensor and corrupted by system noises [4,22]. This observed scene (image) from the sensor may be modeled in the discrete two-dimensional spatial domain as follows:

$$s_i(m, n) = \left\{ r_i(m, n) + v_1(m, n) \right\} \otimes b(m, n) + v_2(m, n) \qquad (1)$$

where $s_i(m, n)$ is the ith detected scene with region of support (ROS) $w_0(m, n)$, $r_i(m, n)$ is the ith target with ROS $w_{r_i}(m, n)$ within the scene, $v_1(m, n)$ is the nonoverlapping noise with ROS$\{w_0(m, n) - w_{r_i}(m, n)\}$ $b(m, n)$ is the point spread function of the environment, $v_2(m, n)$ is the zero mean overlapping additive noise with ROS $w_0(m, n)$, and $\otimes$ denotes a two-dimensional convolution in the spatial domain. It is assumed that $v_2(m, n)$, the model for additive system noise, is Gaussian distributed and uncorrelated with the other components in the scene model and is overlapping the entire scene. The background clutter is modeled by $v_1(m, n)$, which in general, has a nonzero mean and low-pass characteristics. The background clutter is in the same plane as the target and is nonoverlapping because the presence of the target masks the clutter region within the target's region of support. The point spread function, $b(m, n)$, is analogous to the impulse response in system theory. The Fourier transform of $b(m, n)$ or the optical transfer function (OTF) degrades the high-spatial-frequency response of the observation [25]. The presence of atmospheric turbulence and aerosols between the target and the sensor imposes a loss in the detection performance in terms of the probability of detection and probability of recognition [22,26].

6.3 DERIVATION OF THE OPTIMAL DISTORTION-INVARIANT MMSE FILTER

The desired output in this case is a sharp peak at the location of the target with a quiet background to allow unambiguous declaration of its presence [27]. We select the minimum mean squared error (MMSE) as the optimization criterion to design the filtering algorithm [21–23]. The optimal filter aims to minimize the squared difference between a desired output and the filter output.

We aim to design the MMSE filter to be distortion invariant using the total mean-squared-error criterion. The ability to have one distortion invariant filter to detect a target with out-of-plane rotational distortion is accomplished by extending the MMSE filter to be sensitive to target distortions as defined by a training set. We assume that when there is a target in the scene, the occurrence of the different views of the target within the training set is equally likely. The scene with the ith target view present is modeled as follows:

$$
\begin{aligned}
s_i(m, n) = r_i(m, n) \otimes b(m, n) + v_1(m, n)\big[w_0(m, n) - w_{r_i}(m, n)\big] \\
\otimes\, b(m, n) + v_2(m, n)w_0(m, n)
\end{aligned}
\tag{2}
$$

where the training set is $\{r_i(m, n) \text{ for } i = 1, 2, \ldots, J\}$, the corresponding regions of support for the set is $\{w_{r_i}(m, n) \text{ for } i = 1, 2, \ldots, J\}$ and $w_0(m, n)$ is the scene window.

We wish to design a distortion invariant filter [4,7] that will produce a maximum peak value at the correlation output plane while maintaining a quiet background. The mean squared error for the ith target view is

$$
\epsilon_i = E\left(\sum_{m=1}^{M}\sum_{n=1}^{N}|y_i(m, n) - y_d(m, n)|^2\right)
\tag{3}
$$

where $y_i(m, n)$ is the distortion-invariant filter output in response to the ith target view and $y_d(m, n)$ is a Kronecker delta function (the desired output) at the target location. The strategy is to minimize the total mean squared error for the entire training set; that is,

$$
\epsilon = \sum_{i=1}^{J}\epsilon_i = \sum_{i=1}^{J} E\left(\sum_{m=1}^{M}\sum_{n=1}^{N}|y_i(m, n) - y_d(m, n)|^2\right)
\tag{4}
$$

Using Parseval's theorem on Eq. (4), we can express the total mean squared error for the training set in the discrete spatial frequency domain as

$$
\epsilon = \sum_{i=1}^{J} E\left(\frac{1}{MN}\sum_{k=1}^{M}\sum_{l=1}^{N}|Y_i(k, l) - 1|^2\right)
\tag{5}
$$

$$\epsilon = \sum_{i=1}^{J} E\left(\frac{1}{MN}\sum_{k=1}^{M}\sum_{l=1}^{N}|H(k,l)S_i(k,l) - 1|^2\right) \tag{6}$$

$$\epsilon = \sum_{i=1}^{J} E\left(\sum_{k=1}^{M}\sum_{l=1}^{N}\{H(k,l)S_i(k,l)S_i^*(k,l)H^*(k,l) - H(k,l)S_i(k,l)\}\right.$$

$$\left. - S_i^*(k,l)H^*(k,l) + 1\right) \tag{7}$$

Interchanging the order of summation and moving the statistical expectation operations inside the summations, we have

$$\epsilon = \sum_{k=1}^{M}\sum_{l=1}^{N}\sum_{i=1}^{J}\left[\left\{H(k,l)E|S_i(k,l)|^2\right\}H^*(k,l) - H(k,l)E\{S_i(k,l)\}\right.$$

$$\left. - E\{S_i^*(k,l)\}H^*(k,l) + 1\right] \tag{8}$$

Taking the gradient of Eq. (8) with respect to the function $H(k,l)$ and setting the result to zero yields

$$\nabla_H\epsilon = \frac{1}{MN}\sum_{k=1}^{M}\sum_{l=1}^{N}\left[H^*(k,l)E\left\{\sum_{i=1}^{J}|S_i(k,l)|^2\right\} - E\left\{\sum_{i=1}^{J}S_i(k,l)\right\}\right] = 0 \tag{9}$$

Solving Eq. (9) for $H^*(k,l)$ yields the optimal MMSE distortion-invariant filter for the training set denoted as

$$H_{\text{opt-comp}}(k,l) = \frac{\sum_{i=1}^{J} E\{S_i^*(k,l)\}}{\sum_{i=1}^{J} E\{|S_i(k,l)|^2\}} \tag{10}$$

Expanding out the statistical expectations, we have the optimal distortion-invariant MMSE filter:

$$H_{\text{opt-comp}}(k,l) = \left(\sum_{i=1}^{J}[R_i^*(k,l) + \mu_{n_1}W_{c_i}^*(k,l)]B^*(k,l)\right)$$

$$\left\{\sum_{i=1}^{J}\left[|[R_i(k,l) + \mu_{n_1}W_{c_i}(k,l)]B(k,l)|^2 + \Phi_{n_1}(k,l)\right.\right.$$

$$\left.\left.\otimes|W_{ci}(k,l)B(k,l)|^2 + \Phi_{n_2}(k,l) \otimes |W_0(k,l)|^2\right]\right\}^{-1} \tag{11}$$

where $\otimes$ denotes two-dimensional convolution in the spatial frequency domain and $*$ denotes complex conjugation. The expressions $R(k, l)$, $W_c(k, l)$, and $B(k, l)$ are the two-dimensional discrete Fourier transforms of the reference target, its silhouette, and the point spread function, respectively. In the denominator, $\Phi_{n_1}(k, l)$ and $\Phi_{n_2}(k, l)$ are the power spectral densities of the zero mean nonoverlapping (the contribution of the mean background value is accounted for by the terms with μ_{n_1}) and overlapping noises; $W_0(k, l)$ is the Fourier transform of the scene window and μ_{n_1} is the mean value of the background clutter over the entire scene.

The denominator in Eq. (11) is the power spectral density of the non-overlapping background, the disjoint target (and its silhouette), and the overlapping system noise; it is always positive and real. The presence of the silhouette in the numerator and the denominator serve to normalize the effect of the background mean level at the output. The inclusion of the target's region of support in the formulation of the optimal filter is essential for performance improvement [8,10] because it increases the separation between the signal and noise output magnitudes by forcing the output noise to have near-zero mean.

6.3.1 Performance Metrics

We now define and describe the following performance metrics [4,7,28–32]: ROC curve and two peak signal to background noise measures. These metrics are used to quantify and compare performance of various filtering algorithms but, more importantly, the agreement of simulation results and the expected theoretical results.

The ROC Curve

The ROC curve is used to compare the relative performance of different processors. It plots the probability of detection versus the probability of false alarm for a range of threshold values. The desired performance of detection systems is usually specified in terms of the probability of detection (p_d) and the probability of false alarm (p_{fa}) for a given input signal-to-noise ratio. For example, a typical specification may require the following: $p_d = 0.5$ with $p_{fa} < 10^{-4}$ at a range of 10 km.

We now describe a procedure to estimate the probability density functions (not necessarily Gaussian) needed to evaluate the p_d and p_{fa} based on a finite number of experimental trials. The number of independent samples required is of the order of 10^6, which will not be achievable via field measurements and, therefore, we typically employ computer simulations to validate performance specifications. In order to generate the representative ROC curve for a detector, it is necessary to obtain the conditional density functions of the output samples for the noise-only and signal-plus-noise hypotheses. In most cases, the number of noise samples per realization of an input image scene is high so that the sample size is not an issue for generating a representative noise-only histogram. For a typical 256-pixel × 256-pixel correlation output plane, we can easily collect 40,000 noise samples with sufficient standoff distance from the correlation peak to exclude samples that may have some correlation with the peak. On the other hand, because each computationally intensive two-dimensional correlation realization only produces 1 sample point for the target-plus-noise case, it is not unusual to have only 100–200 samples of the correlation peak levels to generate its histogram.

One standard method for estimating the signal-plus-noise conditional density function using a finite set of data is the Parzen estimator [30,31]. It allows a smoothed probability density function to be generated when the number of samples is too small to use the histogram as the estimate of the true probability density function. Consider a one-dimensional random vector $\{x_j, j = 1 \text{ to } N\}$ arising from N statistical realizations. It is assumed that x_j are independent and identically distributed random variables with an unknown probability density function $f(x)$. The goal is to estimate $f(x)$ using the samples $\{x_j\}$. The estimation is called Parzen's approximation of probability density functions; the approximate probability density function is given as

$$\hat{f}(x) = \frac{1}{Nh} \sum_{j=1}^{N} K\left(\frac{x - x_j}{h}\right) \tag{12}$$

where N is the number of independent samples, h is the smoothing interval size, and x is the value at which the probability density function is to be estimated. The typical choice of $K(\cdot)$ is a Gaussian function with zero mean and unit variance [31]:

$$K(v) = \frac{1}{\sqrt{2n}} e^{-v^2/2} \tag{13}$$

Thus, with a few hundred samples, we can generate an approximate PDF to represent the distribution of the signal-plus-noise magnitude. In this chapter, 200 samples were used to approximate the probability density functions under the signal-plus-noise hypothesis for generation of the respective ROC curves. However, the estimated probability density function is quite accurate with only about 100 samples if the smoothing kernel and the smoothing intervals are chosen properly [28,30,31].

Let $\hat{f}(x_0)$ and $\hat{f}(x_1)$ be the estimated probability density functions using Parzen's estimation for the noise-only and the signal-plus-noise hypotheses. We can write the probability of false alarm as

$$p_{\text{fa}} = \int_{\lambda}^{\infty} \hat{f}(x_0) \, dx_0 \qquad (14)$$

where the decision threshold λ is determined by the required probability of false alarm and the corresponding probability of detection is written as

$$P_d = \int_{\lambda}^{\infty} \hat{f}(x_1) \, dx_1 \qquad (15)$$

Thus, we can generate experimental ROC curves for any filtering algorithms, including many non-linear filters that are not mathematically tractable. The minimization of the probability of false alarm and the maximization of the probability of detection are contradictory goals. In order to improve detection performance, we need to maximize the "distance" between two density functions and shrink their spreads (standard deviations). This may be accomplished by proper signal and data processing, to a certain extent, if the system hardware and other operating parameters cannot be improved.

Peak-to-Output Energy and Peak-to-Correlation Intensity

The peak-to-output energy (POE) is defined as the ratio of the square of the expected value of the output signal at the target location to the expected value of the average output energy. The statistical expectation is evaluated over the number of independent runs of the filtering algorithm using different noise realizations. The POE is a good metric to quantify the filter performance using computer simulations where the scene is stationary. The POE is defined as

$$\text{POE} = \frac{\left| E\{y_a(m_0, n_0)\} \right|^2}{E\left\{ (1/MN) \sum_{m=1}^{M} \sum_{n=1}^{N} |y_a(m, n)|^2 \right\}} \qquad (16)$$

The peak-to-correlation intensity (PCI) ratio is also an appropriate metric to quantify the peak sharpness and background uniformity for *each statistical realization* because it is maximized when the background noise is zero and a high level peak is present. The PCI is defined as

$$PCI = \frac{I_p}{(1/MN)\sum_{m=1}^{M}\sum_{n=1}^{N}|y_a(m,n)|^2} \qquad (17)$$

where I_p is the peak intensity, which is defined as the output intensity as the target location. The PCI may be expressed in decibels by taking the base-10 logarithm of Eq. (17) and multiplying the result by 10. It should be noted that the PCI is the same as the peak-to-correlation energy (PCE) [22] if the division by MN is removed from Eq. (17). The maximum value of the PCE is unity and the maximum value of PCI is MN. Thus, the PCI may be interpreted as the peak intensity to average energy per output pixel, making this metric independent of the size of the output correlation plane. The PCI and POE are good measures of the output's compliance to the ideal case of a Kronecker delta function.

6.4 SIMULATION: DESCRIPTION AND RESULTS

In this section, the computer simulations and the results for the optimal filter are presented. The total mean squared error (MSE) for all of the filters are compared to show that the environmental filter with the correct point spread function yields the lowest MSE. The POE [4] is the metric used to quantify the peak sharpness and background noise suppression of the entire output plane. The maximum POE is also achieved when the correct blur compensation is applied. The discrimination of the distortion invariant filter is demonstrated. The ROC curves and target localization accuracy results are two important metrics used to show quantitatively the performance improvement attributable to proper blur compensation in the filter design. For each scenario, the scene is corrupted by noise and blurred and then processed by a bank of filters designed with different point spread functions.

6.4.1 Target Set

The following test images of a T-62 main battle tank from the Center for Imaging Sciences (CIS) database are used to test the distortion-invariant MMSE filter given in Eq. (11). Figure 1 shows a training set of nine T-62 main battle tank images with different out-of-plane rotational distortions. The difference in out-of-plane angle between adjacent images is $10°$. The

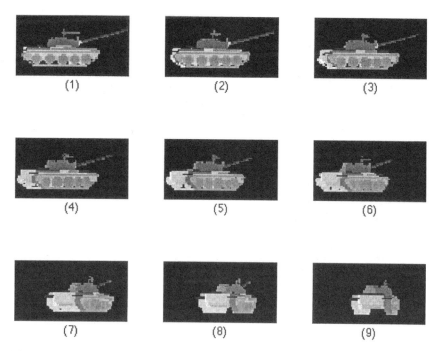

(1) (2) (3)

(4) (5) (6)

(7) (8) (9)

Figure 1 True class training image set used to construct the distortion-invariant MMSE filter. Image numbers 1 through 9 represent the T-62 tank training set from an aspect angle of $0°$ to $80°$ with a $10°$ increment.

training set covers the range of out-of-plane rotation from $0°$ to $80°$ with a $10°$ increment (i.e., $0°, 10° \ldots$, and $80°$). The sizes of the T-62 tank images are very similar for the entire training set. The distortion-invariant filter generated with this training set is not designed to detect the same target pointing in the opposite direction (e.g., the mirror image of training set images). The images shown in Fig. 2 were used to test the false class rejection of the distortion invariant MMSE filter. It is obvious that images 10 and 11 are false class targets. The tank image numbers from 12 through 16 are the same T-62 tank, but they are pointing in the opposite direction to the images used to construct the distortion-invariant filter (they are, in fact, mirror images are the true class targets). The out-of-plane rotation angles for the false class tanks are $358°$, $353°$, $343°$, $334°$, and $327°$. The regions of support and target details are very similar to the true class and, therefore, we expect a filter response to be higher than that of images 10 and 11. We will show that the distortion-invariant filter responds to this

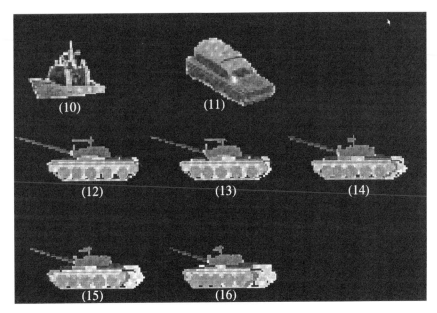

Figure 2 False class images. Images 10 and 11 are obvious false class images. Images 12 through 16 are images of the same T-62 tank shown in 3 but they are going in the opposite direction; note that the guns are pointing left.

true class target even though the gun is pointing in the opposite direction because of the shape similarity. In some applications, the correct determination of the direction in which the target is pointing is an important attribute. Filters that emphasize higher spatial frequencies are more discriminating and may be used for differentiating such small details. In practice, it is often necessary to construct a bank of distortion-invariant filters to cover all possible target headings. Because the filter response degrades when the set of training images is large, the designer must trade-off between detection performance and the size of the training set. Figures 3 shows eight images that are covered by the training set (Fig. 1); these are the nontraining true class images. The target angles for images 18–25 in Fig. 3 are $2°$, $6°$, $7°$, $8°$, $9°$, $17°$, $26°$, and $33°$, respectively. The distortion-invariant filter constructed with the training set shown in Fig. 1 is expected to detect these nontraining images. The detection performance of the distortion-invariant filter using the T-62 tank images in Fig. 3 will be presented.

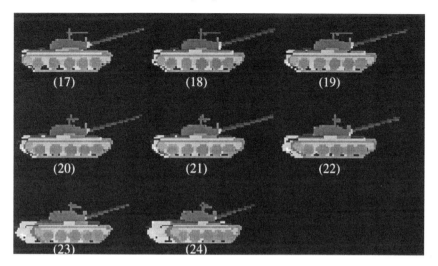

Figure 3 T-62 true class nontraining images covered by the training set.

6.4.2 Description of the Simulation

Two independent sources of noise are used in the scene modeling. The nonoverlapping noise is modeled as Gaussian with a mean of μ_1 and variance σ_1^2. This represents the background against the target. The environmental blurring function is generated using a two-dimensional Gaussian symmetric kernel.

6.4.3 Optimality with Respect to the Design Criterion

The compliance with the design criterion of optimal environmentally compensated filtering for a distortion-invariant filter is discussed in this section. Figure 4 shows the performance of environmental filters designed with a training set of nine training images. The distortion-invariant filter is less sensitive to distortion than the single-reference filter at the cost of lower detection performance [4,7,22]. However, distortion-invariant filters may be the only viable solution for implementing an automatic target classifier on some legacy systems with limited computational resources.

The distortion-invariant filter (composed of nine out-of-plane distorted views of the T-62 tank set) responds with a sharp peak at the location of a target that is within the training set and rejects other targets. We tested

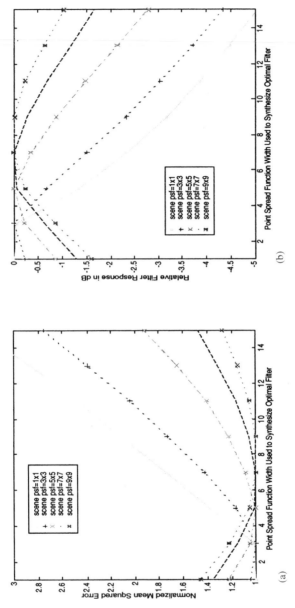

Figure 4 Performance of environmentally compensated distortion-invariant filters for various degraded (blurred) and noisy scenes. The input signal-to-noise ratios are normalized for the different scenes so that the curves are comparable. Eight different distortion-invariant filters were used to process each degraded scene. (a) The total mean squared error for five different scenes; (b) the relative POE for a family of five degraded input scenes.

the distortion-invariant filter's response to each of the targets in the training set. Figure 4a shows (the result for one of the targets in the training set, all others are similar) that the total mean squared error is minimized when the correct point spread function (PSF) is used to synthesize the filter. This confirms that the design criterion is met. The background and system noise statistics are known exactly. The POE and the PCI are also maximized when the filter parameters (PSF and noise statistics) match that of the input scene. The average loss in filter performance is about 1.0 dB (compare with 1.5 dB in the single-reference filter) if the point spread function used to construct the filter is 1×1 or 7×7 when the true environmental blurring is equivalent to a 5×5 point spread function. As in the single-reference case, when a scene has no environmental blurring, the filter response is most sensitive, as manifest in the quickest rate of change away from optimality in both the MSE and POE curves in the case when the scene PSF is equal to 1 (-o- curve).

6.4.4 Discrimination of the Distortion-Invariant Filter

The distortion-invariant filter is designed to register a high correlation peak against a quiet noise floor when the target in the input scene matches one of the images in the training set. The ability of the distortion-invariant filter to reject scene targets that are not in the training class is a measure of its discrimination.

We investigate the performance of the distortion-invariant filter using a subset of the T-62 main battle tank image dataset from Center for Imaging Sciences. This dataset consists of out-of-plane rotational views of a T-62 tank at $1°$ increment from $0°$ to $359°$. Zero degree represents the broadside view of the tank with the gun pointing to the right. The images shown in Fig. 1 is a subset of nine views in which the out-of-plane rotation angle starts from $0°$ to $80°$ with a $10°$ increment; this is the true class training set. A distortion-invariant filter using Eq. (12) using the true class training set is constructed. The distortion-invariant filter parameters included the second-order statistics of the nonoverlapping and overlapping noises. The set of true class training images (Fig. 1), the set of false class images (Fig. 2), and the set of true class nontraining images (Fig. 3) are processed through the distortion-invariant filter. Each image is processed 20 times with a different statistical realization of the noises. The background mean is 0.3, the non-overlapping variance is 0.01, and the overlapping noise variance is 0.01. The POE values, the mean correlation peak values, and its standard deviation for each of the images are tabulated in Table 1. The tabulated values for the correlation peaks are multiplied by a constant of 10^4 to make mental comparison easier.

Table 1 Simulation Results for the Distortion-Invariant
MMSE Filter with T-62 Images

True class training images	Aspect angle	POE	$\overline{peak} \times 10^4$	$\sigma_{peak} \times 10^4$
1	0	495.3	59	2.68
2	10	719.9	71	2.38
3	20	962.6	84	3.01
4	30	836.0	78	2.05
5	40	682.0	70	2.52
6	50	492.0	62	2.05
7	60	555.7	66	3.01
8	70	533.9	65	2.73
9	80	466.9	58	2.75
False class images				
10 car	—	76.2	21	1.52
11 ship	—	73.9	21	1.45
12	358	301.6	46	2.28
13	353	320.8	47	2.03
14	343	274.9	44	1.94
15	334	251.6	44	1.89
16	327	214.6	40	2.28
True class nontraining				
17	2	457.6	56	2.51
18	6	531.4	60	2.52
19	7	580.6	63	2.17
20	8	596.6	63	3.03
21	9	620.8	65	2.86
22	17	758.1	72	2.59
23	26	889.3	82	2.42
24	33	706.7	71	2.44

It is clear from the POE and peak values in Table 1 that the distortion-invariant MMSE filter detects all true class training and nontraining images. The maximum peak level due to a false class image and the minimum due to a true class image is significant enough that misclassification is unlikely. For example, the Fisher's discrimination ratio [28] between images 1 and 10 is 152.1, indicating a sufficiently large separation-to-scatter ratio. Another way of looking at the discrimination is that the ratio of the difference in mean peak values divided by the combined standard deviation is over 10. It should be noted that images 12 through 16 may be classified as false class or as true class nontraining (outside the range of distortion covered by the distortion-invariant filter because they are pointing in the opposite direction than the images in the training set). This means that the detection threshold may be set to accept or reject images 12 through 16 depending on the application. The POE values (average of 20 realizations) for these 24 objects are plotted in Fig. 5.

We next examined the performance of the MMSE distortion-invariant filter (for the T-62 tank training set) when environmental blur is present. The test scene has a nonoverlapping background mean of 0.3 and variance of 0.01; the overlapping noise variance is 0.01. The scene shown is blurred by a 7×7 Gaussian kernel. We test the performance of the compensated and the uncompensated filters against a true class (image 2), a false class (image 11), and a true class nontraining (image 19) image (the image number refers to those shown in Figs. 1–3). The test results are tabulated in Table 2. The compensated filter performs better as expected, as indicated by the POE values (335.7 versus 427.5 for training image 2 and 351.1 versus 443.9 for nontraining image 19). The Fisher's discrimination ratio between true class and false class images 2 and 11 is 241.4 for the uncompensated filter and 463.5 for the compensated filter.

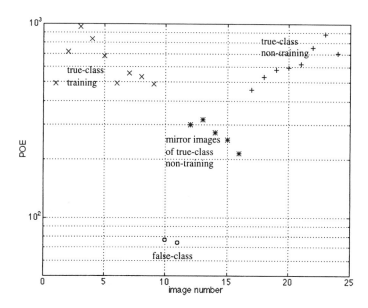

Figure 5 The POE values for $N = 20$ runs. Images 1 through 9 are the true class training images, indicated by $\times$. Images 10 and 11 are obvious false class images, indicated by $\bigcirc$. Images 12 through 16, marked by $*$, are minor images of nontraining images in the true class. The false class images, 17 through 24, are indicated by $+$.

Table 2   Results for the Distortion-Invariant MMSE filter with Environmental Blur (T-62)

Image #	Results for uncompensated filter			Results for compensated filter		
	POE	$\overline{\mathrm{Peak}} \times 10^4$	$\sigma_{\mathrm{peak}} \times 10^4$	POE	$\overline{\mathrm{Peak}} \times 10^4$	$\sigma_{\mathrm{peak}} \times 10^4$
2	335.71	50.0	2.26	427.51	50.0	1.28
11	3.03	4.23	1.89	4.30	4.53	1.68
19	351.13	51.0	2.78	443.88	52.0	2.27

We now consider the discrimination ability of the MMSE distortion-invariant filter in the presence of other true class and false class images in the same scene. Object 1 in Fig. 6 is a true class training image, object 2 is a true class nontraining image, objects 3 and 4 are false class images. The scene in Fig. 7 shows the environmentally corrupted input scene presented to the filter. This scene has a complex (natural) background consisting of foliage, man-made structures, ocean, and sky. The background mean value is 0.4 and the background standard deviation is 0.17. The important point is that the nonoverlapping background is nonhomogeneous so that the filter parameters for the nonoverlapping background is therefore mismatched for most of the regions within the scene. The true class tank objects are both

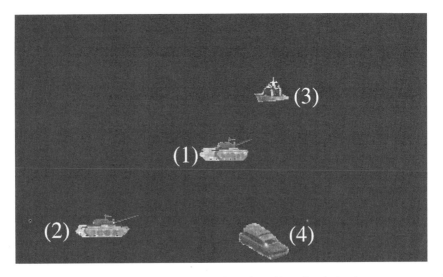

Figure 6   Objects for MMSE distortion-invariant filter discrimination test.

Figure 7 Naturally complex input scene presented to the MMSE distortion-invariant filtering algorithms.

40 × 80 pixels in size. The scene is blurred by a 5 × 5 Gaussian kernel and overlapping noise with variance of 0.01 is added. The input scene shown in Fig. 7 has a scene size equal 260 × 450 pixels. Two MMSE distortion-invariant filters are constructed using the true class training set (Fig. 1); the first filter includes the estimated noise parameters in the design and the second filter includes both the noise parameters and the point spread function. Figures 8a and 8b show the correlator output of these two MMSE distor-

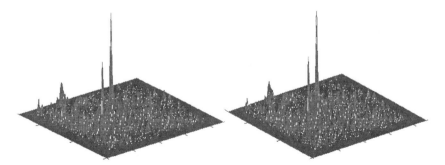

Figure 8 Complex background scenario: (a) correlator outputs for the MMSE distortion-invariant filter with no environmental compensation; (b) with environmental compensation.

tion-invariant filters. Both filters detect the true class T-62 tank objects readily. The compensated filter yields a slightly smoother background in the output correlation plane.

To illustrate the complexity of this scene, we compare it with the case when the nonoverlapping noise is homogeneous. Figure 9 shows the input scene with homogeneous background with the same noise parameters. The correlator outputs for the uncompensated and the compensated filters in response to the input scene shown in Fig. 9 are shown in Figs. 10a and b, respectively. The correlator output planes are much more uniform than those shown in Fig. 10 because the filter noise parameters were matched to the entire scene. The POE for the compensated filter is about 30% better than the uncompensated filter due to the reduction of background noise variance.

We conclude that the MMSE distortion-invariant filter performs well in both natural (inhomogeneous) and homogeneous background noises.

6.4.5 ROC Curve Results

The ROC curve allows the designer to compare different filtering algorithms over the range of operating points to select the required probability

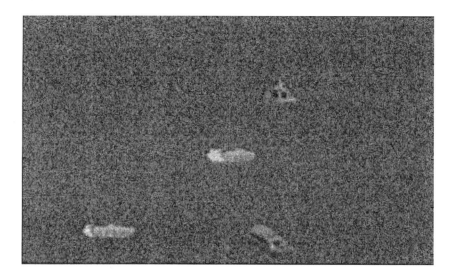

Figure 9   Input scene with homogenous background. Scene is blurred by a 5 × 5 Gaussian kernel. The nonoverlapping mean is 0.4, the standard deviation of the background is 0.17, and the standard deviation of the overlapping noise is 0.1.

Figure 10 Homogeneous background scenario: (a) correlator outputs for the MMSE distortion-invariant filter with no environmental compensation; (b) with environmental blur compensation.

of detection and probability of false alarm for his or her particular application.

We now present the ROC curves for the distortion-invariant filter [Eq. (12)]. The true class T-62 tank images (Fig. 1) are used to construct the uncompensated and the blur-compensated distortion-invariant MMSE filters.

The detection performance of the distortion-invariant MMSE filter is quite good even when the noise level is very high against a true class nontraining image. In this case, image 2 in Fig. 1 is used. The input scene has a background noise mean of 0.5, nonoverlapping noise variance of 0.01, and overlapping noise variance of 0.64. The scene is also blurred by a 7×7 Gaussian kernel. The probability density functions of the blur-compensated (dash-dot curves) and uncompensated filter (solid curves) using a T-62 tank true class nontraining image are shown in Fig. 11a. The blur compensated improves the performance by reducing the variances of the noise-only and the signal-plus-noise probability density functions. The peak values of the blur-compensated output are lowered, but the variance reductions are sufficient to give better peak-to-correlation intensity ratio and a better ROC curve (Fig. 11b).

We now compare the performance between the single reference filter and the nine-image distortion-invariant filter for the T62 main battle tank training set (see Fig. 12). The object to be detected is the T-62 tank image at an out-of-plane angle of 25°, a true class nontraining image. We compare the relative performance between the optimal MMSE filter and the optimal MMSE distortion-invariant filter (nonoverlapping noise mean is 0.5 with variance of 0.01, the overlapping noise variance is 0.64, and the scene is blurred by a 7×7 Gaussian kernel). The uncompensated and the blur-compensated ROC curves for the distortion-invariant MMSE filter are

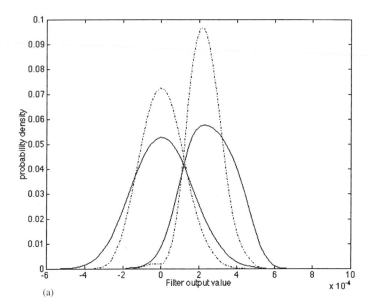

(a)

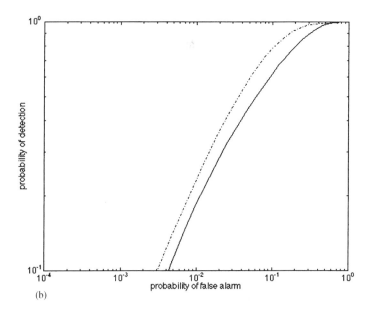

(b)

plotted using the solid-line type and the dash-dot-line type, respectively. The uncompensated and the blur-compensated ROC curves for the single-reference MMSE optimal filter are plotted using the + symbol and the * symbol, respectively. We note that both the compensated and the uncompensated filters perform well against a true class training image. The compensated distortion-invariant filter performs better than the uncompensated single-reference filter for the T-62 tank training set.

The probability density functions of the blur-compensated (dash-dot curves) and uncompensated distortion invariant filter (solid curves) are shown in Fig. 13a. As in the case of the single-reference MMSE filter, the noise standard deviations are reduced, resulting in less overlap of the density functions for the two hypotheses. The mean level of the correlation peak for the blur-compensated filter is reduced, but the detector performance is better because the noise standard deviations are reduced. The input scene that is processed to yield the density functions has a nonoverlapping background noise mean of 0.3, a nonoverlapping noise variance of 0.01, and an overlapping noise variance of 0.49. The scene (target and background) is blurred by a 7×7 Gaussian kernel before adding the overlapping noise. Figure 13b shows the ROC curves associated with the density functions shown in Fig. 13a. The compensated filter has a superior probability of detection over the entire operating range, but it is not as substantial as the single-reference MMSE filter. The probability of detection associated with a selected probability of false alarm of 3.5×10^{-4} is 0.1 for the compensated filter and less than 0.1 for the uncompensated filter.

We assert that a little blur compensation in the filter design is better than none at all if the scene is degraded by the environment. We use three different distortion-invariant filters (insufficiently, perfectly, and uncompensated for environmental blurring) to process the same corrupted scene. We

Figure 11 (a) The probability density functions of the blur-compensated (dash-dot curves) and uncompensated filter (solid curves) using a T-62 tank true class nontraining image. The scene processed to yield the density function have a background noise mean of 0.5, nonoverlapping noise variance of 0.01, and overlapping noise variance of 0.64. The scene is blurred by a 7×7 Gaussian kernel. (b) The ROC curves associated with the density functions shown in (a). The scene processed to yield the density function has a background noise mean of 0.5, nonoverlapping noise variance of 0.01, and overlapping noise variance of 0.64. The scene is blurred by a 7×7 Gaussian kernel. The compensated filter has superior probability of detection over the entire operating range.

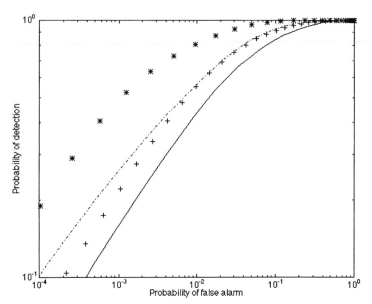

Figure 12 Relative performance of the optimal MMSE filter and the optimal MMSE distortion-invariant filter performance using the same scene (nonoverlapping noise mean is 0.5, variance is 0.01, overlapping noise variance is 0.64, scene is blurred by a 7×7 Gaussian kernel). The target is a true class training T-62 tank. The uncompensated and the blur-compensated ROC curves for the distortion invariant MMSE filter are plotted using the solid-line type and the dash-dot-line type, respectively. The uncompensated and the blur-compensated ROC curves for the single-reference MMSE optimal filter are plotted using the + symbol and the * symbol, respectively.

will examine the case when the filter is designed with too much compensation. Figure 14a shows the probability density functions of the blur-compensated (dash-dot curves) and uncompensated filter (solid curves). The density functions due to a filter designed with a 3×3 Gaussian blur are included ($\bigcirc$ curve). The corrupted input scene processed to yield the output density functions for these three filters have a background noise mean of 0.7, nonoverlapping noise variance of 0.01, and overlapping noise variance of 0.64. The scene is blurred by a 7×7 Gaussian kernel and then overlapping noise is added. Figure 14b plots the ROC curves associated with the density functions. The performance of the distortion-invariant filters is inferior to the single-reference MMSE filter, as expected. However, a little compensation for environmental blur does improve the detection performance, as indicated by the $\bigcirc$ curve in Fig. 14b.

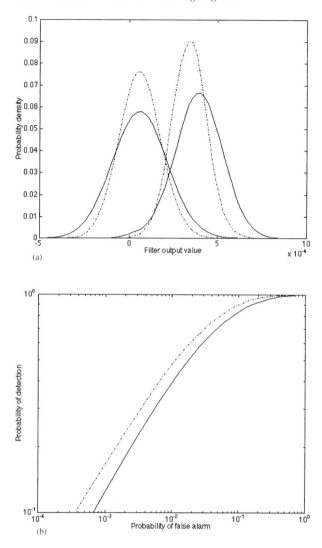

Figure 13   (a) The probability density functions of the blur-compensated (dash-dot curves) and uncompensated filter (solid curves). The scene processed to yield the density function has a background noise mean of 0.3, nonoverlapping noise variance of 0.01, and overlapping noise variance of 0.49. The scene is blurred by a 7×7 Gaussian kernel. (b) The ROC curves associated with the density functions shown in (a). The nonoverlapping noise mean is 0.3, the noise variance is 0.01, and the overlapping noise variance is 0.49. The compensated distortion-invariant filter has slightly superior probability of detection over the entire operating range. The probability of detection associated with probability of false alarm of 3.5×10^{-4} is around 0.1 for both the compensated filter and the uncompensated filter.

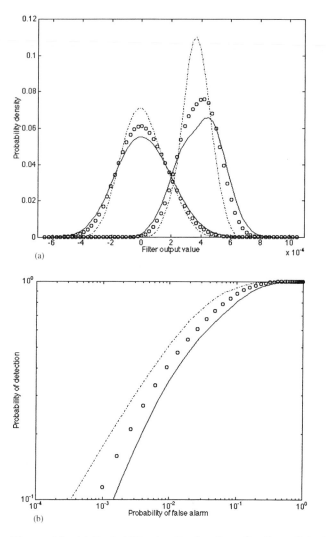

(a)

(b)

Figure 14 (a) Probability density functions for three distortion-invariant filters. The probability density functions of the blur compensated (dash-dot curves) and uncompensated filter (solid curves) are shown. The density functions due to a filter designed with a 3×3 Gaussian blur are included ($\bigcirc$ curve). The scene processed to yield the density function has a background noise mean of 0.7, nonoverlapping noise variance of 0.01, and overlapping noise variance of 0.64. The scene is blurred by a 7×7 Gaussian kernel. (b) ROC curves for the three distortion-invariant filters with output probability density functions shown in (a). The scene processed to yield the density function has a background noise mean of 0.7, nonoverlapping noise variance of 0.01, and overlapping noise variance of 0.64. The scene is blurred by a 7×7 Gaussian kernel. The blur-compensated filter (dash-dot curve) is better than the uncompensated filter (solid curve). The $\bigcirc$ curve denotes the undercompensated filter.

The next distortion-invariant filter ROC is a relatively high signal-to-noise ratio scenario; this is the same scene used in the single-reference case. Figure 15 shows the probability density functions of the blur-compensated (dash-dot curves) and uncompensated filter (solid curves). The ○ curves are the density functions for a filter that is overly compensated (PSF = 9). The scene processed to yield the density functions have a background noise mean of 0.3, nonoverlapping noise variance of 0.04, and overlapping noise variance of 0.25. The scene is blurred by a 5 × 5 Gaussian kernel. It is evident that the distortion-invariant filters do provide usable detection performance. Overcompensation for environmental blur hurts the detection performance as indicated by the ○ curve in Fig. 15b.

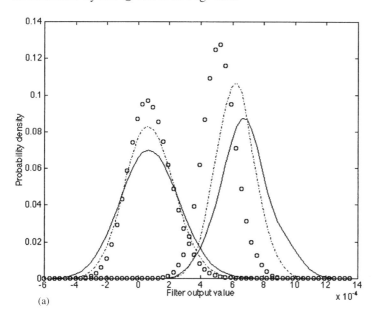

(a)

Figure 15 (a) Probability density functions for three distortion-invariant filters. The probability density functions of the blur-compensated (dash-dot curves), uncompensated filter (solid curves) and the overcompensated filter (○ curve) are shown. The scene processed to yield the density function has a background noise mean of 0.3, nonoverlapping noise variance of 0.04, and overlapping noise variance of 0.25. The scene is blurred by a 5 × 5 Gaussian kernel. (b) ROC curves for the three distortion-invariant filters with output probability density functions shown in (a). The ROC curves for the blur-compensated (dash-dot curves), uncompensated filter (solid curves), and the overcompensated filter (○ curve)'are shown. The scene processed has a background noise mean of 0.3, nonoverlapping noise variance of 0.04, and overlapping noise variance of 0.25. The scene is blurred by a 5 × 5 Gaussian kernel.

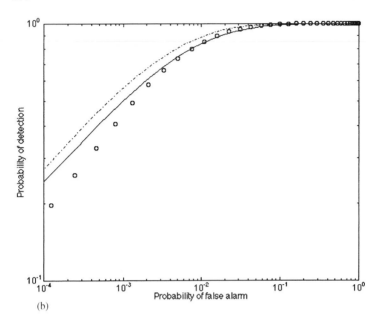

(b)

Figure 15 Continued.

We conclude that distortion-invariant MMSE filters are more versatile but carry a cost of lower detection performance. However, the size of the training set may be traded off against the number of distortion-invariant filters to achieve the desired level of detection/discrimination performance. The application of distortion-invariant filters should be restricted to higher signal-to-noise scenarios. We have shown that proper compensation provides optimal performance for both the single-reference and distortion-invariant MMSE filters.

6.5 SUMMARY

We presented a distortion-invariant filtering algorithm that takes into account the environmental degradation between the target and the observer, the background nonoverlapping noise, the nonstationarity of the scene, nontarget objects, and additive system noise. The optimization criterion is the minimum mean squared error between the desired and actual correlation plane outputs. The detection performance of this algorithm has been validated using simulations and found to be superior to filters that do not take into account the effects of environmental conditions. We use the ROC curve as a metric to quantify filter performance relative to other processors. Much work remains to characterize the performance of this filter with respect to nonstationary clutter, overlapping clutter, multiple-target discrimination, trade-off between detection performance and distortion invariance, filter sensitivity, and robustness.

REFERENCES

1. JL Turin. Introduction to matched filters. IRE Trans Inform Theory IT-6:311–329, 1960.
2. A Vanderlugt. Signal detection by complex filters. IEEE Trans Inform Theory IT-10:139–145, 1964.
3. JW Goodman. Introduction to Fourier Optics. 2nd ed. New York: McGraw-Hill, 1996.
4. B Javidi, J Horner. Real-Time Optical Information Processing. San Diego, CA: Academic Press, 1994.
5. DL Flannery, JL Horner. Fourier optical signal processor. Proc IEEE 77:1511, 1989.
6. P Refregier, V Laude, B Javidi. Nonlinear joint transform correlation: An optimal solution for adaptive image discrimination and input noise robustness. Opt Lett 19(6), 1994.
7. A Mahalanobis. Review of correlation filters and their application for scene matching. In Optoelectronic Devices and Systems for Processing. Critical Reviews of Optical Science Technology Vol. CR 65. Bellingham, WA: SPIE Press, 1996, pp 240–260.
8. JL Horner, PD Gianino. Phase-only matched filtering. Appl Opt 23:812–816, 1984.
9. RC Gonzalez, RE Woods. Digital Image Processing. Reading, MA: Addison-Wesley, 1992.
10. Ph Refregier, J Figue. Optical trade-off filters for pattern recognition and their comparison with Wiener approach. Opt Computer Process 1:245–265, 1991.
11. P Refregier, B Javidi, G Zhang. Minimum square error filter for pattern recognition with nonoverlapping signal and scene noise. J Opt Lett 18:1453–1455, 1993.
12. IEEE Transaction in Image Processing. Special issue on automatic target recognition. 6(1), 1997.
13. HJ Caufield, WT Maloney. Improved discrimination in optical character recognition. Appl Opt 8:2354–2356, 1969.
14. A Mahalanobis. Review of correlation filters and their application for scene matching. In B Javidi, K Johnson, ed. Optoelectronic Devices and Systems for Processing. Critical Reviews of Optical Science Technology Vol. CR 65, Bellingham, WA: SPIE Press, 1996, pp 240–260.

15. A Mahalanobis. Processing of multi-sensor data using correlation filters. Proc SPIE 3466:56–64, 1998.
16. B Javidi, ed. Smart Imaging Systems. Bellingham, WA: SPIE Press, 2001.
17. D Casasent, D Psaltis. Position, rotation, and scale-invariant optical correlation. Appl Opt 15:1795–1799, 1976.
18. D Casasent. Unified synthetic discrimination function computational formulation. Appl Opt 23:1620–1627, 1984.
19. B Javidi, J Wang. Optimum distortion invariant filters for detecting a noisy distorted target in background noise. J Opt Soc Am A 12:2604–2614, 1995.
20. B Javidi, D Painchaud. Distortion invariant pattern recognition using Fourier plane nonlinear filters. Appl Opt 35:318–331, 1996.
21. B Javidi, G Zhang, F Parchekani, P Refregier. Performance of minimum-mean-square-error filters for spatially nonoverlapping target and input-scene noise. Appl Opt 33(35), 1994.
22. F Chan, N Towghi, B Javidi. Distortion-tolerant minimum-mean-squared-error filter for detecting noisy targets in environmental degradation. Opt Eng 39(8):2092–2100, 2000.
23. B Javidi, F Parchekani, G Zhang. Minimum-mean-squared-error filters for detecting a noisy target in background noise. Appl Opt 35(35), 1996.
24. D Sadot, SR Rotman, NS Kopeika. Comparison between high-resolution techniques of atmospherically distorted images. Opt Eng 34(1), 1995.
25. NS Kopeika. A System Engineering Approach to Imaging. Bellingham, WA: SPIE Press, 1998.
26. A Fazlolahi, B Javidi. Error probability of an optimum receiver designed for pattern recognition with nonoverlapping target and scene noise. J Opt Soc Am A 14:1024–1032, 1997.
27. B Javidi, J Wang. Limitations of the classic definition of the signal-to-noise ratio in matched filter based optical pattern recognition. J Appl Opt 31, November 10, 1992.
28. DH Kil, FB Shin. Pattern Recognition and Prediction with Applications to Signal Characterization. Washington, DC: AIP Press, 1996.
29. J Horner. Metrics for assessing pattern-recognition performance. Appl Opt 31(2), 1992.
30. CW Therrien. Decision Estimation and Classification. New York: Wiley, 1989.
31. TY Young, TW Calvert. Classification, Estimation and Pattern Recognition. New York: Elsevier, 1974.
32. K Fukunaga. Statistical Pattern Recognition. 2nd ed. San Diego, CA: Academic Press, 1990, p 59.

7

Electro-Optical Correlators for Three-Dimensional Pattern Recognition

Joseph Rosen
Ben-Gurion University of the Negev, Beer-Sheva, Israel

7.1 INTRODUCTION

Optical pattern recognition is a well-known research field since the pioneer work of VanderLugt [1]. Several books and review articles survey different approaches of this field [2–5]. The optical correlator is the main component in many of the optical image recognition schemes. Spatial correlations can be done fast and in parallel by the use of optics. The functions involved in such correlators are at most two dimensional (2D). However, our real spatial world is three dimensional (3D), and in some applications, 3D objects should be processed along all of their three dimensions. In general, 2D pattern recognition systems cannot determine the exact longitudinal distances between the various targets and cannot map the identified targets in the 3D space. In other words, with a 2D correlator, we cannot be sure which object is in front or behind other objects. For that purpose, we need to extend the correlation from two dimensions to three. The 3D correlation has two advantages over the conventional 2D correlation. First, we employ the information obtained from the 3D shape of the object, including its pattern along its depth dimension. Second, the target's location in the 3D space is exactly identified by the correlator. This essential information might be useful for pattern recognition, image reconstruction, and target tracking systems.

A method for performing 3D optical correlation was proposed by Bamler and Hofer-Alfeis [6]. In their study, the 3D observed scene is first mapped, slice by slice, onto a 2D plane. Then, a conventional 2D optical correlation, with an increased space–bandwidth product, is performed. Karasik [7] modified this concept by means of using a sampling scheme, which reduced the amount of information participating in the correlation process. In these methods, the 3D distribution of the observed scene must be known a priori to the digital computing system, prior to the mapping stage on the 2D plane. In other words, the scene must be processed with intensive digital algorithms to reconstruct the 3D image inside the computer memory before any correlation can be employed. The advantages of the optical processing, namely the directness and the high processing speed, vanish in such schemes.

Recently, three other attempts at 3D optical pattern recognition were reported [8–11]. Although these creative proposals contribute original ideas to the reservoir of image processing techniques, none of them has permitted complete spatial correlation in all three dimensions. Threfore, they have the property of shift invariance only in the transverse plane and not along the longitudinal axis. When the observed object is shifted along the longitudinal axis from the position for which its filter has been designed, the correlation peak either disappears [8,10] or remains at the same location [9,11]. Therefore, these systems are not capable of locating specific objects in a 3D scene.

In 1997, the optical correlator was extended from operation in two dimensions to three [12]. This correlation involves fusing images of objects from a few different points of view and allows objects to be identified and located in 3D space. This process has been demonstrated in a 3D joint transform correlator (JTC), in which a reference and tested objects are observed together from a distance [13,14]. The reference object and the tested objects are projected a few times from different points of view onto a spatial light modulator (SLM), and the projected images are electro-optically processed to yield the desired 3D correlation. This chapter surveys the evolution of the 3D optical correlation developed at Ben-Gurion University (BGU) for the last 4 years.

In the BGU correlator, the system outcome is a direct 3D correlation between realistic 3D objects without the need first to understand and reconstruct the scene by a digital computer. Using optics gives the technique speed advantages. The signal processing can be performed in parallel, using simple devices such as lenses, and at the speed of light. The new scheme removes some of the ambiguity inherent in the current 2D optical correlators. In particular, the device is not confused by objects of different sizes that seem identical only because they are positioned at different distances from the observer. When the observed object is shifted along the longitudinal axis, the correlation peak is proportionally shifted along the same axis and thus permits the object to be located in all the coordinates of the 3D space.

The key concept of the 3D correlation is first to implement electro-optically a 3D Fourier transform (FT) of the observed scene. Then, with the convolution theorem, the desired 3D correlation result is obtained. The electro-optical 3D FT is the main subject of Section 7.2. In the field of 2D optical correlation, there are two basic types of correlators: the VanderLugt correlator (VLC) [1] and the JTC [15]. The 3D extensions of these two types are described in Section 7.3.

7.2 THREE-DIMENSIONAL FOURIER TRANSFORM

According to the convolution theorem, correlation of any dimension can be expressed by two successive FTs. Therefore, the basic concept of the 3D correlation is first to implement optically the 3D FT of the observed scene. In this scheme, 3D objects are scanned from the paraxial point of view and Fourier transformed to the 3D spatial frequency space. The 3D FT is composed from a series of 2D FTs, each of which is performed on a 2D projection of the 3D input function, observed by a camera from a different point of view. The 3D object is observed from various points of view distributed on a finite transverse plane located far from the object. It is assumed that the field of view is wider than the transverse dimension of the input function and that the depth of focus of the camera is longer than the longitudinal dimension of the input objects. There are at least two methods for observing the input scene of the correlator by multiple cameras: the parallel and converging observations. A description of these two attitudes is given in the next two subsections.

7.2.1 Parallel Observation

The first proposed scheme of a parallel observation is shown in Fig. 1. A 3D input object is located in the coordinate system (x_s, y_s, z_s), and contained in the volume $(\Delta x_s, \Delta y_s, \Delta z_s)$. A camera observes the input scene through an imaging lens lcoated a distance L from plane P_1, where P_2 is the transverse plane $z_s = 0$. The lens and the camera are transversely displaced a distance (D_x, D_y) from the z_s axis. All of the observed objects are imaged onto the camera and displayed on a SLM at plane P_2. This SLM is located in the front local plane of a spherical lens with a focal length f and is illuminated by a plane wave. Thus, a 2D FT of the SLM's image is obtained on the lens' back focal plane P_3. The third spatial frequency variable is obtained by displaying a few projections of the plane P_1, each with a different transverse displacement of the camera. To express the relation between the 3D input function and the distributions on plane

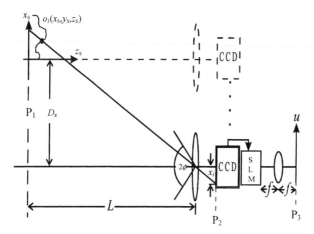

Figure 1 Illustration of the imaging system in the case of parallel observation.

P_3, let us first look at a single point (x_s, y_s, z_s) from the entire input object. The observed point is displayed on the SLM at point (x_i, y_i) and transformed by the lens into a linear phase function $\exp[i2\pi(x_i u + y_i v)/\lambda f]$, where u and v are the coordinates of back focal plane P_3 and λ is the wavelength of the plane wave. Assuming that plane P_1 is displayed on the SLM with a magnification factor M, we can calculate the location of the observed point on the SLM as a function of its location in object space and the amount of camera displacement (D_x, D_y). Using simple geometrical considerations, we can locate the point on the SLM:

$$
\begin{aligned}
x_i &= \frac{M(D_x + x_s)}{1 - z_s/L} \\
y_i &= \frac{M(D_y + y_s)}{1 - z_s/L}
\end{aligned}
\tag{1}
$$

Assuming that $L \gg \Delta z_s$, we approximate (x_i, y_i), by taking only the first two terms of the binomial expansion of $(1 - z_s/L)^{-1}$ as follows:

$$
\begin{aligned}
x_i &\cong M\left(D_x + x_s + \frac{z_s D_x}{L} + \frac{z_s x_s}{L}\right) \\
y_i &\cong M\left(D_y + y_s + \frac{z_s D_y}{L} + \frac{z_s x_s}{L}\right)
\end{aligned}
\tag{2}
$$

The overall field distribution on the rear focal plane that results from a 3D object, $o_1(x_s, y_s, z_s)$, for a given displacement (D_x, D_y) is obtained by integrating over the linear phases contributed by all the object points as follows:

$$O_3(u, v, D_x, D_y) = \int\int\int o_1(x_s, y_s, z_s) \exp\left(i \frac{2\pi}{\lambda f} (x_i u + y_i v)\right) dx_s \, dy_s \, dz_s$$

$$(3)$$

At this point, we assume that $Lf\lambda \gg M\Delta u\Delta z_s\Delta x_s$ and $Lf\lambda \gg M\Delta v$ $\Delta z_s\Delta y_s$, where $(\Delta u, \Delta v)$ are the maximal dimensions of plane P_3. In these assumptions, we require that the maximum phase change contributed by the fourth terms of Eqs. (2) be much less than 1 rad. For a given input, the validity of these assumptions depends on the system parameters, therefore, we can always design the system to satisfy these assumptions. Following the assumptions, the fourth terms in the right-hand side of Eqs. (2) can be neglected, and we approximate the location of each object's point on the SLM as

$$x_i \cong M\left(D_x + x_s + \frac{z_s D_x}{L}\right)$$

$$y_i \cong M\left(D_y + y_s + \frac{z_s D_y}{L}\right)$$

$$(4)$$

Substituting approximations (4) into Eq. (3) yields

$$O_3(u, v, D_x, D_y) = \exp\left(i \frac{2\pi M}{\lambda f} (D_x u + D_y v)\right)$$

$$\times \int\int\int o_1(x_s, y_s, z_s) \exp\left\{i \frac{2\pi M}{\lambda f} \left[x_s u + y_s v + \left(\frac{z_s}{L}\right)\right.\right.$$

$$\left.\left.(D_x u + D_y v)\right]\right\} dx_s \, dy_s \, dz_s$$

$$(5)$$

Equation (5) is a 3D FT, multiplied by a linear phase function, which transforms an object function $o_1(x_s, y_s, z_s)$ into a 3D spatial frequency function $O_3(f_x, f_y, f_z)$, where $f_x = Mu/\lambda f$, $f_y = Mv/\lambda f$, and $f_z = M(D_x u + D_y v)/\lambda f$. We note that f_z is unusually dependent on the transverse spatial frequency variables u and v. The three independent variables in the transformation space are u, v, and the vector (D_x, D_y), but the longitudinal frequency variable f_z is a linear combination of (D_x, D_y) scaled by u and v. Nevertheless, it turns out that this transform [defined by Eq. (5) without the linear phase function before the integral] has features similar to those of the conventional 3D FT. In particular, it satisfies the convolution theorem; therefore, this peculiar transform can be used as a building block in the spatial correlation process.

To recognize the limitations of the system, we first consider the required maximum displacement of the camera $(D_{x,\max}, D_{y,\max})$. Following a conventional Fourier analysis, we know that the maximum camera displacement depends on the longitudinal size of the smallest input element δz_s. Assuming that $D_{x,\max} = D_{y,\max} = D$ and $\Delta u = \Delta v = 2B$, the condition $DB \geq \lambda f L / M \delta z_s$ should be satisfied in order not to lose the information on the smallest longitudinal element. We conclude that the bandwidth of the system in the third dimension f_z depends directly on both the transverse bandwidth and the maximal camera displacement. On the other hand, the condition on the transverse bandwidth is $B \geq \lambda f / M \delta x_s$, where δx_s is the transverse size of the smallest input element. Therefore, the maximal displacement of the camera should follow the condition $D \geq L \delta x_s / \delta z_s$.

The observed object should remain in the field of view of the camera; therefore, another limitation on the camera's displacement is given by the condition $D \leq L \tan \phi - \Delta x_s / 2$, where 2ϕ is the field angle of the imaging system (assume that $\Delta y_s = \Delta x_s$). Shifting the camera beyond this limitation causes the object to disappear from the field of view. On the other hand, the condition $D \gg \Delta x_s$, should be satisfied; otherwise, there is no justification for keeping the third terms on the right-hand side of Eqs. (2) while neglecting the fourth terms.

Although it is convenient to analyze the system in terms of continuous signals, our detected 3D signal is discrete in its all dimensions. This is so because each 2D image is recorded separately as a collection of discrete pixels inside the computer. Therefore, the limitations on the sampling interval along the camera's translation should be considered. Let us assume that the maximual sampling intervals (d_x, d_y) between every two successive camera's shifts satisfy the equation $d_x = d_y = d$ and that the maximal transverse sampling intervals $(\delta u, \delta v)$ in spatial frequency domain satisfy the equation $\delta u = \delta v = b$. The criterion $db \leq \lambda f L / 2 M \Delta z_s$ should be satisfied for a signal to be reconstructed completely, along the z_s direction, from its samples in the spectral domain. When the well-known criterion on the maximal transverse sampling interval $b \leq \lambda f / 2 M \Delta x_s$ is substituted into the longitudinal criterion, we obtain that the maximal sampling interval along the camera's translation should follow the condition $d \leq L \Delta x_s / \Delta z_s$.

7.2.2 Converging Observation

One main drawback of the system proposed so far is the need for cameras with wide fields of view in the extreme points of view. All of the directions of observations for all the points of view are parallel to the same longitudinal axis. Therefore, the field of view must be much wider than the total width of the observed objects (see Fig. 1). Otherwise, one cannot shift the point of view far from the origin without missing parts of the objects in the extreme frames. In the following, the previously proposed system is modified to overcome this drawback.

To overcome the above-mentioned limitation, we examine the converging observation as an alternative configuration. In this architecture, shown in Fig. 2, the detection plane P_2 of each camera is orthogonal to the line that connects the origin points of the object space and the detection plane. In this case, the field angle is independent of the lateral shift; therefore, the field of view can be as narrow as the total object width. Formally, the condition that must be satisfied in this case is only $\tan \phi \geq \Delta x_s/2L$. However, Eqs. (4) are not valid for the converging observation, and we should recalculate the 3D accumulated distribution on plane P_3, as described in the following paragraphs. In order to simplify both the analysis and the experimental implementation, we assume at this point and hereafter that the camera is shifted only on the horizontal plane $y_i = 0$.

To express the relation between the 3D input function and the distribution of the Fourier space for the new configuration, we first look again at a single point (x_s, y_s, z_s) from the entire input object. The observed point is imaged on plane P_2 at point (x_i, y_i). The location of the imaged point (x_i, y_i) depends on the coordinates (x_s, y_s, z_s) and on the angle θ between the longitudinal axis z_s and the direction of observation. According to Fig. 2 and if we assume that plane P_1 is imaged with a magnification factor M, the location of the imaged point on plane P_2 is

$$x_i = M(x_s \cos \theta + z_s \sin \theta), \qquad y_i = M y_s \tag{6}$$

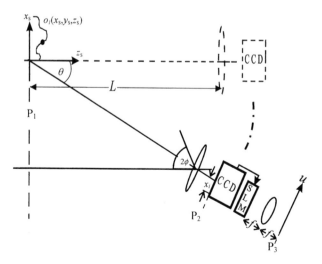

Figure 2 Illustration of the imaging system in the case of converging observation.

Each imaged point (x_i, y_i) is Fourier transformed to a 2D linear phase function. Therefore, the overall complex amplitude on plane P_3 is

$$O_3(u, v, \theta) \propto \int\int o_2(x_i, y_i, \theta) \exp\left(i\frac{2\pi}{\lambda f}(ux_i + vy_i)\right) dx_i \, dy_i \qquad (7)$$

where $o_2(x_i, y_i, \theta)$ is the transparency distribution on plane P_2 at some angle θ, and f is the focal distance of the Fourier lens L_1.

Let us look now at the complex amplitude $O_3(u, v, \theta)$ as a function of the input object. For a single element of size $(\delta x_s, \delta y_s, \delta z_s)$ and brightness $o_1(x_s, y_s, z_s)$, from the entire 3D object function, the complex amplitude on plane P_4 is

$$O_3(u, v, \theta) \propto o_1(x_s, y_s, z_s) \exp\left(i\frac{2\pi}{\lambda f}(ux_i + vy_i)\right) \delta x_s \, \delta y_s \, \delta z_s$$

$$= o_1(x_s, y_s, z_s) \exp\left(i\frac{2\pi M}{\lambda f}(ux_s \cos\theta + vy_s + uz_s \sin\theta)\right)$$

$$\delta x_s \, \delta y_s \, \delta z_s$$

$$\qquad (8)$$

where the final part of Eq. (8) is obtained after substituting Eq. (6) into the first part of Eq. (8). Next, we examine the influence of all points of the object $o_1(x_s, y_s, z_s)$. Because the SLM is illuminated by a coherent plane wave, the electromagnetic field contributed from each image point is summed. The object is three dimensional; therefore, the overall complex amplitude distribution on the Fourier plane P_3 from the entire 3D object $o_1(x_s, y_s, z_s)$ is obtained by integration over the linear phases contributed by all object points, as follows:

$$O_3(u, v, \theta) \propto \int\int\int o_1(x_s, y_s, z_s)$$

$$\exp\left(i\frac{2\pi M}{\lambda f}(ux_s \cos\theta + vy)x + uz_s \sin\theta\right) dx_s \, dy_s \, dz_s \qquad (9)$$

Equation (9) is again similar to a 3D FT, which transforms an object function $o_1(x_s, y_s, z_s)$ into a 3D spatial frequency fucntion $O_3(f_x, f_y, f_z)$, where $f_x = (Mu\cos\theta)/\lambda f$, $f_y = Mv/\lambda f$, and $f_z = (Mu\sin\theta)/\lambda f$. Note that f_x and f_z are both dependent on the transverse variable u and on the angle θ. This dependence distinguishes Eq. (9) from a pure 3D FT, but, apparently, we can obtain 3D spectral fucntions of three independent orthogonal spatial frequency variables with a proper coordinate transform, as discussed in the following paragraph.

Note that, in distant imaging, whereas $L \gg D_x$, the approximations $\cos\theta \approx 1$ and $\sin\theta \approx \theta \approx D_x/L$ are valid. In this case, Eq. (9) is reduced to the form of Eq. (5) without the linear phase function given before the integral. Therefore, in comparison with the parallel observation, the converging observation is less restricted by the distance between the imaging system and the objects. However, in the regime of close imaging, we have to map the function $O_3(u, v, \theta)$ from coordinates (u, v, θ) to $(u\cos\theta, v, u, \sin\theta)$. This mapping is more complicated than the mapping from $(u, v, D_x/L)$ to $(u, v, uD_x/L)$, which is suitable for both cases (i.e., the parallel and the converging observations in the regime of distant imaging). Mapping the function $O_3(u, v, \theta)$ on the 3D spatial spectral space enables us, by an additional 3D FT, to obtain the desired 3D correlation distribution.

7.3 THREE-DIMENSIONAL CORRELATIONS

There are two main types of optical correlator, namely the VLC [1], also known as 4-f correlator, and the JTC [15]. We have been able to extend both configurations, from operation in two dimensions to three. For the 3D JTC, we have proposed two different types of reference function. The first reference function [12–14] is identical to the same object that the system is intended to identify. Thus, a reference object and tested objects are observed together from a distance and projected several times from different points of view on a SLM. The projected images are electro-optically processed, yielding the desired 3D correlation. The second type of 3D JTC has a synthetic reference function [16], which gives the system the property of invariance to some object's geometrical distortions. Both types are described in the following first two subsections.

The extension of the 2D VLC toward the three dimensions has actually yielded a hybrid 3D correlator [17]. It is hybrid in the sense that this correlator combines concepts from the JTC as well as the VLC. It is also hybrid in the other sense that this correlator combines optical and electrical subsystems. In the hybrid correlator, only the tested objects are observed from different points of view. The 3D spatial spectrum of the observed scene is multiplied by a proper filter to yield the desired 3D correlation at the output space. This type of 3D correlator is described in Subsection 7.3.3.

7.3.1 Joint Transform Correlator with Real Reference

The complete process of the 3D JTC is shown schematically in Fig. 3. The 3D input space of the JTC contains a reference object at some point, say the origin, and a few tested objects, denoted together by the function

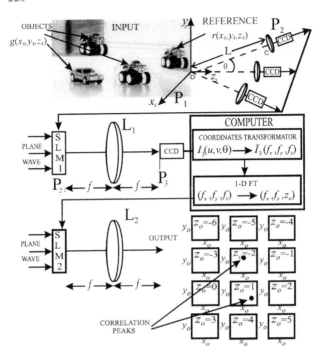

Figure 3 Schematic of the 3D joint transform correlator.

$g(x_s, y_s, z_s)$, and located around some other point, say the point (a, b, c). Therefore, the JTC input function is given by

$$o_1(x_s, y_s, z_s) = r(x_s, y_s, z_s) + g(x_s - a, y_s - b, z_s - c) \tag{10}$$

A camera (or a series of similar cameras) takes pictures of both the scene to be examined and the reference object. If there are several cameras, each is aimed at the same point in the 3D scene from a different direction. With one camera, a series of images is taken sequentially, with the position and point of view changing in between. From each point of view, the camera observes on plane P_1 through an imaging lens located a distance L from plane P_1. The angle between the z axis and the optical axis is denoted by θ. For each θ, the projected function $o_2(x_i, y_i, \theta)$ is displayed on the first spatial light modulator (SLM1), where (x_i, y_i) are the coordinates of plane P_2. The resulting images are Fourier transformed using lens L_1.

Following the analysis of Section 7.2, we can conclude that, for converging observation from the far field, the relation between (x_i, y_i, θ) and (x_s, y_s, z_s) is given by Eqs. (6). It is assumed that the distance L is much longer than the depth of the object function $o_1(x_s, y_s, z_s)$; therefore, the magnification factor is approximately the same constant M for all the object points. Following Eq. (9), the set of 2D intensity images on plane P_2 for various angles θ is

$$I_3(u, v, \theta) = |O_3(u, v, \theta)|^2$$

$$\propto \left| \iiint o_1(x_s, y_s, z_s) \exp\left(i \frac{2\pi M}{\lambda f} (x_s u \cos\theta + y_s v + z_s u \sin\theta) \right) \right.$$

$$\left. dx_s \, dy_s \, dz_s \right|^2$$

$$(11)$$

Substituting Eq. (10) into Eq. (11) yields

$$I_3(u, v, \theta) \propto \left| R(u, v, \theta) + G(u, v, \theta) \right.$$

$$\left. \exp\left(-i \frac{2\pi M}{\lambda f} (ua\cos\theta + vb + uc\sin\theta) \right) \right|^2 \qquad (12)$$

where R and G are the 3D FTs of r and g, respectively, and are defined by

$$R(u, v, \theta) \propto \iiint r(x_s, y_s, z_s)$$

$$\exp\left(i \frac{2\pi M}{\lambda f} (ux_s \cos\theta + vy_s + uz_s \sin\theta) \right) dx_s \, dy_s \, dz_s$$

$$G(u, v, \theta) \propto \iiint g(x_s, y_s, z_s)$$

$$\exp\left(i \frac{2\pi M}{\lambda f} (ux_s \cos\theta + vy_s + uz_s \sin\theta) \right) dx_s \, dy_s \, dz_s \qquad (13)$$

A conventional linear 2D JTC [15] is basically composed of three sequential mathematical operations: a FT of the input scene, calculation of the square magnitude of the FT, and another FT. In our 3D JTC, we add an additional operation of coordinate transformation before the final 3D FT in which every point in the space (u, v, θ) is mapped to the space $(u\cos\theta, v, u\sin\theta)$. It is suggested that such mapping be implemented electronically, either with software, as was done in the present study, or with electronic hardware.

The intensity I_3 is recorded by another camera into the computer in which a coordinate transform from (u, v, θ) to $(u\cos\theta, vu\sin\theta)$ is performed. The obtained function in the new coordinate system is

$$\tilde{I}_3(f_x, f_y, f_z) = \left| \tilde{R}(f_x, f_y, f_z) + \tilde{G}(f_x, f_y, f_z) \right.$$

$$\left. \exp\left(-i \frac{2\pi M}{\lambda f} \left(au\cos\theta + bv + cu\sin\theta \right) \right) \right|^2$$

$$= |\tilde{R}(f_x, f_y, f_z)|^2 + |\tilde{G}(f_x, f_y, f_z)|^2$$

$$+ \tilde{R}(f_x, f_y, f_z)\tilde{G}^*(f_x, f_y, f_z) \tag{14}$$

$$\exp\left(i \frac{2\pi M}{\lambda f} \left(au\cos\theta + bv + cu\sin\theta \right) \right)$$

$$+ \tilde{G}(f_x, f_y, f_z)\tilde{R}^*(f_x, f_y, f_z)$$

$$\exp\left(-i \frac{2\pi M}{\lambda f} \left(au\cos\theta + bv + cu\sin\theta \right) \right)$$

where $\tilde{R}$ and $\tilde{G}$ are the transformed functions R and G in the new spatial frequency coordinates given by $f_x = Mu(\cos\theta)/\lambda f$, $f_y = Mv/\lambda f$, and $f_z = Mu(\sin\theta)/\lambda f$. Without this mapping to the 3D spatial frequency space, we cannot use the convolution theorem and get the desired 3D correlation.

In this stage, the transformed function $\tilde{I}_3(f_x, f_y, f_z)$ is actually (part of) the square absolute magnitude of the 3D FT of the function $o_1(x_s, y_s, z_s)$. Therefore, following the convolution theorem, another 3D FT of $\tilde{I}_3(f_x, f_y, f_z)$, yields the autocorrelation of $o_1(x_s, y_s, z_s)$. The final 3D FT is performed in two steps; first, one-dimensional digital FT from $(u\cos\theta, v, u\sin\theta)$ to $(u\cos\theta, v, z_0)$ and then multiple 2D optical FT's from $(u\cos\theta, v, z_0)$ to (x_0, y_0, z_0). Similar to an ordinary JTC, one of the four terms of this autocorrelation is the requested 3D cross-correlation between the object function $g(x_s, y_s, z_s)$ and the reference function $r(x_s, y_s, z_s)$. After another 3D FT of $\tilde{I}_3(f_x, f_y, f_z)$ the output result is

$$c(x_0, y_0, z_0) = \iiint \tilde{I}_3(f_x, f_y, f_z) \exp[-i2\pi(x_0 f_x + y_0 f_y + z_0 f_z)] \, df_x \, df_y \, df_z$$

$$= r \otimes r + g \otimes g + (r \otimes g) * \delta(x_0 - a, y_0 - b, z_0 - c)$$

$$+ (g \otimes r) * \delta(x_0 + a, y_0 + b, z_0 + c) \tag{15}$$

where the symbol $\otimes$ and the asterisk ($*$) stand for the 3D correlation and the 3D convolution, respectively, and $\delta(\cdot)$ is the Dirac delta function. Similar to an ordinary 2D JTC, the two last terms of Eq. (15) are the cross-correlations between the reference object and the tested objects. The third and fourth correlation terms are centered around the points (a, b, c) and $(-a, -b, -c)$, respectively. The cross-correlation term can be written explicitly as

$$(g \otimes r)(x_o, y_o, z_o) = \int\int\int g(x, y, z) r^*(x - x_o, y - y_o, z - z_o) \, dx \, dy \, dz$$

$$(16)$$

The first two terms of the autocorrelation in Eq. (15) are centered around the origin. Therefore, if one of the distances (a, b, c) is longer than the respective size of the tested function g, then the cross-correlation is spatially separated from the autocorrelation terms and, thus, becomes detectable.

We tested this electro-optical system by computer simulations. In our example, the input space contains four vehicles, as shown in Fig. 3. One of them, the reference, is located on the right-hand side of the scene and is identical to two cars from the left-hand group of the three vehicles used here as the observed objects. Note that the two lower vehicles are located in front of the reference, whereas the upper vehicle is in back of it. The system should recognize the two cars that are identical to the reference and ignore the other vehicle. Three locations of the camera are also shown in Fig. 3. In this experiment, a single camera was shifted along the horizontal arc by 24 equal displacements, 12 for each side. Each projection was recorded with the camera and Fourier transformed. The intensity patterns of all the 2D FTs on plane P_3, designated as $I_3(u, v, \theta)$, were stored in the computer. Note that the reference can first be recorded alone (off-line) without the observed objects. Then, at the stage of target recognition, each reference projection is displayed on SLM1 beside the corresponding object projection. Alternatively, all the reference projections can be created synthetically by some computer algorithm [16] before they are displayed on SLM1 side by side with the object projections. This kind of system is described in Subsection 7.3.2.

Figure 4 presents 12 out of 25 images taken with the camera from every even position along the baseline. In each image, the group of three tested objects is located on the left-hand side, and the reference is separated on the right-hand side. From the set of 25 intensity patterns of 2D FTs, the 3D spectrum $\tilde{I}_3(f_x, f_y, f_z)$ is composed with the coordinate transform from (u, v, θ) to $(u\cos\theta, v, u\sin\theta)$. An additional 3D FT of $\tilde{I}_3(f_x, f_y, f_z)$ yields the required 3D cross-correlation between the reference object and the tested objects in the first diffraction order.

Three-dimensional plots of the output space (x_o, y_o, z_o) around the region of the first diffraction order are shown in Fig. 5. Each 3D plot of Fig. 5 presents the transverse intensity distribution at some z_o. The two strong correlation peaks on planes $z_o = -2$ and $z_o - 1$ indicate the locations of the two recognized vehicles, which are identical to the reference. At every location of the recognized object, and only there, the correlation values were significantly above some predefined threshold. These correlation peaks indicate the existence of the identified targets.

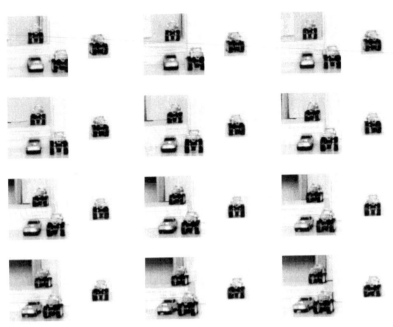

Figure 4   Twelve projections out of 25 of the input scene as observed from different points of view along the baseline.

7.3.2   JTC with Complex Reference

The 3D JTC described in Subsection 7.3.1 suffers from limitations similar to those of the conventional 2D optical correlators, namely sensitivity to geometrical distortions [2,4,5]. Objects from the same class of the reference (i.e., the true class) must appear in the same in-plane and out-plane orientation and be the same size as the reference, to be correctly identified. Otherwise, these objects might be mistakenly classified as belonging to the false class. We describe here a preliminary solution to the problem of sensitivity to distortions. Instead of placing the reference object in the observed scene with the tested objects, we propose a different method of computing and employing the reference function. Our proposed synthetic reference is obtained as a function of the training set, and it is invariant to some distortions determined by this training set. The same 2D synthetic reference function is displayed side by side with each projected image for all of the different points of view. The reference function is designed to recognize every object from the true class, with any distortion defined by the training set, and is capable of locating this object in 3D space.

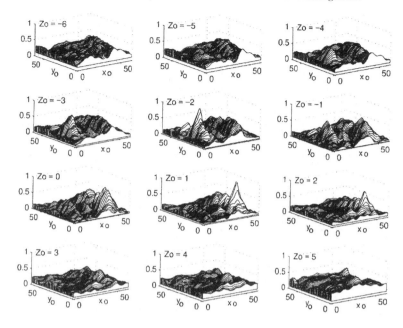

Figure 5 Intensity of the correlation results of the 3D joint transform correlator.

The type of distortions we consider in the present study is object rotations on the *x-z* plane. In other words, the demonstrated system is invariant to any object's rotation, within a limited angular interval, on the *x-z* plane. Our aim is to guarantee that each object from the true class will be identified and located in 3D space no matter what its orientation within the limited angular interval on the *x-z* plane. However, the same concept can be extended toward other kinds of distortions by inclusion of appropriate representatives of these distortions in the training set.

The reference function is actually a synthetic discriminant function (SDF) [5,18], appearing side by side with all object projections. The SDF can be computed by any known optimization algorithm. The fact that the correlation's dimensions have been extended from two to three does not increase the dimensions of the computational problem in our case. Thus, one can choose any off-the-shelf 2D SDF algorithm and implement it in our distortion-invariant 3D correlator. In recent years, many SDF algorithms for distortion-invariant problems have been widely investigated [5]. Among them, the minimum average correlation energy SDF (MACE–SDF) [18] has been successfully demonstrated in a few independent experiments. With the MACE–SDF, one can control the whole correlation plane as well as keep a sharp easily detectable correlation peak. For demonstration purposes of our general concept, we choose the MACE–SDF as the reference function in our first distortion-invariant 3D JTC.

The distortion-invariant 3D JTC is shown in Fig. 6. The cameras record the input scene from different points of view, and each image is displayed on SLM1. Unlike previous versions of the 3D JTC described in Section 7.3.1, here the reference function is not a real object located in the input scene. Instead, plane P_2 is divided into two parts. The various projections of the observed scene, from the various points of view, are displayed, one by one, on one side of plane P_2. On the other side of plane P_2, the same single 2D synthetic reference function constantly appears. In general, the reference function is complex-valued; therefore, one should consider the technique for implementing complex-valued functions in a JTC. In this

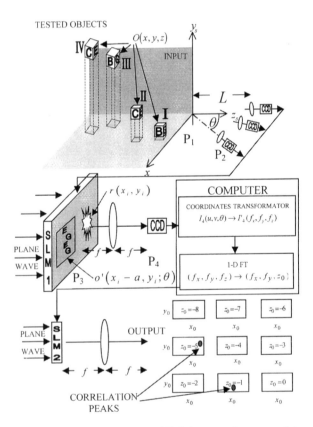

Figure 6 Schematic of the 3D joint transform correlator equipped with the synthetic reference function.

study, the correlator is computer simulated; therefore, we assume that the reference function is complex-valued without any limitations. Other than the new concept of the synthetic reference function, the system is similar to the previous one presented in Section 7.3.1.

In our simulated example, the training set is composed from 42 projections, 21 from the true class and 21 from the false class. The true class contains 21 projections of a cube of $16 \times 16 \times 16$ pixels with the characters B, G, and U on three of its faces. In the central projection ($\theta = 0$), one sees only the letter G. The false class also contains 21 different projections of a cube of the same size, but with the characters C, E, and E on three of its faces. The angular interval of the training set is $100°$, $50°$ to each side, and the angle between any two successive projections in this set is $5°$. These 42 projections of the true and the false classes were used to synthesize the synthetic reference function by the MACE–SDF algorithm.

When the reference function is displayed on plane P_2, the system should be able to recognize and locate true class members, which appear in any orientation within $\pm 50°$ in the *x*-*z* plane. In the test shown in Fig. 6, the input space contains four objects. Two of them, I and III, belong to the true class. Object III is rotated $20°$ from the *z* axis on the *x*-*z* plane. The other two objexts, II and IV, belong to the false class and they should be ignored by the system. In the test stage of our simulation, the point of view was shifted along an arc of $18°$, $9°$ to each side of the *z* axis. The arc's center is in the origin of the *x*-*y*-*z* space, and all of the simulated cameras are directed to this point. The total number of projections in the test stage is 19 in $1°$ increments. Each projection was mapped on plane P_2 side by side with the reference function. Three examples of plane P_2, out of 19, are shown in Fig. 7. Figures 7a, 7b, and 7c show the most extreme projections to the left-hand side of the *z* axis, on the *z* axis, and to the right-hand side of the *z* axis, respectively. The magnitude of the same reference function is shown in the right-hand side of Fig. 7.

All 19 images of plane P_2 were processed by the 3D JTC as described earlier. The output result of this system appears as a collection of correlation planes, each for a different value of the output longitudinal axis z_o. As with every JTC, the expected cross-correlation results between the tested objects and the reference are obtained near the first diffraction order. This region is shown in Fig. 8 for a few values of z_o. The high correlation peaks on planes $z_o = -5$ and $z_o = -1$ indicate the successful identification of the two cubes, I and III, of the true class.

A comparison between two methods of 3D correlation is presented in Fig. 9. In the first method, indicated by triangles, the true class object is correlated with the MACE–SDF for various values of rotation angle in the *x*-*z* plane. In the second method, indicated by rectangles, the same object is

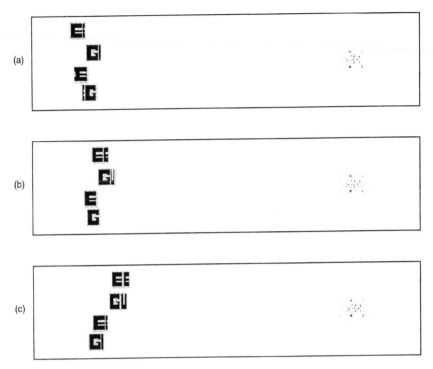

Figure 7 Three images of plane P_2 out of 19 as observed from different points of view.

correlated with the real reference function described in Section 7.3.1. In the range between 0° and 50°, which is the angular range of the training set, the MACE–SDF keeps a stable peak value, whereas the correlation peak of the conventional reference function gradually decreases.

7.3.3 Three-Dimensional Hybrid Correlator

Most of the SLMs, used today as input transparency masks in JTC's, cannot simultaneously provide amplitude and phase modulation with a satisfactory quality. However, many schemes for pattern recognition and other image processing tasks require complex or at least bipolar real reference functions. A possible solution to this problem can be holographic coding at the JTC input plane. In that case, the tested object is sampled by a grating and appears side by side with a computer-generated hologram used as the reference function [19,20]. These systems, however, suffer from a low bandwidth,

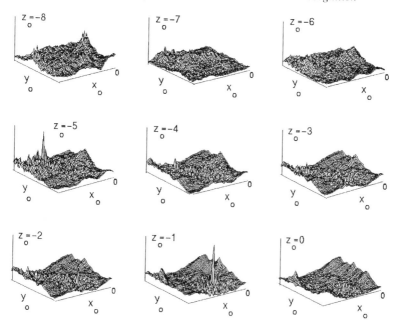

Figure 8 Intensity of the correlation results of the distortion invariant 3D joint transform correlator.

because at most only one third of the available bandwidth actually participates in the process. This drawback causes the loss of the high-resolution information from the observed scene. On the other hand, a VLC equipped with a Fourier hologram can correlate the input image with a general complex reference function. However, the main drawback of the VLC is its inability to process the image's spatial spectrum digitally. When one tries to record the spectrum distribution into a digital processor with any camera, the phase function, which usually contains the objects' shape information is lost. The digital processing is a crucial part of our 3D correlation scheme. Therefore, a new correlator design is required in which the benefits of both classical correlators, the VLC and the JTC, are combined. Such a correlator is described here.

We propose a technique that is similar, but not identical, to one that was employed previously for 2D correlators [21] in which a camera records electronic Fourier holograms of the input scene and transfers them into a computer. Thus, although the hologram distributions are real and positive-valued, the complete complex information of the spatial spectrum is recorded into the computer and can be digitally processed. In addition to

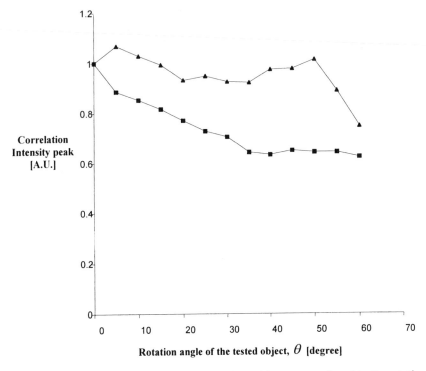

Figure 9 Correlation peak of one true class object versus the object's rotation angle for the conventional reference function (rectangles) and for the SDF reference function (triangles).

being digital processed, the spectrum is multiplied by a filter function in the Fourier plane, and an additional FT of this product yields the desired correlation results. Such a correlator is actually a combination of a JTC (with a point as the reference function) and a VLC (with a spatial filter at the Fourier plane). As a hybrid configuration, it combines the best features of the two types of correlator. Explicitly, it permits an effective complex reference function to be implemented, as is usually possible in VLC. It also lets us perform complicated digital manipulations (in our case, it is a coordinate transformation) of the spatial spectrum of the input function, as is inherently possible with the JTC.

The 3D hybrid correlator is shown in Fig. 10. As previously, a 3D input function $o_1(x_s, y_s, z_s)$, that describes all tested objects in the observed scene is located in coordinate system (x_s, y_s, z_s). The cameras record the input scene from different points of view along an arc on the $(x$-$z)$ plane,

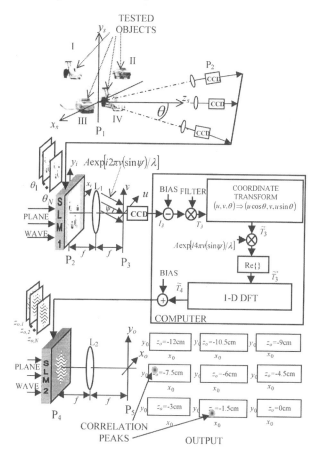

Figure 10 Schematic of the hybrid 3D correlator.

whose center is located at the origin point of (x_s, y_s, z_s). From each point of view, the camera observes plane P_1 through an imaging lens located far from plane P_1. In each point of view, the line OC between the center of th camera's plane and the origin point $(x_s, y_s, z_s) = (0, 0, 0)$ at the x-z plane is orthogonal to the camera's plane. The angle between the z axis and line OC is denoted by θ. For each θ, the projected function $o_2(x_i, y_i, \theta)$ is displayed on SLM1, where (x_i, y_i) are the coordinates of plane P_2.

Next, we consider the intensity distribution on plane P_3 for any angle θ. The complex amplitude coming from SLM1 through lens L_1 interferes with a reference plane wave, hitting the camera plane at an angle ψ from the optical axis on the v-z plane. With the 2D FT relation between planes P_2 and P_3, the intensity on plane P_3 is

$$
\begin{aligned}
I_3(u, v, \theta) &= \left| A \exp\left(i \frac{2\pi}{\lambda} v \sin \psi \right) + O_3(u, v, \theta) \right|^2 \\
&= |A|^2 + |O_3(u, v, \theta)|^2 + A^* O_3(u, v, \theta) \exp\left(-i \frac{2\pi}{\lambda} v \sin \psi \right) \\
&\quad + A O_3^*(u, v, \theta) \exp\left(i \frac{2\pi}{\lambda} v \sin \psi \right)
\end{aligned}
$$

(17)

where A is a constant and $O_3(u, v, \theta)$ is given by

$$
\begin{aligned}
O_3(u, v, \theta) &= \int\int o_2(x_i, y_i; \theta) \exp\left(i \frac{2\pi}{\lambda f} (ux_i + vy_i) \right) dx_i \, dy_i \\
&= \int\int\int o_1(x_s, y_s, z_s) \exp\left(i \frac{2\pi M}{\lambda f} (ux_s \cos \theta + vy_s + uz_s \sin \theta) \right) \\
&\quad dx_s \, dy_s \, dz_s
\end{aligned}
$$

(18)

The last part of Eq. (18) is obtained from the first part due to the same reasons Eq. (8) is obtained from Eq. (6).

Apparently, $I_3(u, v, \theta)$ contains four terms, but only the third term contains the useful spectrum of $o_1(x_s, y_s, z_s)$. We wish to get rid of all unnecessary terms, but, at this stage, we can easily get rid of the bias term by a digital subtraction of the constant term $|A|^2$ from the function $I_3(u, v, \theta)$.

Spatial filtering is the next stage in the correlation process, and the nature of the filter should be determined. Most filters used for pattern recognition in 2D correlators are complex-valued, with nonzero imaginary parts. This is true for hundreds of filters proposed in recent years [5] designed to satisfy many criteria and computed by numerous computational methods. In many cases, the complex filters are coded on real positive computer-generated holograms only to avoid the problem of using a complex-valued optical transparency. However, in our case, the spatial spectrum is already recorded inside the computer; therefore, there is no such constraint on the optical transparency at this stage of the system. We choose for the present demonstration a 3D extension of the 2D phase-only filter (POF) [22]. This choice of filter is made only to illustrate an example of a general complex filter, and the same correlator can be equipped with different filters according to other needs. In fact, the reference function, obtained by the inverse FT of the POF, is just a real, bipolar function. However, the process presented here is valid for general complex reference functions, as well.

The 3D POF of an object function $f(x_s, y_s, z_s)$ is defined as

$$H(u, v; \theta) = \frac{F^*(u, v; \theta)}{|F(u, v; \theta)|} \tag{19}$$

where

$$F(u, v; \theta) = \iiint f(x, y, z)$$
$$\exp\left(i \frac{2\pi M}{\lambda f} (ux\cos\theta + vy + uz\sin\theta)\right) dx\, dy\, dz \tag{20}$$

Multiplying the hologram distributions (less the bias term) by the complex POF function yields

$$T_3(u, v, \theta) = H(u, v, \theta)\left[I_3(u, v, \theta) - |A|^2\right] \tag{21}$$

The product $T_3(u, v, \theta)$ is intended to be Fourier transformed once again to yield the correlation function. At this point, we note that a coordinate transformation should be made before the final FT is performed. That is so because both functions $O_3(u, v, \theta)$ and $F(u, v, \theta)$ have the form of a 3D FT with Fourier coordinates $(u\cos\theta, v, u\sin\theta)$, but the real coordinates of the physical space at plane P_3 are (u, v, θ). Therefore, to get the desired 3D FT, we must transform the function $T_3(u, v, \theta)$ from coordinate system (u, v, θ) to $(u\cos\theta, v, u\sin\theta)$. Following the coordinate transformation $T_3(u, v, \theta)$, denoted now by $\tilde{T}_3(f_x, f_y, f_z)$ becomes

$$\tilde{T}_3(f_x, f_y, f_z) = |\tilde{O}_3(f_x, f_y, f_z)|^2 \tilde{H}(f_x, f_y, f_z)$$
$$+ A^* \tilde{O}_3(f_x, f_y, f_z)\tilde{H}(f_x, f_y, f_z)\exp\left(-i\frac{2\pi}{\lambda} v\sin\psi\right)$$
$$+ A\tilde{O}_3^*(f_x, f_y, f_z)\tilde{H}(f_x, f_y, f_z)\exp\left(i\frac{2\pi}{\lambda} v\sin\psi\right) \tag{22}$$

where $\tilde{O}_3$ and $\tilde{H}$ denote the functions of the image's spectrum and the filter at the new coordinate system, respectively.

We keep in mind that $\tilde{T}_3$ is going to be displayed on SLM2, which is, again, a transparency medium that can get only positive real values. Consequently, only the real part of $\tilde{T}_3$ can be displayed on SLM2. Thus, from the three terms of $\tilde{T}_3$, Re$\{\tilde{T}_3\}$ becomes an expression of six terms, the three of $\tilde{T}_3$, plus their complex conjugate terms. Without an additional processing in $\tilde{T}_3$ given by Eq. (22), the important term of the convolution between $o_1(x_s, y_s, z_s)$ and $h(x_s, y_s, z_s)$ will be obscured by the convolution between $o_1(-x_s, -y_s, -z_s)$ and $h(x_s, y_s, z_s)$. To avoid this overlap between the correlation terms, we multiply $\tilde{T}_3$ by a linear phase function $\exp[i4\pi v(\sin \psi)/\lambda]$. Multiplying $\tilde{T}_3$ by this linear phase function causes a shift of all the correlation terms to one-half of the correlation plane, whereas taking the real part of $\tilde{T}_3$ produces an additional set of correlation terms in the other half of the correlation plane. Thus, the distribution in this stage becomes

$$
\begin{aligned}
\tilde{T}_3'(f_x, f_y, f_z) = {} & 2\mathrm{Re}\left\{\tilde{T}_3(u, v, \theta)\exp\left(\frac{i4\pi v \sin \psi}{\lambda}\right)\right\} \\
= {} & |\tilde{O}_3(f_x, f_y, f_z)|^2 \tilde{H}(f_x, f_y, f_z)\exp\left(\frac{i4\pi v \sin \psi}{\lambda}\right) \\
& + |\tilde{O}_3(f_x, f_y, f_z)|^2 \tilde{H}^*(f_x, f_y, f_z)\exp\left(-\frac{i4\pi v \sin \psi}{\lambda}\right) \\
& + \tilde{O}_3(f_x, f_y, f_z)\tilde{H}(f_x, f_y, f_z)\exp\left(\frac{i2\pi v \sin \psi}{\lambda}\right) \\
& + \tilde{O}_3^*(f_x, f_y, f_z)\tilde{H}^*(f_x, f_y, f_z)\exp\left(-\frac{i2\pi v \sin \psi}{\lambda}\right) \\
& + \tilde{O}_3^*(f_x, f_y, f_z)\tilde{H}(f_x, f_y, f_z)\exp\left(\frac{i6\pi v \sin \psi}{\lambda}\right) \\
& + \tilde{O}_3(f_x, f_y, f_z)\tilde{H}^*(f_x, f_y, f_z)\exp\left(-\frac{i6\pi v \sin \psi}{\lambda}\right)
\end{aligned}
\tag{23}
$$

Another 3D inverse FT of $\tilde{T}_3'$ yields six diffraction orders that are spatially separated from one another along the y_o axis, because of the different linear phase factors that multiply each term. In our setup, this final 3D FT is performed in two steps. First, a one-dimensional digital FT from $u \sin \theta$ to z_o is performed as follows:

$$
\begin{aligned}
\tilde{T}_4(u \cos \theta, v, z_o) = {} & \int \tilde{T}_3'(u \cos \theta, v, u \sin \theta) \\
& \exp\left(-i\,\frac{2\pi M}{\lambda f}\,(z_o u \sin \theta)\right) d(u \sin \theta)
\end{aligned}
\tag{24}
$$

The remaining 2D FTs from $(u \cos \theta, v)$ to (x_o, y_o) are done optically, where each 2D FT is made for a different value of z_o. For any value of z_o, the 2D function $\tilde{T}_4(u \cos \theta, v, z_o)$ plus a proper bias C, to ensure positive values, is displayed on SLM2. Then, with lens L_2, a series of 2D FTs is obtained in the output. The number of 2D transforms in the sequence is actually related to the measure of z_o. When one has only a single SLM, the series of 2D optical FTs is considered sequentially, one at a time. Otherwise, FTs can be treated in parallel.

For our purposes, the interesting term in Eq. (23) is the third one, inasmuch as that term yields the desired convolution between $o_1(x_s, y_s, z_s)$ and $h(x_s, y_s, z_s)$. Because of the spatial separations among all the output terms, we let ourselves ignore all of the terms other than the third one. Some of the higher-order terms may even be eliminated because of the low sampling rate of SLM2. Therefore, correlation results are obtained from an inverse 3D FT of the third term as follows:

$$c(x_o, y_o, z_o) = \int \int \int \tilde{O}_3(u\cos\theta, v, u\sin\theta)\tilde{H}(u\cos\theta, v, u\sin\theta)$$

$$\exp\left(\frac{i2\pi v \sin\psi}{\lambda}\right)$$

$$\times \exp\left(-i\frac{2\pi M}{\lambda f}(x_o u\cos\theta + y_o v + z_o u\sin\theta)\right) \tag{25}$$

$$d(u\cos\theta)\,dv\,d(u\sin\theta)$$

$$= \left[\int\int\int o_1(x, y, z)h(x_o - x, y_o - y, z_o - z)\,dx\,dy\,dz\right]$$

$$* \delta\left(x_o, y_o - \frac{f(\sin\psi)}{M}, z_o\right)$$

The desired correlation result is obtained about the point $(x_o, y_o, z_o) = [0, f(\sin\psi)/M, 0]$.

In the experiment shown in Fig. 10, the observed scene contained four toy cars, I–IV, in various locations. The two race cars, I and IV, were used here as the objects to be recognized, whereas the two patrol cars, II and III, were the false objects. Cars I, II, II, and IV were located at distances of 57.5, 60.5, 50, and 51.5 cm, respectively, from the camera. The distance between the point of origin of the scene and the camera was 50 cm. Three examples of 3 projections from the first experiment are shown in Fig. 11. Figures 11a, 11b, and 11c show the scene as viewed by the camera positioned at θ angles of $-15°$, $0°$, and $15°$, respectively, from the optical axis. The angular increment between successive projections was $1°$ and the angles of $\pm15°$ were the extreme angles of this experiment.

All of the 31 projections were sequentially displayed on SLM1. From SLM1, each projection was optically Fourier transformed by lens L_1 and interfered with the reference plane wave on the camera plane P_3. As a result, 31 electronic Fourier holograms of the various projections were recorded in the computer. Three of the holograms, corresponding to the three projections of Fig. 11, are shown in Fig. 12. From each hologram, first the bias was subtracted and then the functions were multiplied by POFs. Every filter was computed for each projection of the same race

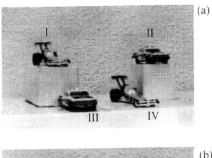

Figure 11   Three images of the scene out of 31, as observed from different points of view at the angles (a) $\theta = -15°$, (b) $\theta = 0°$, and (c) $\theta = 15°$.

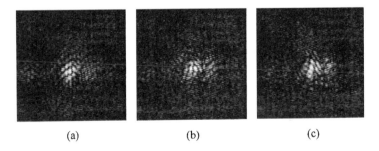

Figure 12   Three electronic Fourier holograms of the three images shown in Fig. 11, as recorded by the CCD at plane P_3.

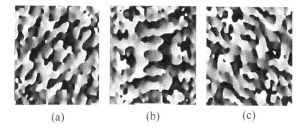

(a)   (b)   (c)

Figure 13   Three examples out of 31 of POFs computed from a single race car observed by the CCD on plane P_2 from the angles (a) $\theta = -15°$ (b) $\theta = 0°$, and (c) $\theta = 15°$.

car position at $(x, y, z) = (0, 0, 0)$, according to Eq. (19). The phase distribution of the computed 256×256 pixel POFs of the extreme left-hand, central, and extreme right-hand projections are shown in Figs. 13a, 13b, and 13c, respectively. After multiplying all of the 31 holograms by the proper complex POFs, we performed the coordinate transformation mentioned earlier. As a result of the coordinate transformation, all of the current data were organized into a 3D complex-valued matrix of size $256 \times 256 \times 67$ pixels. The value 67 is the size of the longitudinal frequency axis approximated by the number $\max(u)\max[\sin(\theta)] = 256\sin(15°)$ according to the coordinate transformation. This value is actually the longitudinal bandwidth of the system and it indicates the resolving power of the system along the longitudinal axis. More precisely, if the size of a pixel on plane P_3 is d, then, according to Eq. (18), the smallest size along the z axis that the system can resolve is $\lambda f / 67 dM$.

To get good separation among the various convolution terms on the output plane, we multiplied the current function by the linear phase function $\exp[i4\pi v(\sin \psi)/\lambda]$. Then, to obtain the final correlation results, we needed to perform a 3D FT on the last matrix. As mentioned earlier, we first performed a digital one-dimensional FT from $(u\cos\theta, v, u\sin\theta)$ to $(u\cos\theta, v, z_o)$. At this point, the real parts of the obtained 2D functions, plus some bias (to ensure a positive-valued matrix), were displayed successively on SLM2, each pattern for a different value of z_o. Nine patterns of the given 67 z_o values are shown in Fig. 14. The correlation results are depicted in Figs. 15 and 16. Figure 15 shows the intensity distribution on output plane P_5, corresponding to the transverse planes with the same z_o values as in Fig. 14. Only the three central diffraction orders are shown here. From Eq. (23), it is clear that each diffraction pattern at any z_o value should be symmetric about the origin. Therefore, we can use the area of the two first diffraction orders as the region in which to look for the recognition peaks.

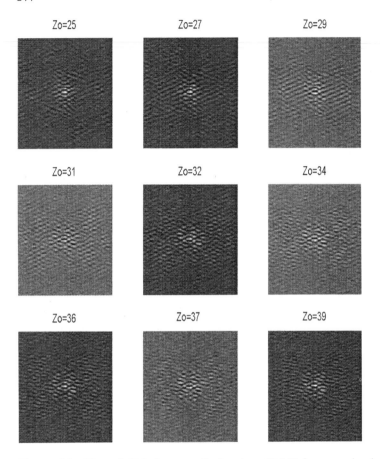

Figure 14   Nine of 67 holograms displayed on SLM2 for several values of longitudinal axis z_o.

The strongest peaks in the two first diffraction orders are found at $z_o = -1.5$ and $z_o = -7.5$ cm. These are the summits of the two recognition peaks of race cars IV and I, respectively. The area of the lower first diffraction order was used as the data for the series of 3D plots shown in Fig. 16a. Each 3D plot shows the intensity distribution on a transverse plane along the z_o axis. Two recognizable peaks appear in the locations of the two true class objects I (at $z_o = -7.5$ cm) and IV (at $z_o = -1.5$ cm), whereas false cars II and III did not grow peaks above the noise level. Every correlation peak has a finite width in all of its three dimensions. In addition to the ordinary dependence of this width on the size of the correlated functions, the width along z_o is inversely dependent on the angular range of the observing cameras, which

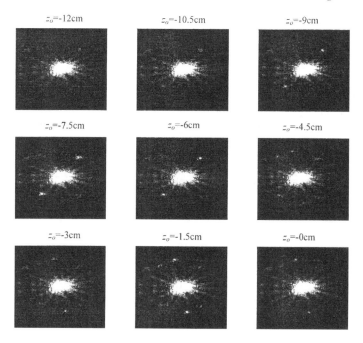

Figure 15 Nine of the output correlation planes obtained by optical Fourier transform of the nine holograms displayed on SLM2.

was $30°$ in this experiment. The other clearly shown peaks in Fig. 16, at $z_o = -6$ cm and $z_o = -9$ cm, are part of the same identification peak whose summit appears at $z_o = -7.5$ cm. Similarly, the peaks at $z_o = 0$ and $z_o = -3$ cm are part of the same identification peak whose summit appears at $z_o = -1.5$ cm. To facilitate observation of the 3D correlation plots, a schematic of the input scene from a top view is included in Fig. 16b.

7.4 CONCLUSIONS

We have shown three possible configurations for 3D correlation between real-world functions. Chronologically, the conventional JTC has been first extended from two to three dimensions. This JTC can locate targets in the entire 3D space, but it is sensitive to geometrical distortions. To overcome this last problem, we have proposed a more sophisticated reference function. This reference is a computer-generated complex function. The complexity of the reference function puts another obstacle on the implementation of the

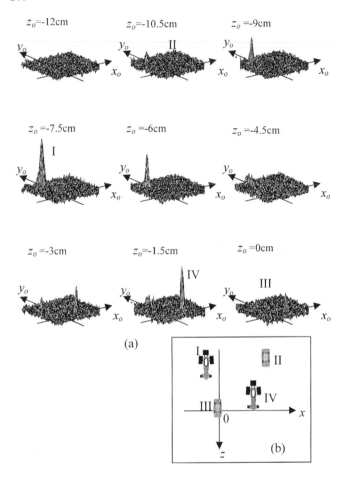

Figure 16 (a) Intensity distribution near the first diffraction order of nine correlation planes, with different values of the longitudinal axis z_o, resulted from the experiment of the 3D hybrid correlator. (b) Top view on the input scene

3D JTC. The 3D hybrid correlator is a solution for this obstacle, and, indeed, for the first time, we have demonstrated a complete experimental 3D correlation with the 3D hybrid correlator.

ACKNOWLEDGMENT

This research was supported by the Israel Science Foundation. Part of this research has been done with Youzhi Li from BGU.

REFERENCES

1. A B VanderLugt. Signal detection by complex spatial filtering. IEEE Trans Inform Theory IT-10:139–145, 1964.
2. J W Goodman. Introduction to Fourier Optics. 2nd ed. New York: McGraw-Hill, 1996, pp 251, 101.
3. F T S Yu and S Jutamulia. Optical Signal Processing, Computing, and Neural Networks. New York: Wiley, 1992, p 34.
4. J Shamir. Optical Systems and Processes. Bellingham, WA: SPIE Pres, 1999.
5. B V K Vijaya Kumar. Tutorial survey of composite filter designs for optical correlators. Appl Opt 31:4773–4801, 1992.
6. R Bamler, J Hofer-Alfeis. Three and four-dimensional filter operations by coherent optics. Opt Acta 29:747–757, 1982.
7. Y B Karasik. Evaluation of three-dimensional convolutions by two-dimensional filtering. Appl Opt 36:7397–7401, 1997.
8. T C Poon, T Kim. Optical image recognition of three-dimensional objects. Appl Opt 38:370–381, 1999.
9. J J Esteve-Taboada, D Mas, J Garcia. Three-dimensional object recognition by Fourier transform proliometry. Appl Opt 38:4760–4765, 1999.
10. B Javidi, E Tajahuerce. Three-dimensional object recognition by use of digital holography. Opt Lett 25:610–612, 2000.
11. J J Esteve-Taboada, J Garcia, C Ferreira. Rotation-invariant optical recognition of three-dimensional objects. Appl Opt 39:5998–6005, 2000.
12. J Rosen. Three-dimensional optical Fourier transform and correlation. Opt Lett 22:964–966, 1997.
13. J Rosen. Three-dimensional electro-optical correlation. J Opt Soc Am A 15:430–436, 1998.
14. J Rosen. Three-dimensional joint transform correlator. Appl Opt 37:7538–7544, 1998.
15. C S Weaver, J W Goodman. A technique for optically convolving two functions. Appl Opt 5:1248–1249, 1996.
16. Y Li, J Rosen. Three-dimensional pattern recognition with a single two-dimensional synthetic reference function. Appl Opt 39:1251–1259, 2000.
17. Y Li, J Rosen. Three-dimensional optical correlators with general complex filters. Appl Opt 39:6561–6572, 2000.
18. A Mahalanobis, B V K Vijaya Kumar, D Casasent. Minimum average correlation energy filters. Appl Opt 26:3633–3640, 1987.
19. D Mendlovic, E Marom, N Konforti. Complex reference-invariant joint-transform correlator. Opt Lett 15:1224–1226, 1990.
20. U Mahlab, J Rosen, J Shamir. Iterative generation of complex reference functions in a joint-transform correlator. Opt Lett 16:330–332, 1991.
21. J Rosen, T Kotzer, J Shamir. Optical implementation of phase extraction pattern recognition. Opt Commun 83:10–14, 1991.
22. J L Horner, P D Gianino. Phase-only matched filtering. Appl Opt 23:812–816, 1984.

8

Three-Dimensional Object Recognition by Means of Digital Holography

Enrique Tajahuerce
Universitat Jaume I, Castellón, Spain

Osamu Matoba
Institute of Industrial Science, University of Tokyo, Tokyo, Japan

Bahram Javidi
University of Connecticut, Storrs, Connecticut

8.1 INTRODUCTION

Optical information processing techniques based in the operation of correlation have proved to be a real alternative to electronic processing in applications such as security, encryption, and pattern recognition [1–12]. Optical techniques provide parallel operation, a high space-bandwidth product, and a large degree of freedom to secure data. These advantages, together with recent advances in electro-optical devices and components, have contributed to the progression of optical correlation methods. However, many of the reported techniques are based in the ability of a lens to perform optically a two-dimensional (2D) Fourier transform operation. Thus, optical pattern recognition technqiues have been commonly restricted to 2D input images.

In the last years, there has been an increasing interest in three-dimensional (3D) optical information processing because of its vast potential applications. In this direction, several ideas have been reported to extend optical techniques to 3D pattern recognition. One basic approach consists in processing a set of different 2D projections of both the 3D scene and the 3D reference object [13]. Recognition is achieved by conventional 2D optical

Fourier transformations performed sequentially over a set of images which have been previously stored by using multiplexing holographic techniques. To avoid optical processing of a large number of 2D perspectives, the collection of 2D projections of the reference object can be used to design an efficient distortion-invariant filter that is used to recognize different 2D views of the 3D input objects [14–16]. In other cases, the 3D object is represented by describing the perspectives with moment invariants such as Fourier–Mellin descriptors [17]. This method allows one to perform 3D object classification using feature space trajectories.

Some studies analyze how to perform 3D convolutions or correlations by planar encoding of 3D images. In this way, conventional Fourier processing techniques can be applied to the resulting 2D encoded image. In one approach, one dimension of the 3D image is sampled and the resulting set of 2D sections of the object are recorded in one plane [18,19]. Other techniques propose different functions to achieve the 3D to 2D mapping [20].

Another class of methods involves the acquisition of different 2D perspectives also, but the recognition is implemented optoelectronically by 3D Fourier transformations [21–23]. In this procedure, a set of 2D perspectives is Fourier transformed optically and converted, with a coordinate transformtion, into data in a 3D frequency space. This information can be processed optoelectronically with 3D correlation techniques to yield the desired 3D recognition. In the practical implementation of the method, the 3D data are processed electronically and 2D Fourier transformations are performed optically. This method has been extended to distortion-invariant pattern recognition with synthetic reference functions [24].

Surface measurement techniques such as Fourier transform profilometry or range imaging methods have also been proposed for 3D shape recognition. In the first approach, a fringe pattern is projected over the 3D reference and input objects [25,26]. The information about the depth and shape of the objects carried out by the distorted grating is obtained by Fourier-transform profilometry and is used to recognize 3D objects in real time with conventional 2D correlations. This method has also been extended to 3D pattern recognition invariant under rotations along the line of sight [27]. Similar ideas have been proposed with other range imaging techniques [28]. In this case, the recognition process is performed with digital 3D correlations.

Holographic methods have been applied to 3D image recognition taking advantage of the ability of a single hologram to record 3D information of the input object. This avoids sequential recording of several 2D perspectives and maintains the phase information [29]. In this direction, one technique has been reported that uses heterodyne scanning interferometry to record a hologram of a 3D reference object. The reference is compared with a 3D input object using the same holographic method. However, this procedure needs further operations to achieve full shift invariance and requires a complex codification of the reference hologram to perform the recognition step [30] Recently a modification by which full 3D shift invariance is possible has been proposed. It uses the Wigner distribution function to analyze the power-fringe-adjusted-filtered correlation output in order to extract the 3D location of the target object [31,32].

Other digital holography techniques provide interesting approaches in order to apply holographic techniques to 3D image processing and pattern recognition. In general, by digital holography, the fully complex information characterizing a Fraunhofer or Fresnel diffraction pattern of an input object is measured, stored, and communicated digitally [33–41]. Holograms are captured by an intensity-recording device, such as a charged-coupled device (CCD) camera, and the input object can be reconstructed by using a digital computer that approximates numerically a diffraction integral. These techniques avoid an analog holographic recording and the corresponding chemical or physical development. The information contained in the digital hologram may also be reconstructed optically after transmission by using computer-generated holograms (CGHs).

In particular, on-axis phase-shifting interferometry permits to record digital holograms with high efficiency. This is a technique well adapted to the space and dynamic range characteristics of the CCD cameras commonly used as intensity detectors [42–44]. This phase measurement method is more precise than using off-axis holography and can also be implemented in real time. Recently, the ability of phase-shift digital holograms to reconstruct in the computer three-dimensional (3D) objects has been reported [45,46]. Also, a technique for optoelectronic encryption of 2D and 3D images has been proposed [47–49].

In this chapter, we describe several methods for 3D image recognition based on phase-shifting digital holography [50,51]. First, the complex amplitude distribution generated by a 3D object at a single plane located in the Fresnel diffraction region is recorded by phase-shifting interferometry. Similarly, a digital hologram of a 3D reference pattern is recorded to be used as correlation filter. In this way, pattern recognition using 3D information can be performed by applying correlation techniques to the digital holograms. In one case, the correlation algorithms are applied directly to the fresnel digital hologram of the reference and input objects [50]. This technique provides high discrimination and, furthermore, allows one to measure small rotations of the reference object. Nevertheless, no shift invariance is obtained along the optical axis. In a second approach, the method is improved, achieving full 3D shift invariance, by applying 3D correlation algorithms to information reconstructed in the object space [51].

In Section 8.2, we briefly review a method to record a Fresnel digital hologram of a 2D object by using phase-shifting interferometry. In Section 8.3, we apply this technique to the case of 3D input objects and we describe our optoelectronic methods for performing 3D image recognition by using fast Fourier-transform correlation algorithms. We also show how a small rotation of the 3D reference object can be measured. In Section 8.4, we describe how to improve the previous method, achieving full 3D shift-invariance. Finally, Section 8.5 summarizes the conclusions.

8.2 DIGITAL HOLOGRAPHY WITH PHASE-SHIFT INTERFEROMETRY

Digital holographic techniques are based in the electronic recording of interference patterns between the diffraction field under study and a reference beam. Among others, there are two basic techniques specially well adapted to information processing applications: off-axis digital holography and phase-shifting interferometry [33–44]. In off-axis digital holography, the object and the reference beam form a very small angle in order to obtain fringe patterns that can be resolved by an intensity detector such as a CCD camera. Only one interferogram is required to measure the amplitude and phase of the diffraction pattern in the computer. Other diffraction patterns generated by the input object can be reconstructed by simulating the reference beam and computing a Fresnel diffraction integral numerically.

In phase-shifting interferometry, the object and the reference beams are directed on-axis toward the detector [42–44]. The method consists of recording several interference patterns, at least three, between the diffraction field generated by the object and a reference beam. The phase of the reference beam is changed continuously or by steps during the recording. This phase shifting can be achieved in different ways by using moving mirrors, diffraction gratings, phase retarders, or liquid-crystal devices [43]. From the different intensity images recorded by the detector, it is possible to obtain in the computer, by applying phase-shifting algorithms, the amplitude and phase of the diffraction pattern generated by the object.

In our experiments, the phase of the reference parallel beam is shifted using phase retardation plates. One approach consists of using two wave plates, a half-wave plate and a quarter-wave plate in the reference beam of a Mach–Zehnder interferometer, as is shown in Fig. 1. The optical system is based on a Mach–Zehnder interferometer. An argon laser beam, expanded and collimated, is divided by a beam splitter into two plane wave fronts traveling in different directions. After reflecting in a mirror, the object beam impinges on the input transparency with transmittance, $t(x, y)$. The light

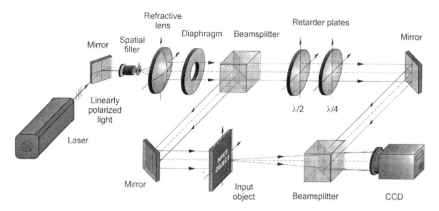

Figure 1 Optical system to record a Fresnel digital hologram of a 2D input object.

diffracted by the input object impinges into the CCD detector through the second beam splitter. Therefore, aside from constant factors, the complex amplitude distribution of the Fresnel diffraction pattern at the output plane, $H_t(x, y)$, is given by

$$H_t(x, y) = \exp\left(i\,\frac{\pi}{\lambda d}\,(x^2 + y^2)\right) \int_{-\infty}^{\infty} \int t(x', y')$$

$$\times \exp\left(i\,\frac{\pi}{\lambda d}\,(x'^2 + y'^2)\right) \exp\left(-i\,\frac{2\pi}{\lambda d}\,(xx' + yy')\right) dx'\,dy' \quad (1)$$

where λ is the wavelength of the laser beam. By measuring the amplitude and phase of $H_t(x, y)$, the amplitude of the input function, $t(x, y)$, may be recovered by computing an inverse Fresnel transform operation either optically or numerically.

The parallel reference beam passes through two phase retarders (one quarter-wave plate and one half-wave plate) and is reflected by a mirror. The system is aligned in such a way that the reference beam generates a plane wave traveling perpendicular to the CCD sensor after reflecting in the second beam splitter. The light provided by the argon laser is linearly polarized. In this way, by suitably orienting the phase retarders, the phase of the reference beam can be shifted to different values, as is shown in Fig. 2. Let us assume that the phase of the parallel beam after the second retarder plate is φ_0 when the fast axis of both plates is aligned with the direction of polarization. In this way, by aligning successively the different slow and fast axes of the phase retarders with the direction of polarization of the incident light, different phase values $\varphi_0 + \alpha_p$ with $\alpha_1 = 0$, $\alpha_2 = -\pi/2$,

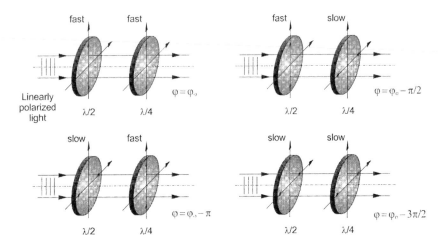

Figure 2 Phase-shifting with retarder plates.

$\alpha_3 = -\pi$, and $\alpha_4 = -3\pi/2$ can be produced. Therefore, the complex field generated by the reference at the output plane can be written as

$$R(x, y; \alpha) = A_R \exp[i(\varphi_R + \alpha_p)] \tag{2}$$

where A_R and φ_R are arbitrary constant values of the amplitude and phase, respectively. The intensity pattern recorded by the linear intensity recording device is then given by

$$I_p(x, y, \alpha_p) = |H_t(x, y) + R(x, y; \alpha_p)|^2 \tag{3}$$

with H_t and R given by Eqs. (1) and (2) respectively. Equation (3) can be written in the following way:

$$I_p(x, y) = [A_H(x, y)]^2 + A_R^2 + 2A_H(x, y)A_R \cos[\phi_H(x, y) - \varphi_R - \alpha_p] \tag{4}$$

From this equation, it is straightforward to show that the complex light field at the output plane can be evaluated by combining the four intensity patterns resulting from the interference of the object beam with the reference beam phase shifted by the previous values of α_p. This technique is known as the four-step method. In this way, we are able to measure the phase

$$\phi_D(x, y) = \phi_H(x, y) - \varphi_R = \arctan\left(\frac{I_4(x, y) - I_2(x, y)}{I_1(x, y) - I_3(x, y)}\right) \tag{5}$$

and the amplitude

$$A_D(x, y) = A_H(x, y)A_R$$
$$= \tfrac{1}{4}\{[I_1(x, y) - I_3(x, y)]^2 + [I_4(x, y) - I_2(x, y)]^2\}^{1/2} \quad (6)$$

As the phase φ_R and amplitude A_R of the reference beam are constant factors they can be substituted by 0 and 1, respectively, in the previous equations. Therefore, aside from constant factors, the complex field diffracted by the 2D input object that is recorded by the digital hologram, $H_t(x, y)$, is just obtained from the combination of the amplitude and phase in Eqs. (5) and (6), respectively {i.e., $H_t(x, y) = A_D(x, y)$ $\exp[i\phi_D(x, y)]$}.

8.3 THREE-DIMENSIONAL PATTERN RECOGNITION WITH DIGITAL HOLOGRAPHY

Phase-shifting digital holography can also be used to record the light diffracted by a 3D input object at a given plane. In this way, it is possible to reconstruct the 3D object digitally in the computer [45,46,49]. Based on this idea, we have proposed a method for the recognition of 3D objects by digital holography [50]. Utilizing phase-shifting interferometry, our procedure electronically records the complex amplitude distribution generated by a 3D object at a single plane located in the Fresnel diffraction region. Similarly, a digital hologram of a 3D reference pattern is recorded to be used as correlation filter. Thus, pattern recognition using 3D information can be performed by applying correlation techniques to the digital holograms. The optical system for digital holographic recording is depicted in Fig. 3. The object beam illuminates the input 3D object, and the reference beam forms an on-axis interference pattern with the light diffracted by the object onto the CCD camera. The 3D object is located at a distance d from the CCD.

By considering the opaque 3D object as an amplitude distribution $O(x, y, z)$ and neglecting secondary diffraction, the amplitude distribution due to the object beam at the CCD, $H_O(x, y)$, within the Fresnel approximation, can be evaluated from the following superposition integral:

$$H_O(x, y) = A_H(x, y)\exp[i\phi_H(x, y)]$$
$$= \frac{1}{i\lambda} \int\int \int_{-\infty}^{\infty} O(x', y'; z) \frac{1}{z} \exp\left(i\frac{2\pi}{\lambda}z\right)$$
$$\exp\left(i\frac{\pi}{\lambda z}[(x - x')^2 + (y - y')^2]\right)dx'\,dy'\,dz \quad (7)$$

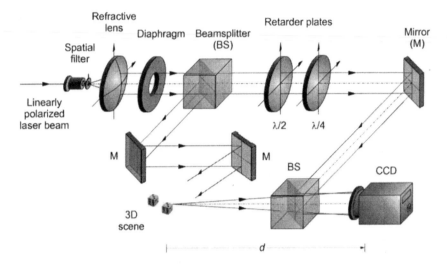

Figure 3   Optical system to record the digital hologram of a 3D object.

As in the previous section, the parallel reference beam passes through two phase retarders, one quarter-wave plate and one half-wave plate, before arriving on axis to the CCD. Again, by aligning successively the different slow and fast axes of the phase retarders with the direction of polarization of the laser beam, different constant phase values $\varphi_R + \alpha_p$ can be produced at the output plane. Therefore, the complex field at the output plane due to the reference beam can be written using Eq. (2).

The intensity pattern recorded by the linear intensity recording device can be written now in the following way:

$$I_p(x, y, \alpha_p) = |H_O(x, y) + R(x, y; \alpha_p)|^2 \tag{8}$$

with H_O given by Eq. (7). Equation (8) can be written as in Eq. (4) but now with $A_H(x, y)$ and $\phi_H(x, y)$ given by Eq. (7). In this way, the complex light field at the output plane can be evaluated with Eqs. (5) and (6). Aside from constant factors, the complex field diffracted by the 3D object that is recorded by the digital hologram, $H_O(x, y)$, is just obtained from the combination of these equations {i.e., $H_O(x, y) = A_D(x, y) \exp[i\phi_D(x, y)]$}.

The reconstruction of an axial view of the 3D input object, at a plane located at a distance d from the output, $U_O(x, y; d)$, can be obtained by computing the following discrete Fresnel integral:

$$U_O(m', n'; d) = \exp\left(-\frac{i\pi}{\lambda d}(\Delta x'^2 m'^2 + \Delta y'^2 n'^2)\right)$$

$$\times \sum_{m'=0}^{N_x-1} \sum_{n'=0}^{N_y-1} H_O(m, n) \exp\left(-\frac{i\pi}{\lambda d}(\Delta x^2 m^2 + \Delta y^2 n^2)\right)$$

$$\exp\left[-i2\pi\left(\frac{mm'}{N_x} + \frac{nn'}{N_y}\right)\right] \tag{9}$$

where variables (m, n) are discrete transversal spatial coordinates in the CCD plane, (m', n') correspond to the object plane coordinates, and N_x and N_y are the number of samples in the x and y directions, respectively. The spatial resolutions at the CCD plane and at the object planes are denoted by $(\Delta x, \Delta y)$ and $(\Delta x', \Delta y')$, respectively. Equation (9) can be computed using a fast Fourier-transform algorithm [52]. It can be shown that the resolutions $\Delta x'$ and $\Delta y'$ along the horizontal and vertical tranversal directions at the object space are given by

$$\Delta x' = \frac{\lambda d}{N_x \Delta x} \quad \text{and} \quad \Delta y' = \frac{\lambda d}{N_y \Delta y} \tag{10}$$

Therefore, in the reconstruction of a view of the 3D input object with Eq. (9), the transversal resolution $(\Delta x', \Delta y')$ changes with the value of the distance d.

As in conventional holography, different regions of the digital hologram record light arising from different perspectives of the 3D object. Thus, we can also reconstruct different views of the 3D object by using different windows in the digital hologram, as is shown in Fig. 4. To reconstruct a

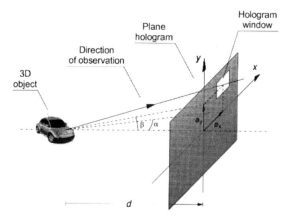

Figure 4 Generation of 3D perspectives from the digital Fresnel hologram.

particular view of the 3D object, first we must consider only a rectangular window, $H_o'(m, n; a_x, a_y)$, in the digital hologram $H_O(m, n)$ centered at the proper location. Parameters a_x and a_y denote the coordinates of the center of the window. The information contained in this region of the hologram corresponds to a direction of observation which subtends angles α and β with the optical axis given by $\alpha = a_x \Delta x/d$ and $\beta = b_x \Delta y/d$. Second, we must consider that the hologram is illuminated by a parallel light beam directed toward the 3D object (i.e., a light beam tilted by angles α and β with respect to the optical axis). In this way, the perspective of the object will remain centered in the process of digital reconstruction of the hologram. Note that the maximum angle of reconstruction is limited only by the size of the detector.

From the above considerations, the rectangular region of the hologram to be used for reconstruction of different angular views of the 3D object must be defined as

$$
H_O'(m, n; a_x, a_y) = H_O(m, n) \, \mathrm{rect}\left(\frac{m - a_x}{b_x}, \frac{n - a_y}{b_y}\right)
$$
$$
\exp\left[i \frac{2\pi}{\lambda d} (\Delta x^2 a_x m + \Delta y^2 a_y n)\right]
\tag{11}
$$

where rect(x, y) is the rectangle function and b_x and b_y denote its transversal size. Now, the discrete complex amplitude distribution $U_O(m', n'; d; \alpha, \beta)$ at a plane located at a distance d and tilted by angles α and β can be computed using the following equation:

$$
U_O(m', n'; d; \alpha, \beta) = \exp\left(\frac{-i\pi}{\lambda d} (\Delta x'^2 m'^2 + \Delta y'^2 n'^2)\right)
$$
$$
\times \sum_{m'=0}^{N-1} \sum_{n'=0}^{N-1} H_O'\left(m, n; \frac{\alpha d}{\Delta x}, \frac{\beta d}{\Delta y}\right)
$$
$$
\times \exp\left(\frac{-i\pi}{\lambda d} (\Delta x^2 m^2 + y^2 n^2)\right)
$$
$$
\exp\left[-i2\pi\left(\frac{m'm}{N_x} + \frac{n'n}{N_y}\right)\right]
\tag{12}
$$

Points on the surface of the object at distances z from the hologram different than d will appear defocused in the reconstructed image. However, the planes of reconstruction can be changed easily in the computer, starting from the same digital hologram. Note also that the field of focus can be increased, diminishing the size of the hologram window at the expense of a reduction of resolution.

An alternative way of reconstructing a given perspective of the 3D input object at a distance d is to use the propagation transfer function method instead of Eq. (12); that is,

$$U_O(m, n; d; \alpha, \beta) = \mathcal{F}^{-1}\left\{\mathcal{F}\left[H'_O\left(m, n; \frac{\alpha d}{\Delta x}, \frac{\beta d}{\Delta y}\right)\right]\right.$$

$$\left.\exp\left[-i\pi\lambda d\left(\frac{u^2}{(\Delta x N_x)^2} + \frac{v^2}{(\Delta y N_y)^2}\right)\right]\right\} \quad (13)$$

where $\mathcal{F}$ denotes the fast Fourier transform and u and v are discrete spatial frequencies. In this approach, the resolution at the input and output planes are the same for any propagation distance d.

In order to perform recognition of 3D objects, we use correlation techniques based on Fourier matched filters starting from the information obtained by digital holography. Let us consider the previous 3D object $O(x, y, z)$ as a reference object to be recognized. A second 3D object, $P(x, y, z)$, is located also at a distance d from the CCD, and its corresponding digital Fresnel hologram is obtained by phase-shift interferometry, $H_P(x, y)$. The correlation between different views of this 3D input object with a given perspective of the reference object $O(x, y, z)$ can be evaluated by extracting partial windows from $H_P(x, y)$ at different positions in the hologram, followed by reconstructing the object at a distance d and computing the correlation with the reconstruction of the reference. However, this approach requires extensive computing operations. Alternatively, using Eq. (13), the correlation intensity of one perspective of the reference $U_O(x, y; d; \alpha, \beta)$ with one different perspective of the input $U_P(x, y; d; \alpha', \beta')$ generated from the respective holograms can be written directly as

$$C_{OP}(x, y; \alpha, \beta; \alpha', \beta') = |\mathcal{F}^{-1}\{\mathcal{F}\{U_O(x, y; d; \alpha, \beta)\}\mathcal{F}^*\{U_P(x, y; d; \alpha', \beta')\}\}|^2$$

$$= |\mathcal{F}^{-1}\{\mathcal{F}\{H'_O(x, y; d; \alpha, \beta)\}\mathcal{F}^*\{H'_P(x, y; d; \alpha', \beta\phi)\}\}|^2$$

$$(14)$$

Thus, by correlating different regions of the two holograms, properly modified by a linear phase factor, we are evaluating the correlation between different perspectives of the 3D objects. It is to be noted, however, that by directly using the information in the Fresnel digital hologram to evaluate the correlation, the technique is not shift invariant under axial displacements. Also, rough objects involve fast fluctuations of the reconstructed phase under translations in a plane orthogonal to the optical axis, thus reducing the transversal shift invariance. Nevertheless, this technique has a high sensitivity under small object motion, and this fact can be useful for measuring a small rotation of a rough 3D object.

We carried out a 3D object recognition experiment using two reproductions of cars with an approximate size of $25 \times 25 \times 45$ mm. They were located at a distance $d = 865$ mm from the CCD detector. The gray-level pictures of Figs. 5a and 5b were obtained by recording a digital hologram of each car model, $H_O(x, y)$ and $H_P(x, y)$, respectively. The reconstruction was performed by simulating Fresnel propagation with Eq. (7), using a fast Fourier-transform algorithm, applied to the entire hologram.

In Fig. 6a, we show a plot of the autocorrelation of the object in Fig. 5a performed by using the information contained in a partial region of its digital hologram. This partial hologram is obtained by placing a window centered at the origin of $H_O(x, y)$. Figure 6b shows the cross-correlation of the 3D object in Fig. 5a with that in Fig. 5b obtained also by digital holography. The reference hologram is the previous digital hologram, and the input hologram is obtained by placing a window with the same size centered at the origin of $H_P(x, y)$. Both plots are normalized to the same value. For comparison, we present the conventional 2D autocorrelation and in Fig. 6c and the 2D cross-correlation in Fig. 6d using the objects in Figs. 5a and 5b, respectively, registered as 2D intensity images.

In Fig. 7, we show the correlation obtained by digital holography between the 3D object in Fig. 2a used as a reference and the same slightly rotated 3D object used as input. We utilized a window centered at the origin

(a)   (b)

Figure 5   Computer reconstructions by digital holography of (a) the reference and (b) the input 3D objects.

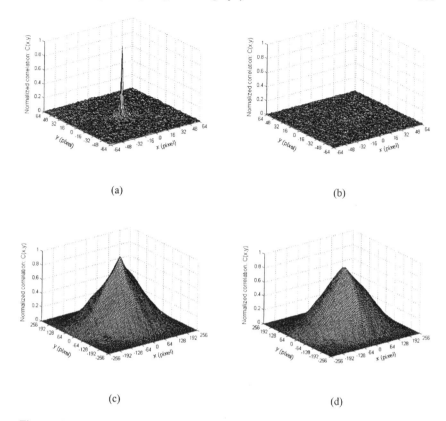

(a)　　　　　　　　　　　　　　(b)

(c)　　　　　　　　　　　　　　(d)

Figure 6 (a) Autocorrelation and (b) cross-correlation of the 3D objects represented in Fig. 5 carried out by digital holography. (c) Conventional 2D autocorrelation and (d) cross-correlation of the irradiance objects.

of the reference hologram and a window with the same size shifted to different positions in the hologram of the input. The information contained in these windows was correlated by using Eq. (14). Figure 7a shows a 3D plot of the maximum peak of the correlation as a function of the displacement (a_x, a_y) of the window in the hologram of the input. A sharp peak is obtained for $a_x = -12, a_y = -2$, corresponding to an angle of view $(\alpha, \beta) = (0.007°, 0.001°)$. Figure 7b shows the correlation for this perspective of the 3D input object. From this result, it is possible to measure the orientation of the input with respect to the reference object.

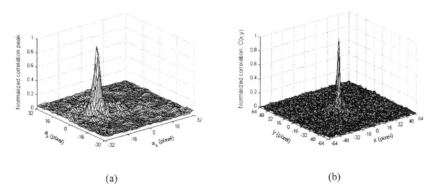

(a)   (b)

Figure 7 Correlation by digital holography of the 3D object in Fig. 5a with a rotated version of the same object: (a) correlation peak versus angle of view and (b) correlation for the angle giving the maximum peak.

8.4 SHIFT-INVARIANT THREE-DIMENSIONAL PATTERN RECOGNITION

It is possible to modify the previous digital holographic technique to achieve shift-invariant 3D pattern recognition [51]. The new method to analyze 3D scenes will allow us to perform 3D correlations between the information obtained from a digital hologram of the 3D reference and a single digital hologram of the input scene. Now, once the holograms have been obtained, we calculate the amplitude and phase of the complex wave front generated by the 3D objects in a set of planes parallel to the output plane within the Fresnel approximation. In other words, in contrast with the technique described in the previous section, instead of using the digital holograms directly, we first evaluate the amplitude distribution of the objects in the 3D object space. This approach permits one to apply 3D correlation techniques to the 3D amplitude distribution generated by the objects. It is also possible to reproduce the complex amplitude distribution in planes tilted with respect the output plane by using partial information from the digital holograms. Therefore, not only the three Cartesian coordinates of the reference in the 3D input space but also the relative out-of-plane rotation of the target with respect to the reference can be obtained. Also, by filtering the phase information, we achieve a system less sensitive to noise and fast fluctuations of the complex distribution produced by the rough surfaces of the objects. With this new method, we are able to achieve full 3D shift-invariant pattern recognition.

As in the previous section, the first step of this optoelectronic procedure to perform 3D pattern recognition is to record a digital hologram of the 3D objects under study. The optical system for digital holographic recording is again that depicted in Fig. 3. By phase-shift interferometry. Eqs. (5) and (6) provide the amplitude $A_H(x, y)$ and phase $\phi_H(x, y)$ of the Fresnel diffraction pattern generated by the 3D scene at the output plane. As before, digital recording of the complex distribution generated by the 3D scene allows us to reconstruct the light field at any plane orthogonal to the output plane, including those planes containing the 3D objects, as shown in Fig. 8. This permits us to reconstruct the 3D light distribution generated by the scene in a volume with a size limited only by the Fresnel approximation and the characteristics of the digital algorithm used to reconstruct the information. Similar 3D objects will generate similar 3D light distributions in volumes located at the same relative distance from the object. Thus, applying 3D correlation techniques to the 3D light distribution volume allow us to recognize the presence and location of a 3D reference in a 3D input scene.

The complex amplitude distribution generated by the 3D object at an arbitrary distance z from the output plane can be evaluated, starting from the digital hologram, by using the following diffraction integral in the Fresnel approximation:

$$U_O(x, y, z) = \frac{1}{i\lambda z} \exp\left(i\frac{2\pi}{\lambda}z\right) \times \int_{-\infty}^{\infty} \int H_O(x', y')$$
$$\exp\left(i\frac{\pi}{\lambda z}[(x - x')^2 + (y - y')^2]\right)dx'\,dy' \tag{15}$$

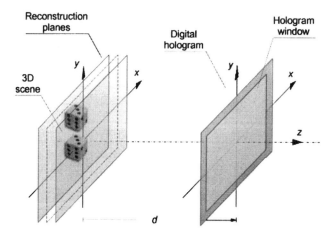

Figure 8 Reconstruction of the amplitude distribution at different planes in the object space from the Fresnel digital hologram of a 3D object.

Substituting Eq. (7) into Eq. (15) and after some mathematical operations, the amplitude distribution at a distance z from the output plane can be written as

$$U_O(x, y, z) = \int_{-\infty}^{\infty} \frac{-1}{\lambda(z' - z)} \exp\left(i\frac{2\pi}{\lambda}(z' - z)\right)$$

$$\exp\left(\frac{i\pi}{\lambda(z' - z)}(x^2 + y^2)\right)$$

$$\times \int_{-\infty}^{\infty} \int O(x', y', z') \exp\left(\frac{i\pi}{\lambda(z - z)}(x'^2 + y'^2)\right)$$

$$\times \exp\left(\frac{i2\pi}{\lambda(z' - z)}(xx' + yy')\right) dx'\, dy'\, dz' \quad (16)$$

For each distance z, Eq. (16) represents a coherent superposition of different Fresnel diffraction patterns. Each diffraction pattern is generated by the light coming from regions of the object located at a fixed distance z' from the output plane and results from a free-space propagation through the distance $\bar{z} = z' - z$. For a reconstruction plane at a fixed distance z, only the regions of the 3D object with a value of z' equal to that of z will appear focused in the reconstruction plane but other areas of the object appear defocused in the same plane. The distribution $U_O(x, y, z)$ constitutes a 3D function that contains information about the location of 3D objects in the 3D scene under consideration.

It is important to remark that the 3D transformation given by Eq. (16) represents a linear shift-invariant operation. A displacement of the 3D input scene, $O'(x, y, z) = O(x - a, y - b, z - c)$, where a, b, and c denote the relative coordinates of the displacement, implies a displacement of the resulting distribution $U_O'(x, y, z) = U_O(x - a, y - b, z - c)$, as can be easily demonstrated substituting $O(x, y, d)$ by $O'(x, y, d)$ in Eq. (7). Thus, application of correlation techniques to $U_O(x, y, z)$ provides information about the presence and position of the reference 3D object. This fact is represented in Fig. 9, where two similar objects are located in two different positions of the 3D space. From Fig. 9, it is clear that the diffraction pattern generated by one of the objects at a given distance $\bar{z}$ from an arbitrary reference point will be the same than that generated by the other input object in a plane located at the same distance. In other words, the diffraction pattern generated by the 3D reference in plane A (see Fig. 9) will be similar to that generated by the 3D object in plane B if this object is similar to the reference, and this can be applied to the rest of planes constituting the diffraction volume.

Let us denote by $R(x, y, z)$ the amplitude distribution of a 3D reference object and by $S(x, y, z)$ that of a different 3D object. The input scene,

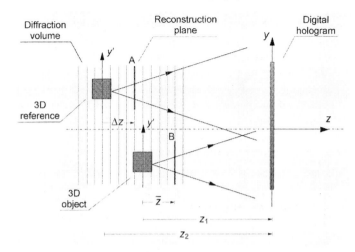

Figure 9 Reconstruction of the irradiance volume for two similar objects. The diffraction patterns generated at planes A and B are also similar.

$O(x, y, z)$, containing both objects, which are located at arbitrary positions, can be described in the following way:

$$O(x, y, z) = R(x - a, y - b, z - c) + S(x - a', y - b', z - d') \qquad (17)$$

where the parameters (a, b, c) and (a', b', c') denote the Cartesian coordinates of the position of the reference $R(x, y, z)$ and the object $S(x, y, z)$, respectively, in the 3D input scene. Due to the shift-invariant property, the 3D correlation between the 3D amplitude distribution $U_O(x, y, z)$ and $U_R(x, y, z)$, both measured by recording respective digital holograms and using Eq. (15), is given by

$$
\begin{aligned}
C_{OR}(x, y, z) &= U_O(x, y, z) * U_R(x, y, z) \\
&= C_{RR}(x, y, z)\delta(x - a, y - b, z - c) \qquad (18) \\
&\quad + C_{SR}(x, y, z)\delta(x - a', y - b', z - d')
\end{aligned}
$$

where $*$ denotes correlation and δ is the Dirac delta function. In Eq. (18), $C_{RR}(x, y, z)$ is the autocorrelation of the 3D amplitude distribution generated by the reference and $C_{SR}(x, y, z)$ is the cross-correlation of the 3D distribution generated by the reference with that generated by the other 3D input object. Due to the autocorrelation operation, a sharp correlation peak is expected to appear, detecting the presence and 3D position of the reference in the input scene. This peak allows us to discriminate the 3D target from other different 3D objects as we show in the experimental results.

Digital reconstruction of the 3D light distribution generated by the input objects can be performed by computing the Fresnel integral in Eq. (15) for different distances z. Let $H_O(m, n)$ be the discrete amplitude distribution of the digital hologram, where m and n are discrete coordinates along the orthogonal directions x and y, respectively. As in the previous section, $x = m\Delta x$ and $y = n\Delta y$, Δx and Δy being the resolution of the CCD detector. The discrete 3D complex amplitude distribution $U_O(m, n, p)$ at a plane orthogonal to the digital hologram, located at a distance $z = \Delta z p$, is given, aside from constant factors, by the following set of Fresnel transformations:

$$
U_O(m, n, p) = \mathcal{F}^{-1}\left\{ \mathcal{F}\{H_O(m, n)\} \exp\left[i\pi\lambda p\Delta z\left(\frac{u^2}{((\Delta x N_x)^2} + \frac{v^2}{(\Delta y N_y)^2} \right) \right] \right\}
$$

(19)

In Eq. (19), Δz is the resolution along the optical axis, N_x and N_y are the number of pixels of the detector along the x and y axis, respectively, and u and v are transversal discrete spatial frequencies along these directions. In evaluating Eq. (19), the transversal resolution remains the same, independent of the propagation distance z, as in Eq. (13), and equal to that in the digital Fresnel hologram plane (Δx, Δy).

Pattern recognition of 3D objects is then performed by evaluating the discrete 3D amplitude distribution generated by the reference, $U_R(m, n, p)$, and the input scene, $U_O(m, n, p)$, from the respective digital holograms using Eq. (19), and computing the correlation in Eq. (18) numerically. This is done by using a discrete Fourier transform approach; that is,

$$
C_{OR}(m, n, p) = \mathcal{F}^{-1}\mathcal{F}\{\{U_O(m, n, p)\}\mathcal{F}^*\{U_R(m, n, p)\}\}
$$

(20)

where now $\mathcal{F}$ denotes the 3D discrete Fourier transformation. It is important to note that to reduce the sensitivity to phase variations (introduced by rough surfaces, by translations of the objects, or by distortions of the object beam), the 3D irradiance distributions $|U_O(m, n, p)|^2$ and $|U_O(m, n, p)|^2$ must be used, instead of the respective 3D complex amplitudes, in Eqs. (18) and (20).

We performed a preliminary experiment on 3D pattern recognition to detect the presence and 3D position of a 3D reference in an input scene. The optical setup in Fig. 3 was constructed again using an Ar laser with wavelength $\lambda = 514.5$ nm. The wave plates were also $\lambda/2$ and $\lambda/4$ phase retarders adapted to the previous wavelength. The reference object was a cubic die with a lateral size equal to 4.6 mm. The center of the die was located at a distance $d_1 = 345$ mm from the output plane. In Fig. 10a, we show a picture of the irradiance distribution at a distance $z = d_1$ mm from the output plane corresponding to the reference object. It was obtained by recording a digital Fresnel hologram, computing the phase and amplitude with Eqs. (5) and (6) and, afterward, evaluating Eq. (19) using a fast Fourier-transform algorithm.

As the input scene we use two similar dice located at difference distances. One was located at the same position as the reference, the other had its center located at a different axial distance from the output plane and was displaced transversally and rotated a small angle around the vertical axis. The 3D amplitude distribution generated by the input scene was obtained by

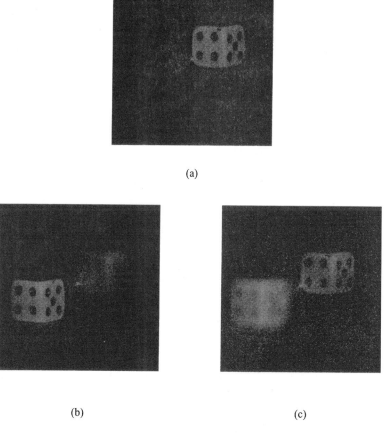

(a)

(b)   (c)

Figure 10 Reconstructions by digital holography of the 3D reference object and the 3D input scene for two focusing distances: (a) reference object, (b) input objects for $z = 315\,\text{mm}$, and (c) input object for $z = 345\,\text{mm}$.

recording a digital hologram and using Eq. (19) to generate a set of 2D distributions with different propagation distances z. This set of 2D functions constitutes the 3D distribution to be used in the recognition step. We show the irradiance distribution generated by the 3D input scene at only two planes: those corresponding to the irradiance at a distance $z = 315\,\text{mm}$ and $z = 345\,\text{mm}$ in Fig. 10b and 10c, respectively. Note that the three 2D distributions in Fig. 10 are just three elements of the set of 2D functions characterizing the 3D reference and input scene.

Before evaluating the correlation in Eq. (20), some previous operations are applied to improve the efficiency of the technique. First, to reduce the computation time, instead of computing the 3D correlation between the two 3D functions generated from the digital holograms, we only consider as the reference function the 2D distribution shown in Fig. 10a. In this way, this 2D reference function can be sequentially correlated in the computer with the different 2D functiosn that characterize the 3D input scene by computing simple 2D correlations. Second, to reduce the sensitivity to phase fluctuation due to the roughness of the object surfaces and other phase effects, the reconstructed phase is filtered and only the irradiance distribution is used in the correlation process. Finally, to increase the discrimination ability, we use a nonlinear filter to perform the correlation. The filter is constructed using only the phase associated to the complex conjugate of the Fourier transform of the reference function. The result, after computing the correlation for different 2D reconstructions of the irradiance distribution associated to the input, constitutes a 3D correlation volume. By analyzing the correlation peaks in this volume, it is possible to determine the 3D position of the reference in the input scene.

In Fig. 11, we show a plot of the correlation between the reference and two different 2D sections of the 3D input amplitude distribution reconstructed from the digital hologram. We selected the same perspective reconstructions that are depicted in Fig. 10, which give us two local maxima of the correlation. Figure 11a corresponds to the correlation between the reference in Fig. 10a with the 2D amplitude distribution in Fig. 10b, whereas Fig. 11b shows the result obtained by correlation of the distributions in Figs. 10a and 10c. Both plots, normalized to the same value, show a clear maximum at the locations of the 3D reference. The maximum in Fig. 11a is lower than that in Fig. 11b due to the small rotation of the 3D object located at this point. The locations of the maxima determines the transversal locations of the objects in the input scene with respect to the original position of the reference. The distance z for which we obtain the maximum value of the correlation peak, $z = 315\,\text{mm}$ and $z = 345\,\text{mm}$, determines the axial position of the two objects. It is important to note that the correlations in Fig. 11 are only two members of the set of correlations achieved for the different distances z.

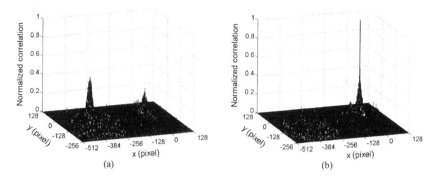

Figure 11 Correlation of the irradiance distribution associated to the reference with that of the input scene for different distances. (a) Correlation of the distribution in Fig. 10a with that in Fig. 10b, and (b) correlation of Fig. 10a with Fig. 10c.

To point out the dependence of the correlation peak with the reconstruction distance, in Fig. 12 we represent the correlation peaks for different propagation distances z. In order to show in the same figure the correlation peak for both objects and different propagation distances, for each distance z we extract a profile of the correlation function along the direction x' in the XY plane that includes both local maxima. This profile is then represented for each distance z in Fig. 12. The plot is then normalized between the minimum and the maximum values of all the correlation peaks. There are two clear local maxima at the two different axial positions of the reference

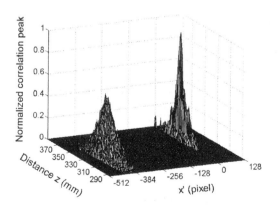

Figure 12 Value of the correlation along the direction x' in the XY plane, that includes both local maxima in Fig.7, for different propagation distances z.

object in the input scene. With the information obtained from this figure, it is possible to localize the 3D reference in the 3D input space. In this way, one of the objects is located at the same position as the original reference object. The second one, similar to the reference but displaced and rotated, is located at 3D coordinates with respect to the original position of the reference given by $(-7.72, -3.35, -30)$ mm.

To analyze the discrimination ability of this technique, we designed a new 3D input scene. In Fig. 13, we show two reconstructions of this input scene in which an object similar to the reference remains in the same position and a second different object, as a false target, has been located with different coordinates. The pictures have been obtained by recording a new digital hologram and reconstructing the amplitude distribution at two different distances. By reconstructing the 2D irradiance distribution as a function of the distance z, as in the previous experiment, it is possible to generate a new 3D input function $U_O'(x, y, z)$ characterizing this second input scene. The 3D correlation of this function with the reference function will again provide information about the presence and position of the reference object in the input scene and discriminating the false target. As in the previous experiment, instead of a real 3D correlation, we perform a set of 2D correlations between the reference distribution in Fig. 10a and different sections of this new 3D input function. The correlation function corresponding to only two different distances is represented in Figs. 14a and 14b. Now, only

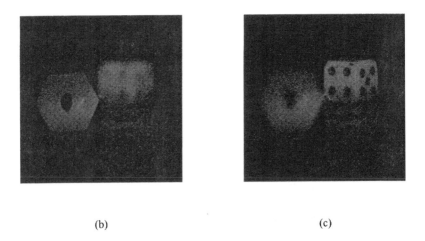

(b)   (c)

Figure 13 Computer reconstructions by digital holography of a second 3D input scene containing a false target for two different focusing distances: (a) $z = 315$ mm and (b) $z = 345$ mm.

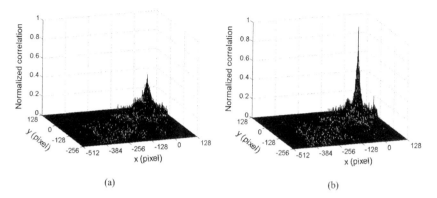

(a)   (b)

Figure 14 Correlation of the irradiance distribution associated to the reference with that of the input scene in Fig. 13 for different distances. (a) Correlation of the distribution in Fig. 10a with that in Fig. 13b, and (b) correlation of Fig. 6a with Fig. 13c.

one maximum in the 3D position corresponding to the location of the reference object is detected, showing the ability of the method to discriminate between different objects.

We would like to note that the object in Fig. 10 closer to the CCD was rotated with respect the reference. In this way, the corresponding correlation peak in Fig. 11a is lower than that associated to the reference in Fig. 11b. The height of this correlation peak can be increased by reconstructing the amplitude distribution generation by the 3D input scene at planes tilted with respect the output plane [51]. This can be done by using partial windows in the corresponding Fresnel digital hologram, as was explained in the previous section (see Fig. 4). This operation allows us to measure the rotation angle of the 3D object under consideration with respect the original reference.

In Fig. 15, we present a diagram of this situation with two objects in the input scene: the refeence and a rotated version of the reference. Let us suppose that the center of the digital hologram faces a given surface of the 3D reference. When the 3D object is tilted a small amount, a different region of the hologram faces perpendicularly to the same surface of the object. By reconstructing the light distribution at different planes, tilted with respect the output plane, using only this region of the digital hologram, we can generate a diffraction volume in which the values of the amplitude generated by the rotated object are similar to those generated by the original reference. Applying correlation techniques now allows us to detect the 3D reference

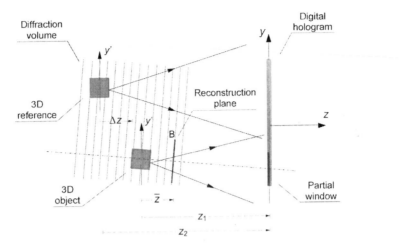

Figure 15   Reconstruction of the irradiance volume with different perspectives. The diffraction patterns at planes A and B generated by the respective objects are similar.

even with out-of-plane rotations. Again, irradiance distributions must be considered, instead of amplitude, to avoid phase effects.

To verify the previous idea, we performed the correlation of the reconstructed reference object in Fig. 10a with reconstructions of different perspectives of the input scene in Fig. 10b. The correlation was evaluated only for a fixed focusing distance, that reconstructing the object located at a distance $z = 315$ mm. The different perspectives were reconstructed by considering only a partial window in the hologram with a horizontal size of 256 pixels and a vertical size equal to the size of the CCD. The window was displaced only in a horizontal direction to obtain a set of different perspectives. As an example, the correlations of two different reconstructed perspectives of the 3D input scene with one of the 3D reference are shown in Fig. 16. In Fig. 16a, the input object was generated with a perspective angle of $-0.9°$ with respect the optical axis, whereas in Fig. 16b, the angle is $+0.9°$. We can compare these results with that obtained in Fig. 11a, where the input object was reconstructed with a view parallel to the optical axis. Note that the correlation peak is higher in Fig. 13a than in Fig. 11a, indicating that the input object was rotated with respect the reference object. In Fig. 17, we show the value of the correlation peak at the location of the object as a function of the lateral displacement of the window used to reconstruct the amplitude distribution (i.e., as a function of the angle of view). A clear maximum appears at a displacement of -544 pixels, corresponding to an angle of view of $0.9°$ with respect to the optical axis.

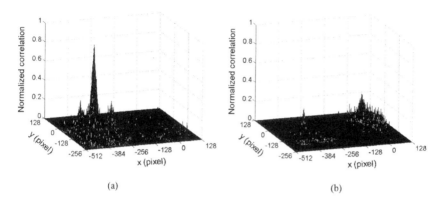

(a)   (b)

Figure 16 Correlation of the irradiance distribution associated to the reference with that of the input scene viewed from different perspective angles: (a) $-0.9°$ and (b) $+0.9°$.

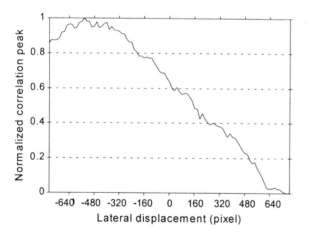

Figure 17 Value of the correlation peak between the reference in Fig. 10a and the input scene, for a fixed distance z, as a function of the lateral position of the window in the input digital hologram.

8.5 CONCLUSIONS

We have described an optoelectronic method to perform 3D object recognition by using digital phase-shifting holography. It is based on the ability of a digital Fresnel hologram to reconstruct different perspectives of a 3D object. The technique allows us to obtain 3D information of the reference and input objects in a single step. Recognition is carried out by a digital matched filter technique applied directly to the holographic information. We have shown that this technique can measure small orientation changes of the 3D object.

Also, we have improved the previous technique in order to achieve full shift invariance and to reduce the sensitivity to noise generated by the rough surfaces of the 3D objects. In this second approach, the correlation operation is applied in the object space to the 3D irradiance distribution generated by the reference and input objects. These 3D irradiance distributions are generated from single digital holograms by reconstructing the amplitude distribution generated by the 3D inputs at different axial distances. The method allows one to perform 3D correlations in order to recognize the presence and 3D position of the reference in the 3D input scene. In the first experiment, the technique has been simplified by evaluating only 2D correlations between 2D sections of the light distribution generated by the 3D reference and the 3D input scene. Also, different perspectives of the 3D input scene can be generated, allowing one to detect the reference even when small out-of-plane rotations have been applied.

The advantages of applying digital holography to pattern recognition have been verified experimentally. By using the first approach, we achieve high discrimination and a method to measure small angular displacements of 3D objects with high accuracy. In the second approach, we obtain a simple method to detect the 3D position of a reference object starting from single digital Fresnel holograms.

REFERENCES

1. A Vander Lught. Signal detection by complex spatial filtering. IEEE Trans Inform Theory IT-10:139–145, 1964.
2. C S Weaver, J W Goodman. A technique for optically convolving two functions. Appl Opt 5:1248–1249, 1966.
3. J L Horner, P D Gianino. Phase-only matched filtering. Appl Opt 23:812–816, 1984.
4. D Psaltis, E G Paek, S S Venkatesh. Optical image correlation with a binary spatial light modulator. Opt Eng 23:698–704, 1984.

5. Ph Réfrégier. Filter design for optical pattern recognition: multicriteria optimization approach. Opt Lett 15:854–856, 1990.

6. B Javidi, J L Horner, eds. Real-time Optical Information Processing. San Diego, CA: Academic Press, 1994.

7. H-Y Li, Y Qiao, D Psaltis. Optical network for real-time face recognition. Appl Opt 32:5026–5035, 1993.

8. B Javidi, J L Horner. Optical pattern recognition for validation and security verification. Opt Eng 33:1752–1756, 1994.

9. Ph Réfrégier, B Javidi. Optical image encryption based on input plane and Fourier plane random encoding. Opt Lett 20:767–769, 1995.

10. B Javidi. Smart Imaging Systems. Bellingham, WA: SPIE Press, 2001.

11. F Goudail, F Bollaro, B Javidi, Ph Réfrégier. Influence of a perturbation in a double phase-encoding system. J Opt Soc Am A 15:2629–2638, 1998.

12. B Javidi. Encrypting information with optical technologies. Phys Today 50(3):27–32, 1997.

13. A Pu, R Denkewalter, D Psaltis. Real-time vehicle navigation using a holographic memory. Opt Eng 36:2737–2746, 1997.

14. D Casasent, G Ravichandran. Advanced distortion-invariant minimum average correlation energy 1MACE2 filters. Appl Opt 31:1109–1116, 1992.

15. B Javidi, J Wang. Optimum distortion invariant filters for detecting a noisy distorted target in background noise. J Opt Soc Am A 12:2604–2614, 1995.

16. B Javidi, D Painchaud. Distortion-invariant pattern recognition with Fourier-plane nonlinear filters. Appl Opt 35:318–331, 1999.

17. Z Ping, Y Sheng, S Deschenes, HH Arsenault. Fourier–Mellin descriptor and interpolated feature space trajectories for three-dimensional object recognition. Opt Eng 39:1260–1266, 2000.

18. J Hofer-Alfeis, R Bamler. Three- and four-dimensional convolution by coherent optical filtering. In W T Rhodes, J R Fienup, B E A Saleh, eds. Transformations in Optical Signal Processing. Bellingham, WA: SPIE Press, 1981, Vol 373, pp 77–87.

19. R Bamler, J Hofer-Alfeis. Three- and four-dimensional filter operations by coherent optics. Opt Acta 29:747–757, 1982.

20. Y B Karasik. Evaluation of three-dimensional convolutions by means of two-dimensional filtering. Appl Opt 36:7397–7401, 1997.

21. J Rosen. Three-dimensional optical Fourier transform and correlation. Opt Lett 22:964–966, 1997.

22. J Rosen. Three-dimensional electro-optical correlation. J Opt Soc Am A 15:430–436, 1998.

23. J Rosen. Three-dimensional joint transform correlator. Appl Opt 37:7538–7544, 1998.

24. Y Li, J Rosen. Three-dimensional pattern recognition with a single two-dimensional synthetic reference function. Appl Opt 39:1251–1259, 2000.

25. J J Esteve-Taboada, D Mas, J García. Three-dimensional object recognition by Fourier transform prolifometry. Appl Opt 38:4760–4765, 1999.

26. N Yoshikawa, T Yatagai. Fringe pattern correlator for three-dimensional object recognition. Opt Lett 25:1424–1426, 2000.

27. J J Esteve-Taboada, D Mas, J García. Rotation-invariant optical recognition of three-dimensional objects. Appl Opt 39:5998–6005, 2000.

28. J Guerrero-Bermúdex, J Meneses, O Gualdrón. Object recognition using three-dimensional correlation of range images. Opt Eng 39:2828–2831, 2000.

29. H J Caulfield, ed. Handbook of Optical Holography. London: Academic Press, 1979.

30. T Poon, T Kim. Optical image recognition of three-dimensional objects. Appl Opt 38:370–381, 1999.

31. T Kim, T C Poon. Extraction of 3-D location of matched 3-D object using power fringe-adjusted filtering and Wigner analysis. Opt Eng 38:2176–2183, 1999.

32. T Kim, T C Poon. Three-dimensional matching by use of phase-only holographic information and the Wigner distribution. J Opt Soc Am A 17:2520, 2000.

33. L Onural, P D Scott. Digital decoding of in-line holograms. Opt Eng 26:1124–1132, 1987.

34. U Schnars. Direct phase determination in hologram interferometry with use of digitally recorded holograms. J Opt Soc Am A 11:2011–2015, 1994.

35. G Pedrini, Y L Zou, H J Tiziani. Digital double-pulsed holographic interferometry for vibration analysis. J Mod Opt 40:367–374, 1995.

36. U Schnars, W P O Jüptner. Direct recording of holograms by a CCD target and numerical reconstruction. Appl Opt 33:179–181, 1994.

37. Y Takaki, H Kawai, H Ohzu. Hybrid holographic microscopy free of conjugate and zero-order images. Appl Opt 38:4990–4996, 1999.

38. J C Marron, K S Schroeder. Three-dimensional lensless imaging using laser frequency diversity. Appl Opt 31:255–262, 1992.

39. U Schnars, T M Kreis, W P O Jüptner. Digital recording and numerical reocnstruction of holograms: reduction of the spatial frequency spectrum. Opt End 35:977–982, 1996.

40. E Cuche, F Bevilacqua, C Depeursinge. Digital holography for quantitative phase-contrast imaging. Opt Lett 24:291–293, 1999.

41. B Javidi, T Nomura. Securing information by means of digital holography. Opt Lett 25:29–30, 2000.

42. J H Bruning, D R Herriott, J E Gallagher, D P Rosenfeld, A D White, D J Brangaccio. Digital wavefront measuring interferometer for testing optical surfaces and lenses. Appl Opt 13:2693–2703, 1974.

43. K Creath. Phase-measurement interferometry techniques. In: E Wolff, ed. Progress in Optics. Amsterdam: North-Holland, 1988, Vol 26, pp 349–393.

44. J Schwider. Advanced evaluation techniques in interferometry. In: E Wolff, ed. Progress in Optics. Amsterdam: North-Holland, 1990, Vol 28, pp 271–359.

45. I Yamaguchi, T Zhang. Phase-shifting digital holography. Opt Lett 22:1268–1270, 1997.

46. T Zhang, I Yamaguchi. Three-dimensional microscopy with phase-shifting digital holography. Opt Lett 23:1221–1223, 1998.
47. E Tajahuerce, O Matoba, S C Verrall, B Javidi. Optoelectronic information encryption using phase-shifting interferometry. Appl Opt 39:2313–2320, 2000.
48. S Lai, M N Neifeld. Digital wavefront reconstruction and its application to image encryption. Opt Commun 178:283–289, 2000.
49. E Tajahuerce, B Javidi. Encryping three-dimensional information with digital holography. Appl Opt 39:6595–6601, 2000.
50. B Javidi, E Tajahuerce. Three-dimensional object recognition using digital holography. Opt Lett 25:610–612, 2000.
51. E Tajahuerce, O Matoba, B Javidi. Shift-invariant three-dimensional object recognition by means of digital holography. Appl Opt 40:3877–3886, 2001.
52. J W Cooley, J W Tukey. An algorithm for the machine calculation of complex Fourier series. Math Comput 19:297–301, 1965.

9

A Distortion-Tolerant Image Recognition Receiver Using a Multihypothesis Method

Sherif Kishk and Bahram Javidi
University of Connecticut, Storrs, Connecticut

9.1 INTRODUCTION

In image recognition, it is desired that the receiver detect a known reference target in the presence of noise [1–7]. Also, the receiver should be able to detect the target even when the target is distorted. Possible sources of distortion are rotation, scaling, and changes in illumination. Numerous methods and algorithms have been developed to detect a distorted version of the target [8–23]. In one such method, a weighted superposition of distorted training true class targets to form a composite matched filter. However, many variations [8–16] of matched filters are proposed to allow distortion tolerance. There is also a vast amount of literature for pattern recognition using neural networks [18] and classification methods [17,18]. However, neural networks are more suitable for target classification rather than detection.

For target detection in the presence of distortions with known statistics such as probability density functions, mean, or variance, algorithms based on optimizing certain cost criteria can be used. Along this line, algorithms and filters have been proposed for detecting a known target under a variety of conditions [19,20]. In Ref 6, a detection algorithm was designed based on multihypothesis testing, but it does not take into account target distortions such as in-plane or out-of-plane rotation. In this chapter, we will use a multiple hypothesis test to locate the distorted targets in the presence of additive noise with unknown parameters. Furthermore, the target is assumed to have unknown constant illumination. Maximum-likelihood estimates of the unknown parameters are obtained, and the likelihood function of the multiple hypothesis is tested to detect the presence of the distorted targets.

The target or a reference signal is detected in the presence of additive noise with unknown statistics and the receiver is designed to be tolerant of the rotation, scaling, and illumination of the target. Computer simulations are presented to evaluate the performance of the receiver for various distorted noisy true class targets with varying illumination and with false class objects.

The chapter is organized as follows. In Section 9.2, an analysis of the algorithm is presented. In Section 9.3, we present computer simulations for various distorted nosiy true class targets with varying illumination and with false class objects. In Section 9.4, the conclusion is discussed.

9.2 ANALYSIS

For simplicity, we will use one-dimensional notation. Let $r(t)$ denote the target value at location t and $s(t)$ be the observed input scene at t. Let us assume that the scene contains an out-of-plane distorted version of the reference signal $r(t)$ and that it is illuminated by an unknown illumination factor a. The algorithm can be generalized to other types of target distortion such as scaling changes. Let $n(t)$ denote the additive white Gaussian noise. Then, if the received scene has a target at position t_j, $s(t)$ can be written as

$$s(t) = ar(t - t_j) + n(t) \tag{1}$$

Let H_j denote the hypothesis that the scene $s(t)$ contains a target r at the location t_j. We can write

$$H_j : s(t) = ar(t - t_j) + n(t) \tag{2}$$

Because we are dealing with discrete data, we can write the hypothesis as

$$H_j : s(t_m) = ar(t_m - t_j) + n(t_m) \tag{3}$$

where t_m is the discrete version of t.

We assume that there are N training true class targetrs that are out-of-plane rotated. These training true class targets will be denoted by $r_i(t)$, where $i = 1, 2, \ldots, N$ is the training target number. Let $H_j(i)$ denote the hypothesis that the scene contains the true class target r_i at location t_j. Thus,

$$H_j(i) : s(t_m) = ar_i(t_m - t_j) + n(t_m) \tag{4}$$

Let H_j denote the hypothesis that the scene contains one of the N training true class targets at location t_j. Then, assuming that all of the N training true class targets are equiprobable and that no more than one target can be at location t_j, an application of Bayes' formula yields

$$P(s(t_m)/H_j) = \frac{1}{N} \sum_{i=1}^{N} \left(\frac{1}{2\pi\sigma^2}\right)^{1/2} \exp\left(-\frac{[s(t_m) - \mu - ar_i(t_m - t_j)]^2}{2\sigma^2}\right) \quad (5)$$

where μ and σ are the mean and the standard deviation, respectively, of the additive noise.

If we assume that there are M samples over the scene $s(t)$ and that the additive noise is white, we obtain the likelihood function of hypothesis H_j:

$$P(s/H_j) = \prod_{m:w(t_m - t_j)=1} \left[\frac{1}{N} \sum_{i=1}^{N} \left(\frac{1}{2\pi\sigma^2}\right)^{1/2} \right. $$
$$\exp\left(-\frac{[s(t_m) - \mu - ar_i(t_m - t_j)]^2}{2\sigma^2}\right)\right]$$
$$\times \prod_{m:w(t_m - t_j)=0} \left[\left(\frac{1}{2\pi\sigma^2}\right)^{1/2} \exp\left(-\frac{[s(t_m) - \mu]^2}{2\sigma^2}\right)\right] \quad (6)$$

where $w(t_m - t_j)$ is a composite target window function; that is, $w(t_m - t_j)$ takes a value of 1 over the support of a target composed of all the training true class targetrs, and 0 outside the support of this composite target.

Because $1/N$ and the exponent $\frac{1}{2}$ do not depend on the hypothetical location t_j, we will drop these terms in subsequent equations. Taking the log of the likelihood, we obtain

$$\log\{P(s/H_j)\} = \sum_{m:w(t_m - t_j)=1} \left\{\log\left[\sum_{i=1}^{N} \left(\frac{1}{2\pi\sigma^2}\right)\right.\right.$$
$$\exp\left(-\frac{[s(t_m) - \mu - a.r_i(t_m - t_j)]^2}{2\sigma^2}\right)\left]\right\}$$
$$+ \sum_{m:w(t_m - t_j)=0} \left\{\log\left[\left(\frac{1}{2\pi\sigma^2}\right) \exp\left(-\frac{[s(t_m) - \mu]^2}{2\sigma^2}\right)\right]\right\} \quad (7)$$

The problem is to find the location of the target in the scene $s(t)$. A reasonable choice for the location of the target is the one that maximizes the a posteriori probability $P(H_j/s)$. If we consider that all of the hypotheses are equiprobable, the test will be reduced to maximizing the conditional probabilities $P(s/H_j)$.

We need to estimate the value of the illumination constant a, and the noise parameters μ and σ^2. To do so, we will maximize the likelihood function with respect to the unknown parameters to find the maximum a posteriori probability "MAP" estimates of these parameters. This is carried out by equating the gradient of the likelihood function with respect to a, μ, and σ^2 to zero:

$$\frac{\partial P(s/H_j)}{\partial a} = 0 \tag{8}$$

$$\frac{\partial P(s/H_j)}{\partial \mu} = 0 \tag{9}$$

$$\frac{\partial P(s/H_j)}{\partial \sigma^2} = 0 \tag{10}$$

From Eq. (8), we obtain

$$\hat{a}_j = \left(\sum_{m=1}^{M} \sum_{i=1}^{N} [s(t_m) - \mu)r_i(t_m - t_j)w(t_m - t_j)] \right) \left(\sum_{i=1}^{N} E_i \right)^{-1} \tag{11}$$

where $\hat{a}_j$ is the estimated illumination constant, and

$$E_i = \sum_{m=1}^{M} [r_i(t_m)w_i(t_m)]^2 \tag{12}$$

is the energy of target r_i.

Because we assume that there is no nonoverlapping noise and that the support of the target is small compared to the scene, to estimate the noise paramters we may use the sample points that are outside of the window. The hypothesis outside of the composite target window $P(s/H_j)_{wo}$ is

$$P(s/H_j)_{wo} = \sum_{m:w(t_m - t_j)=0} \log\left[\left(\frac{1}{2\pi\sigma^2} \right) \exp\left(-\frac{[s(t_m) - \mu]^2}{2\sigma^2} \right) \right] \tag{13}$$

Taking the derivative of Eq. (13) with respect to μ and σ^2 and equating it to zero, we obtain

$$\hat{\mu}_j = \frac{1}{N} \sum_{i=1}^{N} \frac{\sum_{m=1}^{M} [s(t_m)][1 - w(t_m - t_j)]}{M_{io}} \tag{14}$$

where M_{io} is the number of samples outside the composite target window, $\hat{\mu}_j$ is the estimated values of μ, and

$$\hat{\sigma}_j^2 = \frac{1}{N} \sum_{i=1}^{N} \frac{[s(t_m) - \hat{\mu}_j]^2}{M_{io}} \tag{15}$$

where $\hat{a}_j$ is the estimated value of a^2, and using Eqs. (14) and (11), we obtain

$$\hat{a}_j = \left[\sum_{m=1}^{M} \sum_{i=1}^{N} \left(s(t_m) - \frac{1}{N} \sum_{l=1}^{N} \frac{\sum_{n=1}^{M} [s(t_n)][1 - w(t_n - t_j)]}{M_{lo}} \right) \right.$$

$$\left. r_i(t_m - t_j) w(t_m - t_j) \right] \left(\sum_{i=1}^{N} E_i \right) \tag{16}$$

The estimated values for a, μ, and σ^2 will be used in Eq. (6) to obtain the likelihood of finding the target at location t_j.

In the previous analysis, the trained targets are assumed to be equiprobable and the illumination is assumed to be a constant, which are noninformative priors. One may obtain a better estimation for the noise parameters if the illumination is treated as noise with a certain distribution.

9.3 COMPUTER SIMULATION AND DISCUSSION

To test the performance of the receiver presented in this chapter, we show the results of some computer simulations. We used a set of training true class targets with different out-of-plane rotational distortions ranging from $0°$ to $90°$ with increments of $10°$ (i.e., $0°, 10°, \ldots, 90°$). Figure 1 shows the set of training true class targets used in the experiments. The experiments are repeated for different additive noise parameters and different illumination distortions.

We use the receiver operating characteristic (ROC) curve to evaluate the performance of the receiver. The ROC plots the probability of detection (P_d) against the probability of false alarm (P_{fa}) for a range of threshold values and a given input noise. In this test, 500 iterations were used to find the probability of detection and the probability of false alarm. Figure 2 shows the ROC curve using target 1, which has an out-of-plane rotation of $0°$, and the additive noise is white Gaussian. We calculate two ROC curves: one for a noise process that has a standard deviation of 0.7 and the other for a noise process with standard deviation of 1.0.

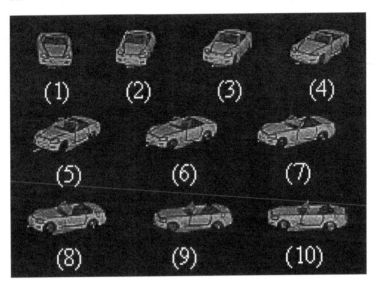

Figure 1   A set of training true targets with different out-of-plane rotational distortions from 0° to 90° with increments of 10°. Zero degrees (target 1) represents the front view of the car.

Figure 3a shows the nontraining true class targets. These are rotated versions of the same car with rotation angles ranges from 21° to 29° with increments of 1°. Figure 3b shows the false object, which is a different car with a 10° out-of-plane rotation angle. In the first experiment, the scene has three targets: one true class training target, one nontrained true class target, and one false object. None of the targets have any illumination distortion. Also, overlapping Gaussian noise having a mean of 0.5 and a standard deviation of 0.7 is added to the entire input scene. Figure 4a shows the scene containing the three targets and each target is labeled with its number, whereas Fig. 4b is the noisy scene having the same three targets. Figure 4c shows the receiver output using Eq. (7) and the estimated illumination distortion using Eqs. (14), (15), and (16). The results shows that the receiver detects the training and the nontraining true class targets while rejecting the false object. We can see that we have peaks at the positions of the training and the nontraining targets, whereas there is no peak at the false object position.

In the second set of experiments, we apply an illumination distortion to a nontraining true class target of 1.3 ($a = 1.3$), whereas the illumination distortion to the training true class target is 0.7. In this experiment, there will be no illumination distortion for the false object. The additive noise used in

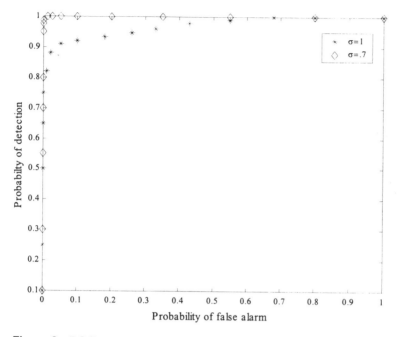

Figure 2 ROC curves for additive noise with standard deviations of $\sigma = 0.7$ and $\sigma = 1$.

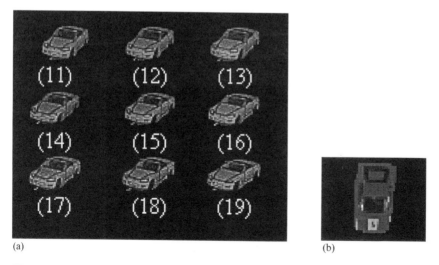

(a)

(b)

Figure 3 (a) The set of nontraining true class targets having an out-of-plane rotation ranging from $21°$ to $29°$ with increments of $1°$. (b) The false class object.

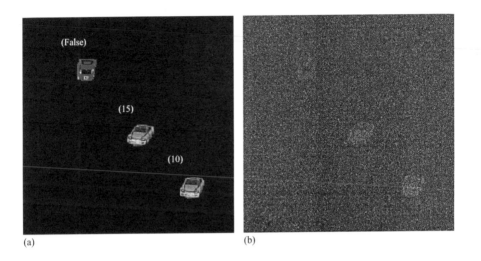

(a)   (b)

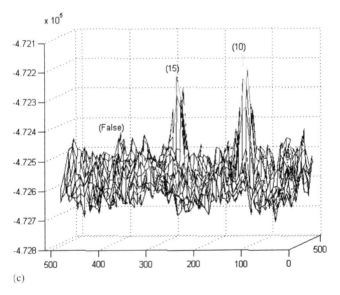

(c)

Figure 4 A scene containing a trained true class target having 10° out-of-plane rotation without illumination distortion, a nontrained true class target having 25° out-of-plane rotation (target 15) without illumination distortion, and a false class object. (b) The same scene with additive noise having a mean of 0.5 and a standard deviation of 0.7. (c) The receiver output.

this experiment is white Gaussian with a mean of 0.5 and a standard deviation of 0.7. Figure 5a shows the scene containing the three targets, Fig. 5b shows the noisy version of the scene, and Fig. 5c shows the output of the filter. As we can see from Fig. 5c, the receiver was able to detect both the training true class and the nontraining true class targets successfully.

In the following experiment, the nontraining true class target illumination distortion is 0.7, the training true class target illumination distortion is 1.3, and the illumination of the false object is not changed. The additive noise used in this experiment is white Gaussian with a mean of 0.5 and a standard deviation of 0.7. Figure 6a shows the scene containing the three targets, Fig. 6b shows the noisy scene, and Fig.6c shows the output of the optimum receiver.

In Fig. 7 we test the system performance when the scene has no targets and contains only noise. Figure 7a shows the scene with the additive noise having standard deviation of 0.7. Figure 7b demonstrates the system output, and as we can see, there is no peak indicating that there is no target in the scene.

In the last experiment, we examined the ability of the receiver to reject a scene containing a false object only in the presence of noise. Figure 8a shows the scene having the false object, Fig. 8b shows the noisy scene, and Fig. 8c shows the receiver output.

Each of the above experiments shows that the receiver can successfully detect the nontraining targets even with high illumination distortion such as the one in Fig. 6. Keeping in mind the very high similarity between the true class targets and the false objects, the results we obtained can be considered very reasonable. For future work, this receiver will be extended to include nonoverlapping clutter noise [24] added around the target.

9.4 CONCLUSION

In this chapter, we developed an algorithm to detect a known target. The algorithm is designed to tolerate out-of-plane rotation of a known reference target and it can also tolerate different types of distortion. The target is buried in additive white stationary noise with unknown statistics. Furthermore, the target is illuminated by an unknown illumination constant. Hypothesis testing is used to locate the target. A set of training images with different out-of-plane rotational distortions ranging from $0°$ to $90°$ with increments of $10°$ was used to obtain a statistical model of the scene. The likelihood function for hypothesis testing is derived based on the composite set of training images. The unknown noise statistics and the unknown illumination on the target are estimated within the context of hypothesis

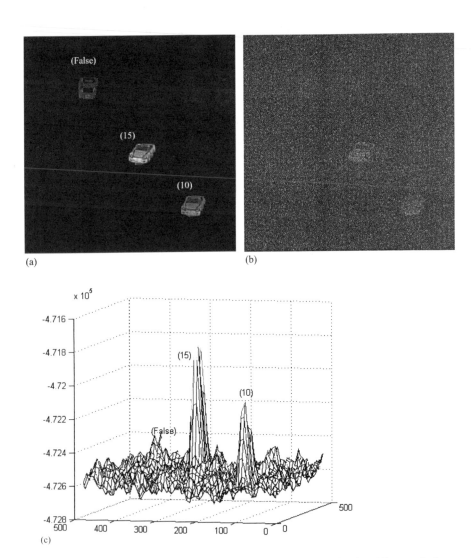

(a)   (b)

(c)

Figure 5 (a) A scene containing a trained true class target having 10° out-of-plane rotation and illumination distortion of 0.7, a nontrained true class target having 25° out-of-plane rotation (target 15) and illumination distortion of 1.3, and a 100% illuminated false class object. (b) The same scene with additive noise having a mean of 0.5 and a standard deviation of 0.7. (c) The receiver output.

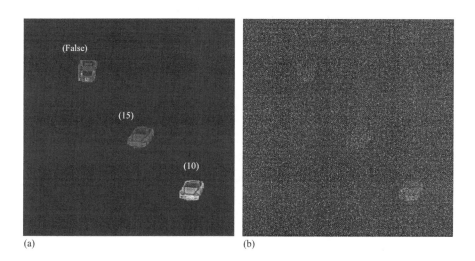

(a)   (b)

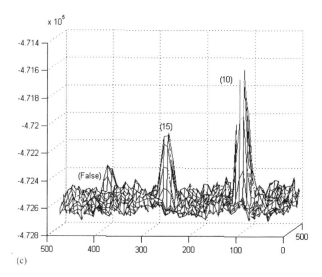

(c)

Figure 6   (a) A scene containing a true class target having 10° out-of-plane rotation and 1.3 illumination distortion, a nontrained true class target having 25° out-of-plane rotation (target 15) and 0.7 illumination distortion, and a 100% illuminated flase class object. (b) The same scene with additive noise having a mean of 0.5 and a standard deviation of 0.7. (c) The receiver output.

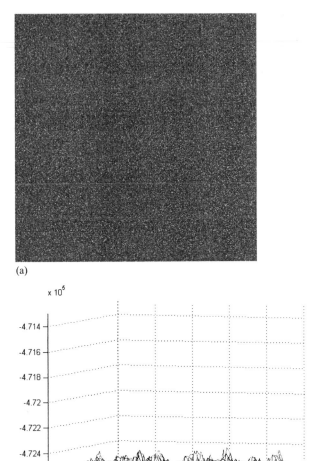

(a)

(b)

Figure 7 (a) The scene with additive noise having a mean of 0.5 and a standard deviation of 0.7. (b) The receiver output.

testing. We carried out a limited number of computer simulations to test the performance of the receiver. These experiments showed that the receiver has a reasonable performance even in highly distorted scenes with very close false objects. We also used a ROC curve to evaluate the performance of the receiver.

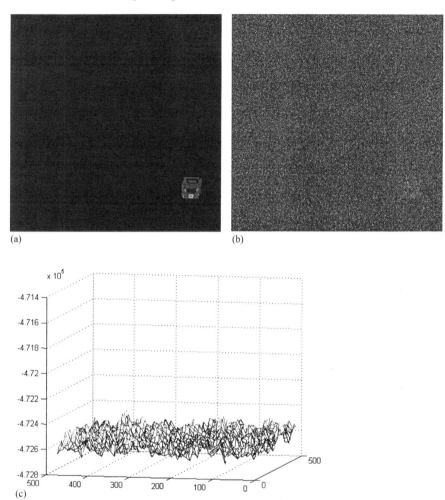

Figure 8 (a) A scene containing a false class object. (b) The same scene with additive noise having a mean of 0.5 and a standard deviation of 0.7. (c) The receiver output

ACKNOWLEDGMENTS

The authors thank Nasser Towghi, Yann Frauel, and Thomas Naughton for their suggestions on the chapter.

REFERENCES

1. R O Duda, P E Hart. Pattern Classification and Scene Analysis. New York: Wiley, 1973.
2. F Fukunaga. Introduction to Statistical Pattern Recognition. New York: Academic Press, 1972.
3. J L Turin. An Introduction to Matched Filters. IRE Trans Inform Theory IT-6:311–329, 1960.
4. J L Horner, P D Gianino. Phase-only matched filtering. Appl Opt 23:812–816, 1984.
5. A VanderLugt. Signal detection by complex spatial filtering. IEEE Trans Inform Theory IT-19:139–145, 1964.
6. B Javidi, P Refregier, P Willett. Optimum receiver design for pattern recognition with nonoverlapping target an scene noise. Opt Lett 18:1660–1664, 1993.
7. A Abu-Naser, N P Galatsanos, M N Wernick, D Schonfeld. Object recognition based on impulse restoration with use of the expectation-maximization algorithm. J Opt Soc Am A 15:2327–2340, 1998.
8. C F Hester, D Casasent. Multi-variant technique for multi class pattern recognition. Appl Opt 19:1758–1761, 1980.
9. R R Kallman. The construction of low noise optical correlation filters. Appl Opt 25:1032–1033, 1986.
10. A Mahalanobis. Review of correlation filters and their application for scene matching. In Optoelectronic Devices and Systems for Processing. Critical Reviews of Optical Science Technology Vol CR 65. Bellingham, WA: SPIE Press, 1996, pp 240–260.
11. A Mahalanobis. Processing of multi-sensor data using correlation filters. Proc SPIE, 3466:56–64, 1998.
12. P Refregier. Filter design for optimal pattern recognition: multicriteria optimization approach. Opt Lett 15:854–856, 1990.
13. L S Jamal-Aldin, R C D Young, C R Chatwin. Synthetic discriminant function filter employing nonlinear space-domain preprocessing on bandpass-filtered images. Appl Opt 37:2051–2062, 1998.
14. Y N Hsu, H H Arsenault. Optical pattern recognition using circular harmonics expansion. Appl Opt 21:4016–4019, 1982.
15. D Casasent, D Psaltis. Position, rotation and scale invariant optical correlations. Appl Opt 15:1795–1799, 1976.
16. P Zi-Liang, E Dalsgaard, Synthetic circular harmonic phase-only filter for shift, rotation, and scaling-invariant correlation. Appl Opt 34:7527–7531, 1995.
17. S Haykin. Neural Networks: A Comprehensive Foundation. Englewood Cliffs, NJ: Prentice-Hall, 1998.
18. W Li, N M Nasrabadi. Invariant object recognition based on network of cascade RCE Nets. Int J Pattern Recogn Artif Intell 7(4):815–829, 1993.
19. B Javidi, J Wang. Optimum distortion-invariant filter for detecting a noisy distorted target in nonoverlapping background noise. J Opt Soc Am A 12:2604–2614, 1995.

20. N Towghi, B Javidi. Generalized optimum receiver for pattern recognition with multiplicative, additive and non-overlapping background noise. J Opt Sco Am A 15:1557–1565, 1998.
21. H J Caulfield, W T Malongey. Improved discrimination in optical character recognition(L), Appl Opt 8:2354–2356, 1969.
22. B Javidi. Smart Imaging Systems. Bellingham, WA: SPIE Press, 2001.
23. H H Arsenault, D Lefebvre. Homomorphic cameo filter for pattern recognition that is invariant with changes in illumination. Opt Lett 25:1567–1569, 2000.
24. B Javidi, J Wang. Limitations of the classic definition of the signal-to-noise ratio in matched filter based optical pattern recognition. J Appl Opt 31, 1992.

10

Correlation Pattern Recognition: An Optimum Approach

Abhijit Mahalanobis
Lockheed Martin, Orlando, Florida

10.1 INTRODUCTION

There has been considerable interest in using correlators for pattern recognition. Correlators are inherently shift invariant, allowing us to locate patterns (such as moving targets) in the input scene merely by locating the correlation peak. Thus, we do not need to segment or register the images prior to correlation, as we have to do in alternate methods for pattern recognition. Accordingly, a wide variety of correlation algorithms and architectures have been investigated (both optical and digital) to deal with the important problem of finding patterns in images.

The matched spatial filter (MSF) is perhaps the most widely known correlation function [1] and is based on a single view of the pattern to be detected. It is optimal [in the sense of yielding maximal signal-to-noise ratio (SNR)] for detecting a completely known pattern in the presence of additive noise. Unfortunately, MSFs are not adequate for many pattern recognition applications because their correlation peak degrades rapidly when the input patterns deviate (sometimes even very slightly) from the reference. These variations in the patterns are often due to common phenomenon such as scale changes and rotations.

One approach to overcoming the distortion sensitivity of matched spatial filters is to use one MSF for every view. However, this leads to the use (and storage) of an enormous number of filters that make such an approach impractical. The alternative is to design *composite filters* that can be optimized to yield better distortion tolerance than the MSFs. Over the years, a plethora of techniques [2] have been developed to obtain various types of composite correlation filter. These filters are derived from a set of representative views of the object known as the *training images*. Hence, in principle, correlation filters can be trained to recognize any object as long as the distortions can be adequately characterized by the training set. As we shall see, the proper selection, registration, and clustering of training images is an important step in the design of composite filters.

In this chapter, our interest is limited to a class of composite filters whose response to distortions can be analytically optimized. In Section 10.2, we discuss the *maximum average correlation height* (MACH) filter [3] that is optimized to detect a target in the presence of both noise and distortions. It has been shown that the MACH filter out performs the MSF when both noise and distortions are taken into account [4]. This is followed in Section 10.3 by a discussion of a higher-order variant of the MACH filter known as the *polynomial correlation filter* (PCF) [5]. Section 10.4 describes the optimal separation of classes using correlation via a quadratic method known as the *distance classifier correlation filter* (DCCF) [6]. Finally, in Section 10.5, we visit the topic of training image alignment in order to get the maximal performance from the correlation filters [7].

10.2 THE MACH FILTER

In this section, we address the detection of a distorted target image (or pattern) in the presence of additive noise. Our objective is to design a correlation filter such that its performance is optimized with respect to not only noise but also distortions. Let $g(m, n)$ denote the correlation surface produced by a filter $h(m, n)$ in response to the input image $f(m, n)$. Strictly speaking, the entire correlation surface $g(m, n)$ is the output of the filter. However, the point $g(0, 0)$ is often referred to as the "filter output" or the correlation "peak." By maximizing the correlation output at the origin, we will be forcing the real peak to be large. With this interpretation, the *peak* filter output is given by

$$g(0, 0) = \sum \sum f(m, n)h(m, n) = \mathbf{f}^T \mathbf{h} \tag{1}$$

where the superscript T denotes the transpose and $\mathbf{f}$ and $\mathbf{h}$ are the vector versions of $f(m, n)$ and $h(m, n)$, respectively.* Typically, for noise toler-

---

*In this chapter, uppercase bold characters denote matrices and lowercase bold characters represent vectors. Lowercase italics indicate images in the space domain and uppercase italics are used to represent their Fourier transforms.

ance, it is desired that this filter output be as immune as possible to the effects of noise.

The key idea for optimizing distortion tolerance is to treat the correlation plane as a new pattern generated by the filter in response to an input image. We start with the notion that the correlation planes are *linearly transformed* versions of the input image, obtained by applying the correlation filter. It may then be argued that if the filter is distortion tolerant, its output will not change much even if the input pattern exhibits some variations. Thus, the emphasis is not only on the correlation peak but also on the entire shape of the correlation surface.

With the above discussion in mind, a metric for distortion is defined as the average variation in images after filtering. If $g_i(m, n)$ is the correlation surface produced in response to the ith training image, we can quantify the variability in these correlation outputs by the average similarity measure (ASM) defined as follows:

$$\text{ASM} = \frac{1}{N} \sum_{i=1}^{N} \sum_m \sum_n [g_i(m, n) - \bar{g}(m, n)]^2 \qquad (2)$$

where $\bar{g}(m, n) = (1/N) \sum_{j=1}^{N} g_j(m, n)$ is the average of the N training image correlation surfaces. ASM is a mean square error measure of distortions (variations) in the correlation surfaces relative to an average shape. In an ideal situation, all correlation surfaces produced by a distortion-invariant filter (in response to a valid input pattern) would be the same and ASM would be zero. In practice, minimizing ASM improves the filter's stability.

We now discuss how to formulate ASM as a performance criterion for filter synthesis. Using Parseval's theorem, ASM can be expressed in the frequency domain as

$$\text{ASM} = \frac{1}{Nd} \sum_{i=1}^{N} \sum_k \sum_l |G_i(k, l) - \bar{G}(k, l)|^2 \qquad (3)$$

where $G_i(k, l)$ and $\bar{G}(k, l)$ are two-dimensional (2D) Fourier transforms of $g_i(m, n)$ and $\bar{g}(m, n)$, respectively. In vector notation,

$$\text{ASM} = \frac{1}{Nd} \sum_{i=1}^{N} |\mathbf{g}_i - \bar{\mathbf{g}}|^2 \qquad (4)$$

Let $\mathbf{m} = (1/N) \sum_{i=1}^{N} \mathbf{x}_i$ represent the average of the training image Fourier transforms. In the following discussion, we treat $\mathbf{M}$ and $\mathbf{X}_i$ as diagonal matrices with the same elements along the main diagonals as in vectors **m** and $\mathbf{x}_i$. Using the frequency-domain relations $\mathbf{g}_i = \mathbf{X}_i^*\mathbf{h}$ and $\bar{\mathbf{g}} = \mathbf{M}^*\mathbf{h}$, the ASM can be rewritten as follows.

$$
\begin{aligned}
\text{ASM} &= \frac{1}{Nd} \sum_{i=1}^{N} |\mathbf{X}_i^*\mathbf{h} - \mathbf{M}^*\mathbf{h}|^2 \\
&= \frac{1}{Nd} \sum_{i=1}^{N} \mathbf{h}^+ (\mathbf{X}_i - \mathbf{M})(\mathbf{X}_i - \mathbf{M})^* \mathbf{h} \\
&= \mathbf{h}^+ \left(\frac{1}{Nd} \sum_{i=1}^{N} (\mathbf{X}_i - \mathbf{M})(\mathbf{X}_i - \mathbf{M})^* \right) \mathbf{h} \\
&= \mathbf{h}^+ \mathbf{S} \mathbf{h}
\end{aligned}
\tag{5}
$$

where the matrix $\mathbf{S} = (1/Nd)\sum_{i=1}^{N}(\mathbf{X}_i - \mathbf{M})(\mathbf{X}_i - \mathbf{M})^*$ is also diagonal, making its inversion relatively easy.

In addition to being distortion tolerant, a correlation filter must yield large peak values to facilitate detection. Toward this end, we maximize the filter's response to the training images on the average. However, unlike traditional *synthetic discriminant functions* (SDFs) [2], no hard constraints are imposed on the filter's response to training images at the origin. Rather, we simply desire that the filter should yield a large peak on the average over the entire training set. This condition is met by maximizing the *average correlation height* (ACH) criterion defined as

$$
\text{ACH} = \frac{1}{N} \sum_{i=1}^{N} \mathbf{x}^+\mathbf{h} = \mathbf{m}^+\mathbf{h}
\tag{6}
$$

Finally, it is, of course, desirable to reduce the effect of noise (and clutter) on the filter's output by minimizing the *output noise variance* (ONV). It has been shown elsewhere [2] that ONV is also a quadratic term given by $\mathbf{h}^+\mathbf{C}\mathbf{h}$, where $\mathbf{C}$ is the noise covariance matrix. For simplicity, we assume a white-noise model, although the actual noise model should be used whenever available. To make ACH large while reducing ASM and ONV, the filter is designed to maximize

$$
\begin{aligned}
J(\mathbf{h}) &= \frac{|\text{ACH}|^2}{\text{ASM} + \text{ONV}} \\
&= \frac{|\mathbf{m}^+\mathbf{h}|^2}{\mathbf{h}^+\mathbf{S}\mathbf{h} + \mathbf{h}^+\mathbf{C}\mathbf{h}} \\
&= \frac{\mathbf{h}^+\mathbf{m}\mathbf{m}^+\mathbf{h}}{\mathbf{h}^+(\mathbf{S} + \mathbf{C})\mathbf{h}}
\end{aligned}
\tag{7}
$$

The optimum filter which maximizes this criterion is the dominant eigenvector of $(S + C)^{-1}mm^+$ or

$$h = \gamma(S + C)^{-1}m \tag{8}$$

where γ is a normalizing scale factor. The filter in Eq. (8) is referred to as the MACH filter and, in some sense, represents one of the most attractive composite correlation filters. The optimality of the MACH filter for handling distortions in the presence of additive noise has been examined in Ref. 4.

10.2.1 Relation to the Filters for Nonoverlapping Noise

In the preceding sections, we have discussed the design of correlation filters assuming that the noise is additive. Often however, cases arise where the noise process (referred to as the background clutter) does not overlap with the pattern to be recognized. In such cases, the noise is said to be *spatially disjoint* from the signal of interest. In this subsection, we will discuss a modified filter design technique that is optimum for recognizing distorted patterns (i.e., more than one view of an object) in the presence of nonoverlapping noise. The model presented here is intentionally simplified for the ease of discussion. Excellent work on this subject has been conducted by Refregier and Javidi et al. The reader is referred to Refs 8 and 9 for further details.

Let us assume that a set of images $x_i(m, n)$, $i = 1, 2, \ldots, N$, must be recognized in the presence of a stationary nonoverlapping (spatially disjoint) noise process $v(m, n)$. The Fourier transform of the images is denoted by $X_i(k, l)$. The noise has a mean value of m_v and power spectral density $S_v(k, l)$. We define a region of support function $w_i(m, n)$ for each of the images [i.e., a window that is unity if (m, n) is a point within $x_i(m, n)$]. The Fourier transform of the support functions is given by $W_i(k, l)$. The received signal can be expressed as

$$y_i(m, n) = x_i(m, n) + v(m, n)[1 - w_i(m, n)] \tag{9}$$

This signal is filtered to produce the output $g_i(m, n)$ [i.e., $g_i(m, n) = y_i(m, n) \otimes h(m, n)$]. We assume without loss of generality that images $x_i(m, n)$ are nominally centered and that the peak output is measured at the center of the correlation plane. The metric to be optimized is *peak-to-correlation energy* (PCE), expressed as

$$PCE = \frac{|E\{y_i(0, 0)\}|^2}{E\{(|y_i(m, n)|)^2\}} \tag{10}$$

It has been shown [9] that the filter which maximizes PCE is

$$H(k, l) = \left(\sum_{i=1}^{N} X_i(k, l) + m_v[\delta(k, l) - W_i(k, l)] \right)$$

$$\times \left[\sum_{i=1}^{N} \left(|X_i(k, l) + m_v[\delta(k, l) - W_i(k, l)]|^2 \right. \right.$$

$$\left. \left. + \frac{1}{2\pi} |\delta(k, l) - W_i(k, l)|^2 * S_v(k, l) \right) \right]^{-1} \tag{11}$$

It is interesting to note the similarities between the optimum filter in Eq. (11) and the MACH filter. The term $m_v[\delta(k, l) - W_i(k, l)]$ has the effect of padding the background of the training images in the space domain with the mean value of the noise (i.e., all points outside the signal but in the image frame are set to m_v). If we absorb this term in the definition of the training image so that $\hat{X}_i(k, l) = X_i(k, l) + m_v[\delta(k, l) - W_i(k, l)]$, we get

$$H(k, l) = \left(\frac{1}{N} \sum_{i=1}^{N} \hat{X}_i(k, l) \right)$$

$$\times \left[\frac{1}{N} \sum_{i=1}^{N} |\hat{X}_i(k, l)|^2 + \left(\frac{1}{N} \sum_{i=1}^{N} \frac{1}{2\pi} |\delta(k, l) - W_i(k, l)|^2 \right) \right.$$

$$\left. * S_v(k, l) \right]^{-1}$$

$$= \frac{M(k, l)}{D(k, l) + C_v(k, l)} \tag{12}$$

where $M(k, l) = (1/N) \sum_{i=1}^{N} \hat{X}_i(k, l)$ is the average of the padded training images, $D(k, l) = (1/N) \sum_{i=1}^{N} |\hat{X}_i(k, l)^2|$ is their average power spectral density, and

$$C_v(k, l) = \left(\frac{1}{N} \sum_{i=1}^{N} \frac{1}{2\pi} |\delta(k, l) - W_i(k, l)|^2 \right) * S_v(k, l) \tag{13}$$

is a modified power spectral density for the noise. Thus, the optimum filter for detecting targets in nonoverlapping noise is very similar to the MACH filter when the training images are padded with the expected mean of the background noise. An alternative to padding the training images with a constant is to set their average value to zero. This corresponds to the case where the background noise and the signals to be detected are expected to be at the same mean level. The modified noise power spectrum in Eq. (13) can be easily obtained in the space domain by multiplying the autocorrelation function of the noise process by the average autocorrelation function of the *complement* of the training image support functions.

10.3 POLYNOMIAL CORRELATION FILTERS

The MACH filter and other correlation techniques mentioned in Section 10.2 generally perform linear operations on the data. In this section, we describe an architecture for obtaining higher-order correlations and for simultaneously correlating multiple sources of data. The fundamental difference in the approach described here and traditional methods for correlation filters is that we treat the output as a nonlinear function of the input. Thus, if $\mathbf{x}$ represents the input image in vector notation, then the vector $\mathbf{g_x}$, which represents the output correlation plane, is expressed as a polynomial function of $\mathbf{x}$ as

$$\mathbf{g}_N = \sum_{i=1}^{N} \mathbf{A}_i f_i[\mathbf{x}] \tag{14}$$

where $f[\cdot]$ is *any nonlinear function* of $\mathbf{x}$. It should be noted that the expression in Eq. (14) represents quantities in the space domain and that the nonlinearity is applied to the pixels in the image. We refer to the form in Eq. (14) as the polynomial correlation filter (or PCF). To ensure that the output is *shift invariant*, all of the coefficient matrices $\mathbf{A}_i$ are required to be Toeplitz. Then, it can be shown that each term in the polynomial can be computed as a linear shift-invariant filtering operating; that is,

$$\mathbf{A}_i f(\mathbf{x}) \equiv h_i(m, n) \otimes f_i[x(m, n)] \tag{15}$$

or that filtering $f_i[x(m, n)]$ by $h_i(n, n)$ is equivalent to multiplying $f_i[\mathbf{x}]$ by $\mathbf{A}_i$. The output of the polynomial correlation filter can be mathematically expressed as

$$g_x(m, n) = \sum_{i=1}^{N} h_i(m, n) \otimes f_i[x(m, n)] \tag{16}$$

The corresponding structure of the filter is shown in Fig. 1.

10.3.1 Derivation of the Solution

The objective is to find the filters $h_i(m, n)$ such that structure shown in Fig. 1 optimizes a performance criterion of choice. Experience has shown that for

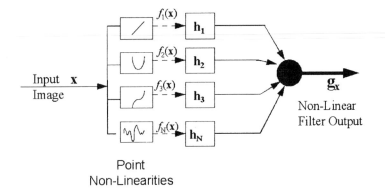

Point
Non-Linearities

Figure 1   The *N*th-order polynomial correlation filter form.

correlation purposes, a useful approach is to maximize the MACH performance criterion

$$J(\mathbf{h}) = \frac{|\mathbf{m}^+\mathbf{h}|^2}{\mathbf{h}^+\mathbf{B}\mathbf{h}} \tag{17}$$

where **h** is the filter vector in the frequency domain, **B** is a diagonal matrix related to a spectral quantity (such as ACE, ASM, and ONV or some combination of these), and **m** is the mean image vector, also in the frequency domain. Of course, the premise is that a higher-order (non-linear) solution will yield much higher values of $J(\mathbf{h})$ than the corresponding linear solutions.

To simplify discussions, we first discuss the derivation of a second-order filter. We also assume, for now, that the nonlinearity is simply raising the pixels to an integer power (specifically 2) to obtain the second term of the polynomial. In this case, the PCF output is given by

$$g(m, n) = x(m, n) \otimes h_1(m, n) + x^2(m, n) \otimes h_2(m, n) \tag{18}$$

The expression for $J(\mathbf{h})$ is obtained by deriving the numerator and the denominator of Eq. (17). In vector notation, the average intensity of the correlation peak for a second-order filter is

$$|\text{Average peak}|^2 = |\mathbf{h}_1^+\mathbf{m}^1|^2 + |\mathbf{h}_2^+\mathbf{m}^2|^2 + 2\mathbf{h}_1^+\mathbf{m}^1(\mathbf{m}^2)^+\mathbf{h}_2 \tag{19}$$

where $\mathbf{h}_1$ and $\mathbf{h}_2$ are vector representations of the filters associated with the first and second terms of the polynomial and

$$\mathbf{m}^k = \frac{1}{L} \sum_{i=1}^{L} \mathbf{x}_i^k \tag{20}$$

is the average of the Fourier transform of the nonlinearly altered training images x_i^k, $1 \leq i \leq L$. For illustration purposes, the denominator of the performance criterion in Eq. (17) is chosen to be the ASM metric while noting that it can easily be any other quadratic form such as ONV or ACE or any combination thereof. The ASM for the second-order nonlinear filter is given by

$$\text{ASM} = \frac{1}{L} \sum_{i=1}^{L} |\mathbf{h}_1^* \mathbf{X}_i^1 + \mathbf{h}_2^* \mathbf{X}_i^2 - \mathbf{h}_1^* \mathbf{M}^1 - \mathbf{h}_2^* \mathbf{M}^2|^2 \qquad (21)$$

where $\mathbf{X}_i^k$, $1 \leq i \leq L$, is a diagonal matrix whose diagonal terms are the same as the vector $\mathbf{x}^k - i$, and $\mathbf{M}^k$ is their average (also a diagonal matrix). After algebraic manipulations, it can be shown that the expression for ASM is

$$\text{ASM} = \mathbf{h}_1^+ \mathbf{S}_{11} \mathbf{h}_1 + \mathbf{h}_2^+ \mathbf{S}_{22} \mathbf{h}_2 + \mathbf{h}_1^+ \mathbf{S}_{12} \mathbf{h}_2 + \mathbf{h}_2^+ \mathbf{S}_{21} \mathbf{h}_1 \qquad (22)$$

where

$$\mathbf{S}_{kl} = \frac{1}{L} \sum_{i=1}^{L} \mathbf{X}_i^k (\mathbf{X}_i^l)^* - \mathbf{M}^k (\mathbf{M}^l)^*, \qquad 1 \leq k, l \leq 2 \qquad (23)$$

are all diagonal matrices. Defining the block vectors and matrices

$$\mathbf{h} = \begin{bmatrix} \mathbf{h}_1 \\ \mathbf{h}_2 \end{bmatrix}, \quad \mathbf{m} = \begin{bmatrix} \mathbf{m}^1 \\ \mathbf{m}^2 \end{bmatrix}, \quad \text{and} \quad \mathbf{S} = \begin{bmatrix} \mathbf{S}_{11} & \mathbf{S}_{12} \\ \mathbf{S}_{21} & \mathbf{S}_{22} \end{bmatrix} \qquad (24)$$

the expression for $J(\mathbf{h})$ for the second-order filter can be succintly expressed as

$$\begin{aligned} J(\mathbf{h}) &= \frac{|\text{Average peak}|}{\text{ASM}} \\ &= \frac{|\mathbf{h}_1^+ \mathbf{m}^1|^2 + |\mathbf{h}_2^+ \mathbf{m}^2|^2 + 2\mathbf{h}_1^+ \mathbf{m}^1 (\mathbf{m}^2)^+ \mathbf{h}_2}{\mathbf{h}_1^+ \mathbf{S}_{11} \mathbf{h}_1 + \mathbf{h}_2^+ \mathbf{S}_{22} \mathbf{h}_2 + \mathbf{h}_1^+ \mathbf{S}_{12} \mathbf{h}_2 + \mathbf{h}_2^+ \mathbf{S}_{21} \mathbf{h}_1} \\ &= \frac{|\mathbf{m}^+ \mathbf{h}|^2}{\mathbf{h}^+ \mathbf{S} \mathbf{h}} \end{aligned} \qquad (25)$$

The well-known solution which maximizes $J(\mathbf{h})$ is given by

$$\mathbf{h} = \mathbf{S}^{-1} \mathbf{m} \qquad (26)$$

Using the definitions in Eq. (24), the solution for the two filters of the second-order polynomial are

$$\begin{bmatrix} \mathbf{h}_1 \\ \mathbf{h}_2 \end{bmatrix} = \begin{bmatrix} \mathbf{S}_{11} & \mathbf{S}_{12} \\ \mathbf{S}_{21} & \mathbf{S}_{22} \end{bmatrix}^{-1} \begin{bmatrix} \mathbf{m}^1 \\ \mathbf{m}^2 \end{bmatrix} \qquad (27)$$

The inverse of the block diagonal matrix can be symbolically computed, yielding

$$
\begin{bmatrix} \mathbf{h}_1 \\ \mathbf{h}_2 \end{bmatrix} = \begin{bmatrix} [|\mathbf{S}_{12}|^2 - \mathbf{S}_{11}\mathbf{S}_{22}]^{-1}(\mathbf{S}_{12}\mathbf{m}^2 - \mathbf{S}_{22}\mathbf{m}^1) \\ [|\mathbf{S}_{12}|^2 - \mathbf{S}_{11}\mathbf{S}_{22}]^{-1}(\mathbf{S}_{21}\mathbf{m}^1 - \mathbf{S}_{11}\mathbf{m}^2) \end{bmatrix}
\tag{28}
$$

It should be noted that all matrices are diagonal and the inverse operations can be performed without any difficulty. The solution in Eq. (28) can be easily extended to the general Nth-order case. Following the same analysis as for the second-order case, the Nth-order solution is given by

$$
\begin{bmatrix} \mathbf{h}_1 \\ \mathbf{h}_2 \\ \vdots \\ \mathbf{h}_N \end{bmatrix} = \begin{bmatrix} \mathbf{S}_{11} & \mathbf{S}_{12} & \cdots & \mathbf{S}_{1N} \\ \mathbf{S}_{21} & \mathbf{S}_{22} & \cdots & \mathbf{S}_{2N} \\ \vdots & \vdots & \vdots & \vdots \\ \mathbf{S}_{N1} & \mathbf{S}_{N2} & \cdots & \mathbf{S}_{NN} \end{bmatrix}^{-1} \begin{bmatrix} \mathbf{m}^1 \\ \mathbf{m}^2 \\ \vdots \\ \mathbf{m}^N \end{bmatrix}
\tag{29}
$$

The block matrix to be inverted in Eq. (29) can be very large depending on the size of the images. However, because all $\mathbf{S}_{kl}$ are diagonal and $\mathbf{S}_{kl} = (\mathbf{S}_{lk})^*$, the inverse can be efficiently computed using a recursive formula for inverting block matrices [10].

The PCF algorithm can be used to simultaneously correlate data from different sensors. The different terms of the polynomial do not have to be from the same sensor or versions of the same data. Rather, we may view the sensor imaging process and its transfer function itself as the nonlinear mapping function. The concept is illustrated in Fig. 2, in

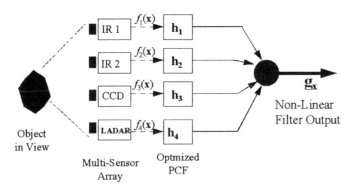

Figure 2   Polynomial correlation architecture for multi-sensor fusion.

which it is shown that data from different sensors (say, for instance, various infrared bands, CCD, and a LADAR) may be directly injected into the architecture, resulting in a fused correlation output. This is equivalent to data fusion in the algorithm. Again, the analysis and the form of the solution remain the same as that Eq. (29).

10.4 DISTANCE CLASSIFIER CORRELATION FILTERS

The correlation filters described in the previous sections were presented as detectors whose response to patterns of interest is carefully controlled by the various optimization techniques. Because these filters are trained only on one class at a time, their response to other classes is not explicitly known. In this section, we address the issue of multiclass discrimination (classification) by explicitly and simultaneously including all of the classes to be separated in the training process. The process further differs from the methods in Sections 10.2 and 10.3 in that we present a different interpretation of correlation filters as a method of applying a *transformation* to the data.

It is well known that the correlation can be viewed as a linear transformation. Specifically, the filtering process can be mathematically expressed as multiplication by a diagonal matrix in the frequency domain. The distance of a vector $\mathbf{x}$ to a reference $\mathbf{m}_k$ under a linear transform $\mathbf{H}$ is given

$$d_k = |\mathbf{Hx} - \mathbf{Hm}_k|^2 = (\mathbf{x} - \mathbf{m}_k)^+ \mathbf{H}^+ \mathbf{H}(\mathbf{x} - \mathbf{m}_k) \tag{30}$$

The filtering process transforms the input images to new images. For the correlation filter to be useful as a transform, we require that the images of the different classes become as different as possible after filtering. Then, distances can be computed between the filtered input image and the references of the different classes that have been also transformed in the same manner. The input is assigned to the class to which the distance is the smallest. The emphasis is shifted from just one point (i.e., the correlation peak) to comparing the entire shape of the correlation plane. These facts, along with the simplifying properties of linear systems, leads to an intriguing realization of a distance classifier with a correlation filter twist.

In this section, we are concerned with finding a solution for the transform matrix $\mathbf{H}$ in Eq. (30). It is assumed that the training images are segmented (although test images are expected to appear unsegmented) and appropriately centered. An image $x(m, n)$ with d pixels can be expressed as a d-dimensional column vector $\mathbf{x}$ or as a $d \times d$ diagonal matrix $\mathbf{X}$ with the elements of $\mathbf{x}$ as its diagonal elements (i.e., Diagonal$\{\mathbf{X}\} = \mathbf{x}$).

Sometimes, the same quantity may be expressed both as a vector, say **m**, and as a diagonal matrix **M**. This implies that **Hm** and **Mh** are equivalent. Again, quantities in the frequency domain are expressed in uppercase italics. Thus, the DFT of $x(m, n)$ is denoted by $X(k, l)$. All analysis presented in this section is carried out in the frequency domain.

As noted in the beginning of this section, the distance classifier uses a global transform **H** to maximally separate the classes while making them as compact as possible. Figure 3 depicts schematically the basic idea using a 3-class example, where $\mathbf{m}_1$, $\mathbf{m}_2$, and $\mathbf{m}_3$ represent the class centers (obtained by averaging the Fourier transforms of the corresponding training images) and **z** represents an unknown input to be classified. The transformation matrix **H** is designed to make the classes distinct by moving the class centers apart while shrinking the boundaries around each class so that **z** can be more accurately identified with its correct class (class 3 in the figure because d_3 is the smallest distance).

The general C class distance classifier problem is formulated by stating that we require the transformed images to be as different as possible for each of the classes. At the same time, the classes should become as *compact* as possible under the transformation matrix. Let $\mathbf{x}_{ik}$ be the d-dimensional column vector containing the Fourier transform (FT) of the ith image of the kth class, $1 \le i \le N$ and $1 \le k \le C$. We assume without loss of generality that each class has N training images. Let $\mathbf{m}_k$ be the mean FT of class k; that is,

$$\mathbf{m}_k = \frac{1}{N} \sum_{i=1}^{N} \mathbf{x}_{ik}, \qquad 1 \le k \le C \tag{31}$$

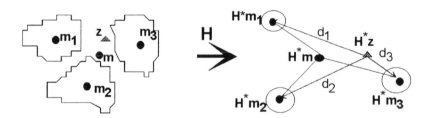

Figure 3 Transformation by **H** increases interclass distance while making each class more compact to simultaneously improve distortion tolerance and discrimination.

Under the transform $\mathbf{H}$, the difference between the means of any two classes (say k and l) is

$$\mathbf{v}_{kl} = \mathbf{H}^*(\mathbf{m}_k - \mathbf{m}_l) \tag{32}$$

Taking the expectation of the elements of $\mathbf{v}_k$ over all frequencies yields

$$\bar{v}_{kl} = E_i\{\mathbf{v}_k(i)\} \cong \frac{1}{d}\,\mathbf{h}^+(\mathbf{m}_k - \mathbf{m}_l) \tag{33}$$

The quantity in Eq. (33) is a measure of *spectral separation* between classes k and l over all frequencies. The sign of $\bar{v}_{kl}$ is not important for the classes to be separated. Therefore, we want to make $|\bar{v}_{kl}|^2$ large. Taking all possible pairs of classes into consideration, we define the *average spectral separation* (ASS) criterion as

$$
\begin{aligned}
A(\mathbf{h}) &= \frac{1}{C^2}\sum_{k=1}^{C}\sum_{l=1}^{C}|\bar{v}_{kl}|^2 \\
&= \frac{1}{C^2}\sum_{k=1}^{C}\sum_{l=1}^{C}|\mathbf{m}_k^+\mathbf{h} - \mathbf{m}_L^+\mathbf{h}|^2 \\
&= \frac{1}{C^2}\sum_{k=1}^{C}\sum_{l=1}^{C}\mathbf{h}^+(\mathbf{m}_l - \mathbf{m}_k)(\mathbf{m}_l - \mathbf{m}_k)^+\mathbf{h}
\end{aligned}
\tag{34}
$$

After some algebraic manipulations, the expression for $A(\mathbf{h})$ can be simplified to

$$A(\mathbf{h}) = \mathbf{h}^+\left[\frac{1}{C}\sum_{k=1}^{C}(\mathbf{m} - \mathbf{m}_k)(\mathbf{m} - \mathbf{m}_k)^+\right]\mathbf{h} = \mathbf{h}^+\mathbf{T}\mathbf{h} \tag{35}$$

where $\mathbf{T} = (1/C)\sum_{k=1}^{C}(\mathbf{m} - \mathbf{m}_k)(\mathbf{m} - \mathbf{m}_k)^+$ is a $d \times d$ nondiagonal matrix of rank $\leq (C - 1)$ and $\mathbf{m} = (1/C)\sum_{k=1}^{C}\mathbf{m}_k$ is the mean of the entire dataset. If $A(\mathbf{h})$ is maximized by the choice of $\mathbf{h}$, the average spectral content of the classes will differ greatly and they will become well separated. At the same time, to improve distortion tolerance within a class, we want to minimize the criterion for compactness given by

$$B(\mathbf{h}) = \frac{1}{C}\sum_{k=1}^{C}\frac{1}{N}\sum_{i=1}^{N}\mathbf{h}^+[\mathbf{X}_{ik} - \mathbf{M}_k][\mathbf{X}_{ik} - \mathbf{M}_k]^*\mathbf{h} = \mathbf{h}^+\mathbf{S}\mathbf{h} \tag{36}$$

We recognize that the term $\mathbf{h}^+\mathbf{S}\mathbf{h}$ is the same as ASM calculated over all classes and, therefore, $B(\mathbf{h})$ is a measure of average class compactness after transformation by $\mathbf{H}$. Our objectives of maximizing $A(\mathbf{h})$ and minimizing $B(\mathbf{h})$ are met by maximizing the ratio of $A(\mathbf{h})$ and $B(\mathbf{h})$; that is, we maximize

$$J(\mathbf{h}) = \frac{A(\mathbf{h})}{B(\mathbf{h})} = \frac{\mathbf{h}^+ \mathbf{T} \mathbf{h}}{\mathbf{h}^+ \mathbf{S} \mathbf{h}} \tag{37}$$

with respect to $\mathbf{h}$. The optimum solution is the dominant eigenvector of $\mathbf{S}^{-1}\mathbf{T}$. We refer to the optimum $\mathbf{h}$ as the distance classifier correlation filter (DCCF).

For testing purposes, the distance to be computed between the transformed input and the ideal reference for class k is

$$d_k = |\mathbf{H}^* \mathbf{z} - \mathbf{H}^* \mathbf{m}_k|^2 = p + b_k - (\mathbf{z}^+ \mathbf{h}_k + \mathbf{h}_k^+ \mathbf{z}), \qquad 1 \le k \le C \tag{38}$$

where $\mathbf{z}$ is the input image, $p = |\mathbf{H}^* \mathbf{z}|^2$ is the transformed input image energy, $b_k = |\mathbf{H}^* \mathbf{m}_k|^2$ is the energy of the transformed kth class mean, and $\mathbf{h}_k = \mathbf{H}\mathbf{H}^* \mathbf{m}_k$ is viewed as the *effective* filter for class k. For images that are real in the space domain, the expression for d_k simplifies to

$$d_k = |\mathbf{H}^* \mathbf{z} - \mathbf{H}^* \mathbf{m}_k|^2 = p + b_k - 2\mathbf{z}^+ \mathbf{h}_k \tag{39}$$

In general, the target may be anywhere in the input image. For shift-invariant distance calculation, we are interested in the smallest value of d_k over all possible shifts of the target with respect to the class references (i.e., the best possible match between the input and the reference for class k). In Eq. (39), because p and b_k are both positive and independent of the position of the target, the smallest value of d_k over all shifts is obtained when the third term (i.e., $\mathbf{z}^+ \mathbf{h}_k$) is as large as possible. Therefore, this term is chosen as the *peak value* in the full space-domain cross-correlation of $\mathbf{z}$ and $\mathbf{h}_k$.

Because there are only C classes to which distances must be computed, we require C such filters. It should be noted that for a given transform $\mathbf{H}$, all d_k, $1 \le k \le C$, have the same term p which could be dropped if the only objective were to find the class to which the distance is the smallest.

10.5 TRAINING IMAGE REGISTRATION TECHNIQUES

Although the science of composite correlation filter design has matured considerably, not much attention has been paid to the registration of training images until now. In this section, we illustrate the benefits of optimum registration by extending the MACH filter and DCCF algorithms discussed in the previous sections. Specifically, we formalize the impact of within-class and between-class alignment of the training set on the performance of the filters and then describe efficient procedures to implement the registration process.

10.5.1 Registration of In-Class Training Images

The problem of registering training images within a class has not been explicitly addressed in previous work on correlation filter synthesis. The issue is often simplified by assuming that images are centered by a user at a predefined aimpoint on the target. In other instances, data generated by a computer model or collected under controlled conditions (e.g., using a turntable) permit us to assume that images represent the object about some axis of rotation; therefore, further registration is not required. In this section, we will discuss the registration of images to directly improve the performance of the correlation filters by adjusting their position to reduce a metric for in-class variance. Let $\mathbf{x}_1, \mathbf{x}_2, \ldots, \mathbf{x}_N$ be column vectors that represent the N training images of a particular class in the space domain. The average image of this set is

$$\mathbf{m}_N = \frac{1}{N} \sum_{i=1}^{N} \mathbf{x}_i \tag{40}$$

where the subscript N indicates that the statistic is estimated using N samples. A *mean square error* (MSE) metric for in-class variance for the same set is defined as

$$\sigma_N^2 = \frac{1}{N} \sum_{i=1}^{N} (\mathbf{x}_i - \mathbf{m}_N)^T (\mathbf{x}_i - \mathbf{m}_N) = \frac{1}{N} \sum_{i=1}^{N} \mathbf{x}_i^T \mathbf{x} - \mathbf{m}_N^T \mathbf{m}_N \tag{41}$$

The above metric can be interpreted as a measure for class compactness, and training images can be registered to minimize its value. We shall use the method of induction to develop the registration algorithm as well as to prove its optimality. Assuming that N images are "perfectly" registered in the minimum MSE sense [i.e., σ_N^2 in Eq. (41) is as small as possible], the question is, how should the $(N + 1)$st image be registered with this set so that σ_{N+1}^2 is also as small as possible?

We now derive the expression for the in-class variance when the $(N + 1)$st image is added to the set. This is given by

$$\sigma_{N+1}^2 = \frac{1}{N+1} \sum_{i=1}^{N+1} \mathbf{x}_i^T \mathbf{x}_i - \mathbf{m}_{N+1}^T \mathbf{m}_{N+1}$$

$$= \frac{1}{N+1} \sum_{i=1}^{N+1} \mathbf{x}_i^T \mathbf{x}_i - \frac{1}{(N+1)^2} \left(\sum_{i=1}^{N} \mathbf{x}_i + \mathbf{x}_{N+1} \right)^T \left(\sum_{i=1}^{N} \mathbf{x}_i + \mathbf{x}_{N+1} \right) \tag{42}$$

Multiplying both sides of Eq. (42) by $(N + 1)^2$ and, after some manipulations, we obtain

$$
\begin{aligned}
(N + 1)^2 \sigma_{N+1}^2 &= (N + 1) \sum_{i=1}^{N} \mathbf{x}_i^T \mathbf{x}_i + (N + 1)\mathbf{x}_{N+1}^T \mathbf{x}_{N+1} \\
&\quad - (N^2 \mathbf{m}_N^T \mathbf{m}_N + 2N \mathbf{m}_N^T \mathbf{x}_{N+1} + \mathbf{x}_{N+1}^T \mathbf{x}_{N+1}) \\
&= \left(N \sum_{i=1}^{N} \mathbf{x}_i^T \mathbf{x}_i - N^2 \mathbf{m}_N^T \mathbf{m}_N \right) + \sum_{i=1}^{N} \mathbf{x}_i^T \mathbf{x}_i \\
&\quad + (N + 1)\mathbf{x}_{N+1}^T \mathbf{x}_{N+1} - \mathbf{x}_{N+1}^T \mathbf{x}_{N+1} - 2N \mathbf{m}_N^T \mathbf{x}_{N+1} \qquad (43)
\end{aligned}
$$

Using Eqs. (41) and (43), we get

$$
\sigma_{N+1}^2 = \frac{N^2}{(N + 1)^2} \sigma_N^2 + \frac{N}{(N + 1)^2} \left(\frac{1}{N} \sum_{i=1}^{N} \mathbf{x}_i^T \mathbf{x}_i + \mathbf{x}_{N+1}^T \mathbf{x}_{N+1} - 2\mathbf{m}_N^T \mathbf{x}_{N+1} \right)
$$

$$(44)$$

Clearly, given that σ_N^2 is as small as it can be, in order to make σ_{N+1}^2 as small as possible after the inclusion of the $(N + 1)$st image, the expression in the large parentheses must be minimized. The first term, $(1/N)\sum_{i=1}^{N} \mathbf{x}_i^T \mathbf{x}_i$, does not impact this process and can be ignored. The second term, $\mathbf{x}_{N+1}^T \mathbf{x}_{N+1}$ [i.e., the energy in the $(N + 1)$st image] depends on the selection of the $(N + 1)$st image, but not on its position relative to the other training images. This term could also be dropped if all of the training images are normalized to unity. However, the third term (i.e., the cross-product between $\mathbf{x}_{N+1}$ and $\mathbf{m}_N$, the average of the N previously registered images) depends on the position of the new training image relative to the average of the others in the set. Because this term occurs with a negative sign, σ_{N+1}^2 is minimum when $\mathbf{m}_N^T \mathbf{x}_{N+1}$ is maximum. This can be ensured by computing the full two-dimensional cross-correlation of $\mathbf{x}_{N+1}$ and the average image $\mathbf{m}_N$, and then shifting $\mathbf{x}_{N+1}$ in the space domain such that the peak cross-correlation value occurs at the origin.

The *optimum* algorithm for registering a set of training images can now be developed as follows. We first compute the full cross-correlations of all possible pairs of training images and select the pair (i.e., $N = 2$) that yields the smallest MSE as defined in Eq. (41). This ensures that the starting value of within class variance (σ_2^2) is as small as possible. Assume that this occurs for the two images $\mathbf{x}_a$ and $\mathbf{x}_b$. We then shift $\mathbf{x}_a$ relative to $\mathbf{x}_b$ such the peak cross-correlation value occurs at the origin. The positions of these two images are now fixed for the remainder of the process. The average of the two *registered* images is defined as $\mathbf{m}_2 = (\mathbf{x}_a + \mathbf{x}_b)/2$. At each step thereafter, we select and register the next training image [i.e., $(N + 1)$st image, $N \geq 2$] to minimize the function

$$\Psi_{N+1} = \mathbf{x}_{N+1}^T \mathbf{x}_{N+1} - 2\mathbf{m}_N^T \mathbf{x}_{N+1} \tag{45}$$

Thus, after selecting the initial pair, the remaining training images (excluding $\mathbf{x}_a$ and $\mathbf{x}_b$) are fully cross-correlated with $\mathbf{m}_2$. The image $\mathbf{x}_C$, which yields the smallest value of Ψ_3, is selected for registration. This is done by shifting $\mathbf{x}_C$ relative to $\mathbf{m}_2$ such that the peak vlaue of their cross-correlation occurs at the origin. Once again, this ensures that the change in variance due the inclusion of the third image (i.e., the change from σ_2^2 to σ_3^2) is as algebraically small as possible. The average image is now updated to $\mathbf{m}_3 = (\mathbf{x}_a + \mathbf{x}_b + \mathbf{x}_c)/3$. The process is repeated with the aim of selecting and registering the training image at each step whose inclusion causes the least increase in the within-class variance. Based on the relationship in Eq. (44), the method can be thus guaranteed to minimize σ_N^2 and register the set of N training images such that the class is represented as compactly as possible. We believe this will directly benefit the distortion tolerance and discrimination properties of the correlation filters.

10.5.2 Optimization of Between-Class Separation for DCCF Synthesis

In this section, we will limit the discussion of between-class registration to the DCCF algorithm, although other multiclass correlation filter algorithms could benefit from the same approach. As discussed in Section 10.3, the optimum DCCF transform maximizes a performance metric $J(\mathbf{h})$ that is the ratio of the *average spectral separation* between the classes and the *average similarity measure* within each class. The implementation of the DCCF algorithm requires the distances to be calculated using correlation in a shift-invariant manner. For any given test image, the algorithm actually computes the *best match* (i.e., smallest distance) to all classes over all possible shifts. Thus, the best match to the correct class has to compete with the best possible matches to all other classes. Therefore, it is advisable to synthesize the DCCFs after registering the class references to reflect the worst possible confusion between them. In other words, we want to first find the relative spatial separation between the classes so that the resulting $J(\mathbf{h})$ is as adversely affected (i.e., small) as possible. Then, we find the transform vector $\mathbf{h}$ that maximizes this minimum $J(\mathbf{h})$. We first illustrate this concept for the two class case and then extend the results to the general N-class case.

Two-Class Case

As discussed in Section 10.3, the basic DCCF algorithm is designed to discriminate between two classes by optimizing the performance criterion

$$J(\mathbf{h}) = \frac{\mathbf{h}^+(\mathbf{m}_1 - \mathbf{m}_2)(\mathbf{m}_1 - \mathbf{m}_2)^+\mathbf{h}}{\mathbf{h}^+\mathbf{S}\mathbf{h}} \tag{46}$$

The frequeny-domain expression for the optimum transform (filter) which maximizes J(h) is given by $\mathbf{h} = \mathbf{S}^{-1}(\mathbf{m}_1 - \mathbf{m}_2)$. Using this solution, it is easy to show that the performance criterion can be simplified to

$$J(\mathbf{h}) = (\mathbf{m}_1 - \mathbf{m}_2)^+\mathbf{S}^{-1}(\mathbf{m}_1 - \mathbf{m}_2) \tag{47}$$

Now, let us discuss how the relative spatial position of $\mathbf{m}_1$ and $\mathbf{m}_2$ can be adjusted prior to computing $\mathbf{h}$ so that $J(\mathbf{h})$ is at its smallest value corresponding to the worst possible confusion between the classes. Expanding the expression for $J(\mathbf{h})$, we obtain

$$J(\mathbf{h}) = \mathbf{m}_1^+\mathbf{S}^{-1}\mathbf{m}_1 + \mathbf{m}_2^+\mathbf{S}^{-1}\mathbf{m}_2 - 2\mathbf{m}_1^+\mathbf{S}^{-1}\mathbf{m}_2 \tag{48}$$

Shifting the center of an image in space domain causes its Fourier transform to be multiplied by a linear phase complex exponential factor. The first two terms in the above equation are independent of phase because $\mathbf{S}$ is diagonal. Thus, the first two terms are unaffected by image shifts. The third term, however, depends on the relative linear phase difference between $\mathbf{m}_1$ and $\mathbf{m}_2$ (or, equivalently, their relative separation in the space domain). Although we will expand the concept in the frequency domain for the general N-class problem, it is easier to solve the problem in the space domain for the two-class case.

To obtain the smallest $J(\mathbf{h})$ which reflects the most overlap between the average class references, we want the third term, $\mathbf{m}_1^+\mathbf{S}^{-1}\mathbf{m}_2$, to be as large as possible. To accomplish this, $\mathbf{m}_1^+\mathbf{S}^{-1}\mathbf{m}_2$ must be equal to the maximum space-domain value in the full cross-correlation between $\mathbf{m}_1$ and $\mathbf{S}^{-1}\mathbf{m}_2$. Defining $\mathbf{h}_2 = \mathbf{S}^{-1}\mathbf{m}_2$, this can be mathematically stated as

$$\max\{\mathbf{m}_1^+\mathbf{S}^{-1}\mathbf{m}_2\} = \max\{\mathbf{m}_1^+\mathbf{h}_2\} = \max_{i,j}\left\{\sum_k\sum_l m_1[k,l]h_2[k+i,l+j]\right\}$$

$$= \max_{i,j}\{g_{12}(i,j)\} \tag{49}$$

where $m_1(i,j)$ and $h_2(i,j)$ are space-domain (shifted by i,j) image format representations of $\mathbf{m}_1$ and $\mathbf{h}_2$, respectively, and $g_{12}(i,j)$ denotes their two-dimensional correlation. Now, $\mathbf{m}_1^+\mathbf{S}^{-1}\mathbf{m}_2$, is, by definition, the value at the origin of the cross-correlation function [i.e., $\mathbf{m}_1^+\mathbf{S}^{-1}\mathbf{m}_2 = g_{12}(0,0)$].

Normally, there is no reason to expect that the largest value will always occur at the origin. However, this condition can be forced by first calculating the full cross-correlation surface $g_{12}(i, j)$, locating the largest value, and then shifting $\mathbf{m}_1$ relative to $\mathbf{m}_2$ such that it now occurs at the origin. By doing so, we have adjusted the spatial location of $\mathbf{m}_1$ relative to $\mathbf{m}_2$ to ensure that the third term in Eq. (48) is as large as possible, which guarantees that the corresponding optimum solution $\mathbf{h} = \mathbf{S}^{-1}(\mathbf{m}_1 - \mathbf{m}_2)$ maximizes $J(\mathbf{h})$ for the toughest comparison between the classes under shift-invariant conditions.

Multiclass Case

To extend the concept of adjusting the phases of the class means to the general C class, let us define $\mathbf{m} = (1/C \sum_{k=1}^{C} \mathbf{m}_k$ as the overall mean of the data space. It was shown in Section 10.3 that the DCCF performance criterion

$$J(\mathbf{h}) = \frac{\mathbf{h}^+ \mathbf{T} \mathbf{h}}{\mathbf{h}^+ \mathbf{S} \mathbf{h}} \tag{50}$$

is maximized by choosing $\mathbf{h}$ to be the dominant eigenvector of $\mathbf{S}^{-1}\mathbf{T}$ with corresponding eigenvalue $\lambda_{\max}$. Recall that $\mathbf{T}$ is an outer product matrix of rank $C - 1$ and can be defined as $\mathbf{T} = \mathbf{E}\mathbf{E}^+$, where $\mathbf{E} = [\mathbf{e}_1 \quad \mathbf{e}_2 \quad \cdots \quad \mathbf{e}_C]$ is a $d \times C$ matrix whose kth column is the vector $\mathbf{e}_k = \mathbf{m} - \mathbf{m}_k$.

Once again, our objective will be to register the different class means such that $J(\mathbf{h})$ is as small as possible and then find $\mathbf{h}$ to maximize the worst-case $J(\mathbf{h})$. It should be noted that the ASM term (a within-class parameter) in the denominator of Eq. (50) will not be affected by the relative spatial position of the class means. This is due to the fact that $\mathbf{S}$ is the sum of the spectral variance matrices of each class $\mathbf{S}_i$ that do not depend on the between-class statistics. Therefore, minimizing $J(\mathbf{h})$ by adjusting the spatial location of the class means has the direct effect of making the numerator in Eq. (50) as small as possible, representing the class separation at its worst. We now develop the algorithm for adjusting the relative position of the class means such that $J(\mathbf{h})$ is minimized.

It can be easily shown that $\mathbf{h}$ must be of the form

$$\mathbf{h} = \mathbf{S}^{-1}\mathbf{E}\mathbf{a} \tag{51}$$

where $\mathbf{a}$ is a $C \times 1$ weight vector. Because $\mathbf{h}$ must satisfy $\mathbf{S}^{-1}\mathbf{T}\mathbf{h} = \lambda_{\max} \cdot \mathbf{h}$, we obtain

$$\mathbf{S}^{-1}\mathbf{E}\mathbf{E}^+\mathbf{S}^{-1}\mathbf{E}\mathbf{a} = \lambda_{\max} \cdot \mathbf{S}^{-1}\mathbf{E}\mathbf{a} \tag{51}$$

which is equivalent to

$$\mathbf{E}^+\mathbf{S}^{-1}\mathbf{Ea} = \lambda_{max}\mathbf{a} \tag{53}$$

which proves that $\mathbf{a}$ must be the dominant eigenvector of $\mathbf{E}^+\mathbf{S}^{-1}\mathbf{E}$ with λ_{max} as the corresponding eigenvalue. Because this is a relatively small $C \times C$ matrix, its eigenvalues and eigenvectors can be numerically computed without difficulty.

Using the results in Eqs. (51) and (53), it is possible to show that ASM [the term for distortion tolerance and the denominator of $J(\mathbf{h})$] can be simplified to

$$\begin{aligned} ASM = \mathbf{h}^+\mathbf{Sh} &= \mathbf{a}^+\mathbf{E}^+\mathbf{S}^{-1}\mathbf{Ea} \\ &= \lambda_{max} \cdot \mathbf{a}^+\mathbf{a} \\ &= \lambda_{max} \end{aligned} \tag{54}$$

where $\mathbf{a}^+\mathbf{a} = 1$ because eigenvectors are normalized. Likewise, the criterion for class separation [which is the numerator of $J(\mathbf{h})$] can be simplified to

$$\begin{aligned} \mathbf{hTh} = \mathbf{a}^+\mathbf{E}^+\mathbf{S}^{-1}\mathbf{EE}^+\mathbf{S}^{-1}\mathbf{Ea} \\ &= \mathbf{a}^+ \cdot \lambda_{max} \cdot \lambda_{max} \cdot \mathbf{a} \\ &= \lambda_{max}^2 \end{aligned} \tag{55}$$

Of course, these are consistent with the known fact that

$$\begin{aligned} J(\mathbf{h}) &= \frac{\mathbf{h}^+\mathbf{Th}}{\mathbf{h}^+\mathbf{Sh}} \\ &= \frac{\lambda_{max}^2}{\lambda_{max}} \\ &= \lambda_{max} \end{aligned} \tag{56}$$

The objective now is to adjust the relative position of the class references such that λ_{max} [and hence $J(\mathbf{h})$] is at its smallest value so that the optimum solution for $\mathbf{h}$ maximizes the worst-case class separation. Due to Eq. (56), this has the added appeal of reducing ASM and improving distortion tolerance.

We now discuss an iterative approach to minimizing λ_{max} for the general C class case. Let $\phi_i(k, l)$ be the frequency-domain linear phase functions to be applied to the class means to optimize their spatial positions. For the ith class, the linear phase change in the frequency domain due to spatial shifts $[a_i, b_i]$ is given by

$$\phi_i(k, l) = \exp\left(-2\pi j \frac{ka_i + lb_i}{d - 1}\right) \tag{57}$$

We define a $C \times 2$ matrix of the shift variables as

$$\Theta = \begin{bmatrix} a_1 & b_1 \\ a_2 & b_2 \\ \vdots & \vdots \\ a_C & b_C \end{bmatrix} \tag{58}$$

These variables can be iteratively optimized to minimize $\lambda_{\max}$ using numerical search routines that are available in several commercial software packages. At each iteration, we recompute the class means using

$$m_i(k, l)\phi_i(k, l) = m_i(k, l)\exp\left(-2\pi j \frac{ka_i + lb_i}{d - 1}\right) \tag{59}$$

and also update the overall mean of the dataset. We then recompute the elements of the matrix in Eq. (53); that is,

$$\mathbf{E}^+ \mathbf{S}^{-1} \mathbf{E} = \begin{bmatrix} \mathbf{e}_1^+ \mathbf{S}^{-1} \mathbf{e}_1 & \mathbf{e}_1^+ \mathbf{S}^{-1} \mathbf{e}_2 & \cdots & \mathbf{e}_1^+ \mathbf{S}^{-1} \mathbf{e}_C \\ \mathbf{e}_2^+ \mathbf{S}^{-1} \mathbf{e}_1 & \mathbf{e}_2^+ \mathbf{S}^{-1} \mathbf{e}_2 & \cdots & \mathbf{e}_2^+ \mathbf{S}^{-1} \mathbf{e}_{2C} \\ \vdots & \vdots & \ddots & \vdots \\ \mathbf{e}_C^+ \mathbf{S}^{-1} \mathbf{e}_1 & \mathbf{e}_C^+ \mathbf{S}^{-1} \mathbf{e}_2 & \cdots & \mathbf{e}_C^+ \mathbf{S}^{-1} \mathbf{e}_C \end{bmatrix} \tag{60}$$

where $\mathbf{e}_i = \mathbf{m}_i - \mathbf{m}$. Because this is a $C \times C$ matrix, the eigenvalues are easy to calculate numerically. To minimize the largest eigenvalue, we simply set $\lambda_{\max}$ to be the cost function of the numerical minimization routines. The iterations proceed by adjusting the elements of Θ until $\lambda_{\max}$ has been optimized as desired.

If an iterative technique is not viable (either due to software or throughput limitations, or if the number of classes is very large), a computationally simpler alternative is to minimize the trace of the matrix in Eq. (60), which will minimize an upper bound on the largest eigenvalue. Although this may not produce the exact solution obtained by direct minimization, a simple noniterative algorithm can be formulated for minimizing the trace criterion. In fact, the algorithm is very similar to that described earlier for registering in-class images and is developed as follows. The trace is given by

$$\text{Trace}\{\mathbf{E}^+\mathbf{SE}\} = \sum_{i=1}^{C} \mathbf{e}_i^+ \mathbf{S}^{-1} \mathbf{e}_i$$

$$= \sum_{i=1}^{C} (\mathbf{m}_i - \mathbf{m})^+ \mathbf{S}^{-1} (\mathbf{m}_i - \mathbf{m}) \tag{61}$$

Careful examination reveals that this is very similar in form to the MSE metric in Eq. (41). Here, $\mathbf{m}$ still represents the overall mean of the images to be registered and $\mathbf{m}_i$ can be likened to the image vectors $\mathbf{x}_i$. As explained earlier, $\mathbf{S}$ is not affected by the relative position of the class means and, therefore, is easily taken into account. With slight modification, the equivalent of Eq. (45) can be expressed as

$$\Psi_{k+1} = \mathbf{m}_{k+1}^+ \mathbf{S}^{-1} \mathbf{m}_{k+1} - 2\mathbf{m}_k^+ \mathbf{S}^{-1} \mathbf{m}_{k+1}, \qquad 1 \le k \le C$$

In other words, assuming that a partial trace based on k class means is already minimum, the $(k + 1)$st class mean is selected and shifted to yield the minimum value of Ψ_{k+1} and, hence, the smallest increase in the trace. The algorithm proceeds exactly as described in Section 10.5.1 by correlating $\mathbf{m}_{k+1}$ (the image to be registered) with $\mathbf{m}_k$ (i.e., the average based on the k images) and then shifting it as necessary to ensure that the peak value of the correlation occurs at the center. The only additional computation is the premultiplication by $\mathbf{S}^{-1}$ in the frequency domain during the correlation operation. Unlike the in-class training image case, the class means cannot be normalized to unity and, therefore, the term $\mathbf{m}_{k+1}^+ \mathbf{S}^{-1} \mathbf{m}_{k+1}$ cannot be dropped.

10.6   A SAMPLE CASE STUDY

In this section, we illustrate the performance of the algorithms presented in this chapter using the 3-class public domain MSTAR data [11]. This set contains 1-ft-resolution SAR images of a BTR70, T72, and a BMP. The training images are at a $15°$ depression angle, whereas the test set is a $17°$ depression. The MACH filters and DCCFs were synthesized to recognize the targets over a range of $45°$. Thus, a set of eight such filters is required to recognize the targets over the full $360°$ azimuth range. The correlation filters were first synthesized without any registration. The test images were assigned to the class to which the distance is the smallest, provided the ratio of the minimum distance to the average of the distances to all the classes is below a threshold of 0.9. In addition, it was also required that corresponding MACH filter yields a correlation peak with a *peak-to-sidelobe ratio* (PSR)* of at least 5.0. The PSR is a measure of the peak sharpness.

---

*PSR = (peak value − mean)/standard deviation.

Figure 4 shows the mean images of BTR70 target based on 22 images between 135° and 180°. The image in Fig. 4a is the result of averaging these images without registration. The MSE as defined in Eq. (41) was found to be 2.03. The image in Fig. 4b is the average of the same training images after the registration algorithm was applied. It is interesting to note that the two mean images look substantially different and that the definition of features in the target signature (such as self-shadows, turrets, and dominant reflectors) are better preserved in the average of the registered training images. Quantitatively speaking, the MSE for the latter case was 1.68, which represents a reduction of 17% in the variance of the set. This type of improvement was noted in all three targets over all clusters of aspect angles.

Figure 5 shows the improvements in the performance of the MACH filter when the in-class registration algorithm is used. The plot in Fig. 5a is the response of the MACH filter to the mean image when the training images are not registered. The peak is noticeably broad and PSR is only 5.75. The plot on the right is the correlation of the average training image and the MACH filter when optimum registration is used. The peak structure is noticeably improved and the PSR is found to be 9.0.

We now describe the improvements observed by optimizing the position of the class means to account for the worst-case between-class separation. The numerical iterative method for minimizing the largest eigenvalue [see Eq. (53)] was implemented using the FMINU function in Matlab^{TM}. The values of λ_{max} before and after optimization are shown in Table 1. The images of the three class means for cluster 4 are shown in Fig. 6 to illustrate their relative position.

In Fig. 6, it can be seen that the optimization process causes the class means to be shifted by a few pixels in order to maximize between-class

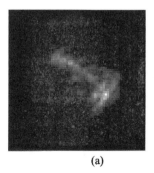 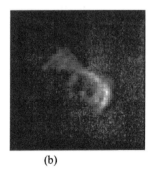

(a)   (b)

Figure 4   Average of BTR70 training images between 135° and 180° (a) without registration and (b) after registration.

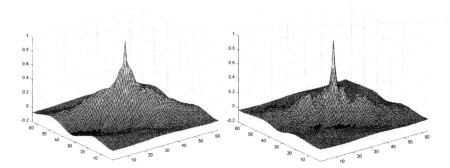

Figure 5 The performance of the MACH filter is significantly improved when the training images are registered. As shown in (a), the average of the unregistered training images produces a broad peak with a PSR of only 5.75. After registration, the peak in (b) is considerably sharper with a PSR of 9.0.

overlap. The solid gray centerlines have been marked on the image to make visual comparisons easier. Given that the DCCF will compute distances over all possible shifts, these images reflect the worst-case potential for confusion. The optimum DCCF solution based on these "position-optimized" images exhibits an enhanced ability to correctly classify a test image as it is compared to all classes over all possible shifts.

Finally, we compare the confusion matrices for the 3-class problem obtained using the MACH and DCCF algorithms with and without the

Table 1 Comparison of λ_{max} Values Before and After Numerical Optimization to Adjust the Relative Position of Class Means to Accommodate the Worst-Case Class Separation

Cluster No. (aspect angles)	Value of λ_{max} before optimization	Value of λ_{max} after optimization
1 (0°–45°)	479	389
2 (45°–90°)	430	344
3 (90°–135°)	356	339
4 (135°–180°)	672	544
5 (180°–225°)	548	412
6 (225°–270°)	458	322
7 (270°–315°)	371	322
8 (315°–360°)	557	454

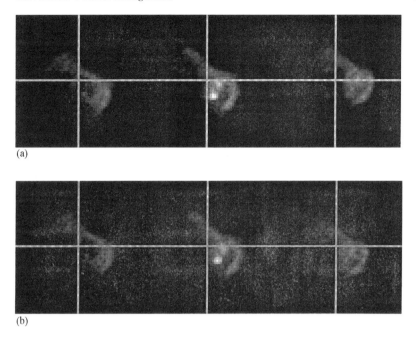

(a)

(b)

Figure 6 Relative position of class mean images of the BTR70, T72, and BMP for cluster number 4 before optimization (a) and after optimization (b) to account for maximum possible confusion.

registration/position optimization algorithms described in this chapter. Table 2 shows the confusion matrix before optimization. Although the results are fairly good, the significant improvements achieved after optimization are also shown in Table 2.

Table 2 Results of Resting the 3-Class ATR Using the MSTAR Dataset (a) Without Optimization and (b) After Registration of Training Images and Optimizing the Relative Position of Class Means

Truth/decision	BTR70	T72	BMP
BTR 70	98.5		1.5
T72	1.1	94.2	4.7
BMP	3.1	1	95.9

We observe that prior to optimization, the correct recognition rate is $P_c = 96.2\%$ and the error rate is $p_e = 3.8\%$. Although this performance is quite good, after optimization the P_c increases to 97.9% and the error rate is reduced to $P_e = 2.7\%$. This change in P_e of 1.1% corresponds to a 29% improvement (or reduction) in the error rate that is achieved by the optimization process. All preprocessing steps and thresholds were kept at the same value in making these comparisons. It should be noted that if the thresholds were raised, many of the error cases could be rejected by the classifier.

10.7 SUMMARY

In this chapter, we have reviewed some recently developed correlation filter design algorithms. Correlation is a powerful tool for pattern recognition, but the performance depends on the filter optimization process. We presented several optimum filters, based on linear and nonlinear applications of correlation. The MACH filter works well as a detector to find targets of a particular class in images with background and noise. The extension of the MACH filter known as the PCF is useful not only for obtaining higher-order correlations but also for simultaneously correlating multiple sources of data, thus achieving fusion in the correlation process itself. The DCCF is a quadratic classifier based on correlation that is very useful for separating multiple classes. Typically, it is advantageous to use the MACH filter for detecting regions of interest and using the DCCFs to classify the detected patterns. We also addressed another important issue in the design of correlation filters, namely the registration of training images. The effectiveness of the MACH filter and the DCCF algorithm is significantly improved by the optimum registration algorithms. Another important benefit of the automatic registration tecnique is that it avoids the need for manual registration that is both tedious and hard to repeat exactly. Finally, we discussed a case study using commonly available SAR images to illustrate the use and performance of the algorithms.

REFERENCES

1. A VanderLugt. Signal detection by complex spatial spatial filtering. IEEE Trans Inform Theory IT-10:139–145, 1964.
2. B V K Vijaya Kumar. Tutorial survey of composite filter designs for optical correlators. Appl Opt 31:4773–4801, 1992.

3. A Mahalanobis, B V K Vijaya Kumar, S R F Sims, J Epperson. Unconstrained correlation filters. Appl Opt 33:3751–3759, 1994.

4. A Mahalanobis, B V K Vijaya Kumar. On the optimality of the MACH filter for detection of targets in noise. Opt Eng 36(10):2642–2648, 1997.

5. A Mahalanobis, B V K Vijaya Kumar. Polynomial Filters for Higher Order Correlation and Multi-Input Information Fusion. Euro American Workshop on Optoelectronic Information Process. SPIE Optical Engineering Press, Barcelona, 1997, pp 221–231.

6. A Mahalanobis, B V K Vijaya Kumar, S R F Sims. Distance classifier correlation filters for multi-class target recognition. Appl Opt 35:3127–3133, 1996.

7. A Mahalanobis, B V K Vijaya Kumar, R T Frankot. Optimization of intraclass and between class registration for correlation filter synthesis. Appl Opt, June 2000.

8. B Javidi, J Wang. Design of filters to detect a noise target in nonoverlapping background noise. J Opt Soc Am A 10:2604–2612, 1994.

9. B Javidi, A Fazlollahi, P Willett, Ph Refregier. Performance of an optimum receiver designed for pattern recognition with non-overlapping target and scene noise. Appl Opt 34(20):3858–3868, 1995.

10. T Kailath, A H Sayed, eds. Fast Reliable Algorithms for Matrices with Structure. Siam, Philadelphia, 1999.

11. Three Class Public MSTAR Dataset (CD Version). Released by DARPA by Veda Inc, the Dayton Group, 1997; for more information see webpage https://www.mbvlab.wpafb.af.mil/public/sdms/main.htm.

11
Optimum Nonlinear Filter for Detecting Noisy Distorted Targets

Seung Hyun Hong and Bahram Javidi
University of Connecticut, Storrs, Connecticut

11.1 INTRODUCTION

Pattern and image recognition techniques have been widely used in military, scientific research, and business areas to process images and patterns. Their applications include the processing of satellite images, object recognition, and identification of fingerprints and retinas. An image or a pattern can be recognized using prior knowledge or the statistical information extracted from the image or the pattern. There are many ways of solving pattern recognition problems. Among the commonly used approaches discussed in this chapter are the correlation approaches [1–7], Bayesian receiver [8,9] and artificial neural networks [10].

The correlation approaches of solving pattern or image recognition problems are used to obtain a high correlation peak between the reference signal and an input scene at the position of the reference signal. The correlation process can be easily implemented in the frequency domain using Fourier transforms, and because it also can be implemented optically (in full or in part), the processing is done very rapidly. Numerous correlation filters [11–28] have been derived based on various metrics.

The matched filter [5,6] was designed to maximize the signal-to-noise ratio (SNR) in the presence of a wide-sense stationary noise that is additive and overlapping with the target. However, the conventional matched filter may not perform well in the presence of nonoverlapping background noise [4,13,28], or in the case of detecting distorted targets such as scaled and/or rotated targets. There are approaches to overcome the detection problem in the presence of nonoverlapping background noise [4,13,18,24,28]. Those approaches consist of optimizing some criteria (such as the probability of detection error or mean square error) to design the filters for the noise model with background clutters. To obtain the distortion tolerance capability [14,16,22,27], the filter must recognize the patterns and objects viewed from various angles and perspectives; thus, a training dataset of reference targets is needed. Several composite filter designs have been proposed to perform distortion-tolerant pattern recognition [17,20,22,27]. Such filters include the filter using the circular harmonic components of the reference signals equal-correlation-peak (ECP)–synthetic-discriminant-function (SDF) filter and minimum-average-correlation-energy (MACE)–SDF filter. The filter using circular harmonic components was proposed for the detection of specific rotated targets. The ECP-SDP filters and MACE-SDF filters are combinations of the reference targets, and the weights for this combination are chosen to have the equal correlation output peak for all training targets at the target location.

One proposed metric is the optimization of the filter to have the discriminating capability without knowing a priori information of the false object and background [15,23]. This nonlinear filter is obtained by optimizing the discriminating capability and by minimizing the output energy due to the additive noise in the input scene and the energy of correlation output in response to the input scene with the false objects or nonoverlapping background noise around the target. However, it does not have the ability to detect distorted targets. In this section, we extend this filter to include the distortion-tolerant capability by constraining all of the training data so that their outputs have certain output correlation peaks. This optimum filter will detect known targets (training targets) and detect the unknown targets that were not trained. Because this filter is energy normalized for both noise and input scene, the correlation output has the sharp peak at the location of the target and low-noise floor elsewhere.

To show the detection performance of the filters, many metrics are used [19,25]. Such metrics include correlation output peak intensity, correlation output peak-to-sidelobe ratio (PSR), peak-to-output energy (POE) ratio, receiver operating characteristics (ROC) curves [29], and output SNR. The POE is a metric to measure the sharpness of the correlation output at the target location. The ROC curves are used to compare the relative performance of different processors. Output SNR is used to measure the variability of the output correlation in response to the noise in the input scene of the filter.

The rest of this chapter is organized as follows. In Section 11.2, we set up and solve the minimization problem that leads to the derivation of the optimum nonlinear distortion-tolerant filter. In Section 11.3, we carry out the performance tests of the derived filter by computer simulations. In Section 11.3.1, the computer simulation setup is described and the correlation output of the derived filter is demonstrated with additive noise and nonoverlapping background noise. In Section 11.3.2, we carry out the performance test. The POE, ROC curves, and the output SNR are revisited and the performance of the filter with various standard deviations of the noise are discussed in the subsections of Section 11.3.2. In Section 11.4, we summarize our work.

11.2 DERIVATION OF OPTIMUM NONLINEAR COMPOSITE FILTER

In this section, we set up the mathematical model of the pattern recognition problem. The filter is designed to have the same correlation output when the input to the filter is one of the reference targets. With this constraint, the filter is to have minimal energy due to the input scene and input noise.

11.2.1 Modeling

Let $r_i(x, y)$ denote one of the reference targets to be detected, where $i = 1, 2, \ldots, T$. T is the size of the reference target set and $n(x, y)$ is assumed to be an additive white Gaussian noise with zero mean and is overlapping over the entire scene. Then, the input scene $s(x, y)$ is

$$s(x, y) = \sum_{i=1}^{T} v_i r_i(x - \tau_{xi}, y - \tau_{yi}) + n(x, y) \tag{1}$$

where τ_{xi} and τ_{yi} are the random positions of the ith reference target on the x axis and y axis, respectively and v_i is a binary random variable which takes on a value of 0 or 1, and v_i indicates whether the target $r_i(x, y)$ is present in the scene. For simplicity in the analysis, the probabilities that v_i takes on value 0 or 1 are the same for all i; that is, $p(v_i = 1) = 1/T$ and $p(v_i = 0) = 1 - 1/T$. The correlation output of the filter is

$$o(x, y) = \sum_{\tau_x=0}^{L-1} \sum_{\tau_y=0}^{M-1} h(x + \tau_x, y + \tau_y)^* s(\tau_x, \tau_y) \tag{2}$$

where, $h(x, y)$ is distortion-tolerant filter, the asterisk denotes complex conjugate, and L and M are the number of sample points in x axis and y axis, respectively. The filter is designed so that when the input to the filter is one of the reference targets, $r_i(x, y)$, which we assumed to be located at the origin, then the output of the filter is

$$o_i(0, 0) = \sum_{x=0}^{L-1} \sum_{y=0}^{M-1} h(x, y)^* r_i(x, y) = C_i \tag{3}$$

where C_i is a positive real constant. Equation (3) can be stated in a Fourier-domain expression

$$o_i(0, 0) = o_i(x, y)|_{\substack{x=0 \\ y=0}} = \frac{1}{LM} \sum_{u=0}^{L-1} \sum_{v=0}^{M-1} H(u, v)^*$$

$$R_i(u, v)e^{j2\pi uxt/L}e^{j2\pi uvyt/M}\Big|_{\substack{x=0 \\ y=0}}$$

$$= \frac{1}{LM} \sum_{u=0}^{L-1} \sum_{v=0}^{M-1} H(y, v) * R_i(u, v) \tag{4}$$

where $H(u, v)$ and $R_i(u, v)$ are the Fourier transforms of $h(x, y)$ and $r_i(x, y)$, respectively. Therefore, from Eqs. (3) and (4), we can get

$$\sum_{u=0}^{L-1} \sum_{v=0}^{M-1} H(u, v)^* R_i(u, v) = LMC_i \tag{5}$$

Equation (5) is the constraint imposed on the filter.

To obtain noise robustness, we minimize the output energy due to the noise. The mean value of the output energy due to the noise as a Fourier-domain expression is

$$E\left[\frac{1}{LM} \sum_{u=0}^{L-1} \sum_{v=0}^{M-1} |H(u, v)|^2 |N(u, v)|^2\right]$$

$$= \frac{1}{LM} \sum_{u=0}^{L-1} \sum_{v=0}^{M-1} |H(u, v)|^2 E|N(u, v)|^2 \tag{6}$$

where E is the expectation operator and $N(u, v)$ is the Fourier transform of $n(x, y)$.

To obtain the discriminating capability, the output energy due to the input scene is minimized. The output energy due to the input scene as a Fourier-domain expression is expressed as

$$\frac{1}{LM} \sum_{u=0}^{L-1} \sum_{v=0}^{M-1} |H(u, v)|^2 |S(u, v)|^2 \tag{7}$$

where $S(u, v)$ is the Fourier transform of $s(x, y)$. Thus, we minimize a linear combination of the output energy due to the noise and the output energy due to the input scene under the filter constraint. We minimize

$$\frac{w_n}{LM} \sum_{u=0}^{L-1} \sum_{v=0}^{M-1} |H(u, v)|^2 E|N(u, v)|^2 + \frac{w_d}{LM} \sum_{u=0}^{L-1} \sum_{v=0}^{M-1} |H(u, v)|^2 |S(u, v)|^2$$

(8)

where w_n and w_d are the positive weights of the noise robustness capability and discriminating capability, respectively.

11.2.2 Derivation of Optimum Nonlinear Composite Filter with Distortion Tolerant Capability

Let $a_{uv} + jb_{uv}$ and $c_{iuv} + jd_{iuv}$ be the elements of $H(u, v)$ and $R_i(u, v)$, respectively, and

$$D(u, v) = \frac{w_n E|N(u, v)|^2 + w_d |S(u, v)|^2}{LM}$$

With these notations, the constraint can be written as

$$\sum_{u=0}^{L-1} \sum_{v=0}^{M-1} H(u, v)^* R_i(u, v) = \sum_{u=0}^{L-1} \sum_{v=0}^{M-1} (a_{uv} - jb_{uv})(c_{iuv} + jd_{iuv})$$

$$= \sum_{u=0}^{L-1} \sum_{v=0}^{M-1} \{(a_{uv}c_{iuv} + b_{uv}d_{iuv})$$

$$+ j(a_{uv}d_{iuv} - b_{uv}c_{iuv})\} = LMC_i$$

(9)

Because LMC_i is a real constant, we can separate the real and imaginary parts and obtain the following set of constraints in real form:

$$\sum_{u=0}^{L-1} \sum_{v=0}^{M-1} (a_{uv}c_{iuv} + b_{uv}d_{iuv}) = LMC_i \qquad \text{for } i = 1, 2, \ldots, T$$

(10a)

$$\sum_{u=0}^{L-1} \sum_{v=0}^{M-1} (a_{uv}d_{iuv} - b_{uv}c_{iuv}) = 0 \qquad \text{for } i = 1, 2, \ldots, T$$

(10b)

Thus, the problem is to minimize

$$
\frac{w_n}{LM} \sum_{u=0}^{L-1} \sum_{v=0}^{M-1} |H(u, v)|^2 E|N(u, v)|^2 + \frac{w_d}{LM} \sum_{u=0}^{L-1} \sum_{v=0}^{M-1} |H(u, v)|^2 |S(u, v)|^2
$$

$$
= \sum_{u=0}^{L-1} \sum_{v=0}^{M-1} (a_{uv}^2 + b_{uv}^2) D(u, v) \tag{11}
$$

under the $2T$ constraints given by Eqs. (10a) and (10b). We use the Lagrange multiplier method [30] to solve this minimization problem. Let

$$
J \equiv \sum_{u=0}^{L-1} \sum_{v=0}^{M-1} (a_{uv}^2 + b_{uv}^2) D(u, v)
$$

$$
+ \sum_{i=1}^{T} \lambda_{1i} \left(LMC_i - \sum_{u=0}^{L-1} \sum_{v=0}^{M-1} a_{uv} c_{iuv} - \sum_{u=0}^{L-1} \sum_{v=0}^{M-1} b_{uv} d_{iuv} \right)
$$

$$
+ \sum_{i=1}^{T} \lambda_{2i} \left(0 - \sum_{u=0}^{L-1} \sum_{v=0}^{M-1} a_{uv} d_{iuv} + \sum_{u=0}^{L-1} \sum_{v=0}^{M-1} b_{uv} c_{iuv} \right) \tag{12}
$$

where, λ_{1i} and λ_{2i} are Lagrange multipliers.

We must find values for a_{uv}, b_{uv}, λ_{1i}, and λ_{2i} that satisfy Eqs. (10a) and (10b) and the following two equations:

$$
\frac{\partial J}{\partial a_{uv}} = 2a_{uv} D(u, v) - \sum_{i=1}^{T} \lambda_{1i} c_{iuv} - \sum_{i=1}^{T} \lambda_{2i} d_{iuv} = 0 \tag{13a}
$$

$$
\frac{\partial J}{\partial b_{uv}} = 2b_{uv} D(u, v) - \sum_{i=1}^{T} \lambda_{1i} d_{iuv} + \sum_{i=1}^{T} \lambda_{2i} c_{iuv} = 0 \tag{13b}
$$

Note that

$$
\frac{\partial^2 J}{\partial a_{uv} \partial a_{ij}} = \frac{\partial^2 J}{\partial b_{uv} \partial b_{ij}} = 2D(u, v) \delta_{uv,ij}
$$

where $\delta_{uv,ij}$ is the Kronecker delta; that is,

$$
\delta_{uv,ij} = \begin{cases} 1, & uv = ij \\ 0, & uv \neq ij \end{cases}
$$

Therefore, J has a minimum value with respect to a_{uv} and b_{uv}. Solving Eqs. (13a) and (13b), we obtain values for a_{uv} and b_{uv} that minimize J and satisfy the required constraints:

$$a_{uv} = \frac{\sum_{i=1}^{T} (\lambda_{1i} c_{iuv} + \lambda_{2i} d_{iuv})}{2D(u, v)} \tag{14a}$$

$$b_{uv} = \frac{\sum_{i=1}^{T} (\lambda_{1i} d_{iuv} - \lambda_{2i} c_{iuv})}{2D(u, v)} \tag{14b}$$

where λ_{1i} and λ_{2i} must satisfy the constraints. In order to obtain λ_{1i} and λ_{2i}, we substitue a_{uv} and b_{uv} given by Eqs. (14a) and (14b) into Eqs. (10a) and (10b) and obtain

$$\sum_{u=0}^{L-1} \sum_{v=0}^{M-1} \frac{1}{2D(u, v)} \sum_{i=1}^{T} \left[\lambda_{1i} (c_{iuv} c_{puv} + d_{iuv} d_{puv}) + \lambda_{2i} (d_{iuv} c_{puv} - c_{ivu} d_{puv}) \right]$$
$$= LMC_p \tag{15a}$$

$$\sum_{u=0}^{L-1} \sum_{v=0}^{M-1} \frac{1}{2D(u, v)} \sum_{i=1}^{T} \left[\lambda_{1i} (c_{iuv} d_{puv} - d_{iuv} c_{puv}) + \lambda_{2i} (d_{iuv} d_{puv} + c_{iuv} c_{puv}) \right]$$
$$= 0 \tag{15b}$$

for $p = 1, 2, \ldots, T$. We introduce the following additional notations to complete the derivation:

$$\boldsymbol{\lambda}_1 \equiv [\lambda_{11} \quad \lambda_{12} \quad \cdots \quad \lambda_{1T}]^t \tag{16}$$
$$\boldsymbol{\lambda}_2 \equiv [\lambda_{21} \quad \lambda_{22} \quad \cdots \quad \lambda_{2T}]^t \tag{17}$$
$$\mathbf{C} \equiv [C_1 \quad C_2 \quad \cdots \quad C_T]^t \tag{18}$$

$$A_{q,r} \equiv \sum_{u=0}^{L-1} \sum_{v=0}^{M-1} \frac{\mathrm{Re}(R_q(u, v)) \mathrm{Re}(R_r(u, v)) + \mathrm{Im}(R_q(u, v)) \mathrm{Im}(R_r(u, v))}{2D(u, v)}$$
$$= \sum_{u=0}^{L-1} \sum_{v=0}^{M-1} \frac{c_{quv} c_{ruv} + d_{quv} d_{ruv}}{2D(u, v)} \tag{19}$$

$$B_{q,r} \equiv \sum_{u=0}^{L-1} \sum_{v=0}^{M-1} \frac{\mathrm{Im}(R_q(u, v)) \mathrm{Re}(R_r(u, v)) - \mathrm{Re}(R_q(u, v)) \mathrm{Im}(R_r(u, v))}{2D(u, v)}$$
$$= \sum_{u=0}^{L-1} \sum_{v=0}^{M-1} \frac{d_{quv} c_{ruv} - c_{quv} d_{ruv}}{2D(u, v)} \tag{20}$$

where the superscript t is the transpose, and $\mathrm{Re}(\cdot)$ and $\mathrm{Im}(\cdot)$ denote the real and imaginary parts, respectively. Let $\mathbf{A}$ and $\mathbf{B}$ be $T \times T$ matrices whose elements at (q, r) are $A_{q,r}$ and $B_{q,r}$, respectively. Note that **A** and **B** are symmetric. Equations (15a) and (15b) can be written as

$$\lambda_1^t \mathbf{A} + \lambda_2^t \mathbf{B} = LM\mathbf{C}^t \tag{21a}$$

$$-\lambda_1^t \mathbf{B} + \lambda_2^t \mathbf{A} = \mathbf{0}^t \tag{21b}$$

Because **A** and **B** are symmetric, we may assume **A** and $(\mathbf{A} + \mathbf{BA}^{-1}\mathbf{B})$ to be nonsingular. Otherwise, when they are singular, they can be adjusted to be nonsingular by a slight change of one of the elements. Solving Eqs. (21a) and (21b), we obtain

$$\lambda_1^t = LM\mathbf{C}^t(\mathbf{A} + \mathbf{BA}^{-1}\mathbf{B})^{-1} \tag{22a}$$

$$\lambda_2^t = LM\mathbf{C}^t(\mathbf{A} + \mathbf{BA}^{-1}\mathbf{B})^{-1}\mathbf{BA}^{-1} \tag{22b}$$

Using Eqs. (14a) and (14b), we obtain the element of the distortion tolerant filter $H(u, v)$:

$$a_{uv} + jb_{uv} = \frac{1}{2D(u, v)} \sum_{i=1}^{T} \{\lambda_{1i}(c_{iuv} + jd_{iuv}) + \lambda_{2i}(d_{iuv} - jc_{iuv})\}$$

$$= \frac{1}{2D(u, v)} \sum_{i=1}^{T} (\lambda_{1i} - j\lambda_{2i})(c_{iuv} + jd_{iuv}) \tag{23}$$

Therefore, the optimum nonlinear distortion tolerant filter $H(u, v)$ is

$$H(u, v) = \frac{LM \sum\limits_{i=1}^{T} (\lambda_{1i} - j\lambda_{2i})R_i(u, v)}{2\{w_n E|N(u, v)|^2 + w_d|S(u, v)|^2\}} \tag{24}$$

Because $n(x, y)$ is a zero-mean white Gaussian noise, $E|N(u, v)|^2 = LM\sigma^2$, where σ^2 is the variance of the additive noise and λ_{1i} and λ_{2i} are obtained using Eqs. (22a) and (22b).

When $T = 1$, **A** and **B** are no longer matrices. Furthermore, $\mathbf{B} = 0$ and

$$\mathbf{A} = \sum_{u=0}^{L-1} \sum_{v=0}^{M-1} \frac{|R_1(u, v)|^2}{2D(u, v)}$$

from Eq. (20) and Eq. (19), respectively. Also, $\lambda_1 = LMC_1\mathbf{A}^{-1}$ and $\lambda_2 = 0$ as obtained from Eqs. (22a) and (22b). Thus, in this case, the optimum nonlinear distortion-tolerant filter becomes

$$H(u, v) = \frac{LM\lambda_1 R_1(u, v)}{2\{w_n E|N(u, v)|^2 + w_d|S(u, v)|^2\}}$$

If we set $w_n = w_d = LM\lambda_1/2$, then we obtain the same optimum nonlinear filter of Ref. 15, which is

$$H(u, v) = \frac{R(u, v)}{E|N(u, v)|^2 + |S(u, v)|^2}$$

Therefore, the newly derived optimum nonlinear distortion-tolerant composite filter is a generalized optimum nonlinear filter in Ref. 15. In the special case of an optimum trade-off filter with the same weights for noise robustness and discrimination peak sharpness, the filter of Ref. 15 can be obtained in the general Bayesian theoretical framework. It is also obtained by the MAP (maximum a posteriori) estimation approach with an inverted gamma probability distribution for the noise spectral density [9,30].

11.3 COMPUTER SIMULATIONS AND PERFORMANCE EVALUATION

We now demonstrate the performance of the derived optimum nonlinear distortion-tolerant filter with numerical simulations performed on an image with a size of 128 × 128 pixels and 256 gray levels. In the computer simulations, we set all C_i's equal to 1 to have the same weight for all training targets. We have chosen both w_n and w_d in $D(u, v)$ as LM (128 × 128). The ability to detect the target is demonstrated by computer simulations for the input scene with overlapping additive and nonoverlapping background noise. The discriminating capability of the filter is measured by POE (i.e., a ratio of the peak to the noise floor). The higher the POE, the more discriminative the filter is. The detecting capability can be also measured by ROC curves, which are the curves of probability of detection (P_d) versus probability of false alarm (P_{fa}). The output SNR of the filter is obtained to measure the noise robustness of the filter. Even though the optimum filter derived in Section 11.2 is not optimized in terms of POE, we can expect that POE will be acceptable, because the filter is optimized in terms of correlation output energy due to the input scene, which is the denominator of the POE.

11.3.1 Description of the Simulations and Correlation Output

In this subsection, we performed computer simulations. Targets to be detected are two helicopters. One of them is a training target and the other one is a nontraining distorted (in-plane-rotated) target. The input scene also has a false object to be rejected. Two targets and one false object are the same size and normalized to have a maximum pixel value of 1.

In the simulation, we used two different sets of reference targets. The first training reference dataset contains 18 reference targets rotated by 20° increments from 0° to 360°, and the second training reference dataset contains 10 reference targets rotated by 10° increments from 0° to 90°. The original reference image is a helicopter with a size of 15 × 25 pixels and it is normalized to have a maximum pixel value of 1. Figure 1a shows an image that contains one of the training reference targets (with 0° rotation) in the upper right corner, a 50° rotated nontraining target in the top center, and a false object in the bottom left. The false object is a tank normalized so that its maximum pixel value is 1. The size of the tank is also 15 × 25 pixels. Figure 1b

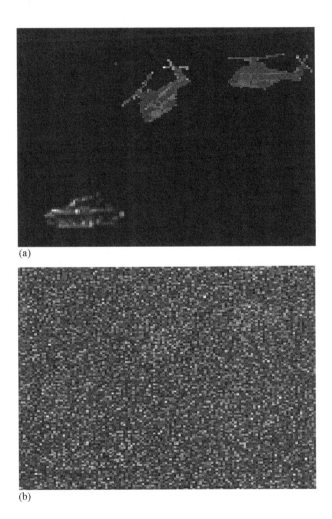

(a)

(b)

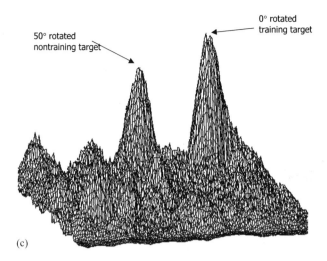

50° rotated nontraining target

0° rotated training target

(c)

Figure 1 (a) Targets used in the simulation. The target from the training dataset, the 50° rotated target from the nontraining dataset, and the false object are 15 × 25 pixels in size and they are normalized so that they have a unity maximum value. (b) Input scene with two targets and one false object, as shown in (a). These are buried in zero-mean additive white Gaussian noise with $\sigma = 0.7$. (c) Correlation output intensity of the optimum nonlinear distortion-tolerant composite filter for the input scene in (b). The filter has been constructed with the first set of reference targets.

is the input scene with one of the training targets, a nontraining target, and a false object buried in the additive noise. The additive noise is a zero-mean white Gaussian noise with standard deviation $\sigma = 0.7$. The computer simulations are performed with this input scene, the filter given by Eq. (24) and λ_{1i} and λ_{2i} given by Eq. (22a) and Eq. (22b), respectively. The correlation output intensity of the composite filter with input scene in Fig. 1b is shown in Fig. 1c. The filter is constructed with the first set of reference targets. The correlation output shows a significant peak for the training target and also a large peak for the nontraining 50° rotated target (which is the worst possible nontraining target between the two rotated training targets at 40° and 60°). Because the correlation output for the false object is significantly lower than the peaks of the training and nontraining targets, we can eliminate the false peaks at the output plane by thresholding the correlation output. Figure 2a is an input scene that contains a training reference target that is 0° rotated in the upper right corner, a 50° rotated nontraining target in the top center, and a false object in the bottom left. It contains a nonoverlapping background colored noise with zero mean, standard deviation of 0.1, and bandwidth of 10. This scene is buried in additive noise of standard deviation 0.7. Again, the com-

(a)

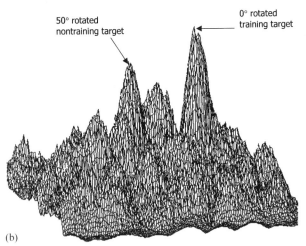

(b)

Figure 2 (a) Input scene with two targets and one false object, as appeared in Fig. 1a. This scene has nonoverlapping background colored noise that is zero mean and has a standard deviation of 0.1 and a bandwidth of 10. These are buried in zero-mean additive white Gaussian noise with $\sigma = 0.7$. (b) Correlation output intensity of the optimum nonlinear distortion-tolerant composite filter for the input scene (a). The filter has been constructed with the first set of reference targets.

puter simulation shows a high peak at the location of the 0° rotated training reference target and also a high peak at the nontraining 50° rotated target. There are some peaks with low intensities due to the background and additive noise. The correlation output intensity is shown in Fig. 2b. By thresholding, we can detect the training target as well as the nontraining target. These simulation results with overlapping additive noise and nonoverlapping background noise show that the optimum nonlinear composite filter provides noise robustness and discriminating capability, as well as good distortion tolerance. Figure 3 shows the correlation output intensity for the filter using the second set of reference targets with the same additive noise that is used in Fig. 1b. The nontraining target at the middle of the top is rotated 5°. The output peak is sharper than in the case of Fig. 1c.

11.3.2 Performance Evaluations

In order to evaluate the statistical performance of the filter designed in Section 11.2, we perform 100 Monte Carlo runs to obtain the average peak-to-output energy ratio, the receiver operating characteristics curves, and the output signal-to-noise ratio. These metrics are used to measure the discriminating capability, the detection performance, and the noise robustness of the filters.

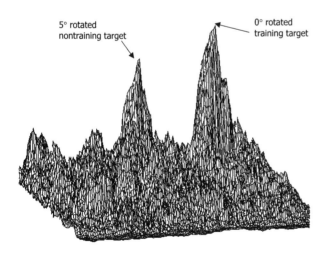

5° rotated
nontraining target

0° rotated
training target

Figure 3 Correlation output intensity of the optimum nonlinear distortion-tolerant composite filter with two targets and one false object, as in Fig. 1a with the same additive noise. However, the nontraining target in the top center has been rotated by 5° and the filter has been constructed with the second set of reference targets.

Peak-to-Output Energy Ratio

The POE is defined as the ratio of the square of the expected value of the output signal at the target location to the expected value of the average output signal energy:

$$
\text{POE} = |E[\,y(0,0)]|^2 \left(E\left\{ \frac{1}{LM} \sum_{x=0}^{L-1} \sum_{y=0}^{M-1} |\,y(x,y)|^2 \right\} \right)^{-1} \tag{25}
$$

where, $y(0,0)$ is the value of the output signal at the target location. The POE is a measure of the discriminating capabilities of the filter, namely high output at the target location and a low-output-noise floor. The higher the POE, the better discriminating capability the filter has. We conducted 100 Monte Carlo runs for each POE curve to obtain the statistical average of the measurements. Figure 4 shows POE curves versus the standard deviation of the additive white Gaussian noise ranging from 0.1 to 0.7. The curve with

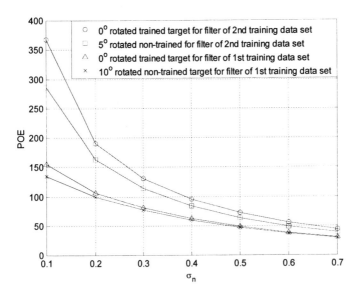

Figure 4   POE curves of the correlation output. Plot with circles is the POE of the $0°$ rotated training target for the filter using the second set of reference targets. Plot with squares is the POE of the $5°$ rotated nontraining target for the same filter using the second set of reference targets. Plots marked with triangles and x symbols are the POE curves of the $0°$ rotated training target and the $10°$ rotated nontraining target, respectively, for the filter using the first set of reference targets. The range of the standard deviations of the additive white Gaussian noise is $[0.1, 0.7]$ with step size 0.1.

circles is the POE curve of the correlation outputs for the $0°$ rotated training target. The curve with squares is the POE curve of the correlation outputs for the $5°$ rotated nontraining target. The two curves marked with triangles and x symbols are the POE curves of the correlation outputs for a $0°$ rotated training target and a $10°$ rotated nontraining target, respectively. The first two curves (marked with circles and squares) are obtained when the filter is constructed with the second set of reference targets; the last two curves are obtained for the filter using the first set of reference targets. The POE of the training target is higher than that of the nontraining target, and the POE of the filter with the training dataset of small increments in rotation angle is higher than that of the filter with the training dataset of larger increments in rotation angle. Naturally, as the noise power increases, the value of POE decreases. Figure 5 shows POE curves versus the standard deviation of the additive white Gaussian noise ranging from 0.1 to 0.7 with background

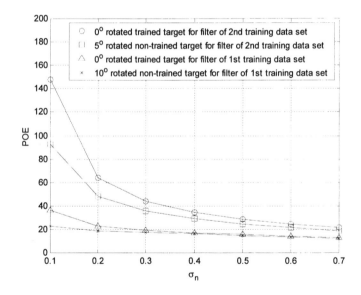

Figure 5 POE curves of the correlation output. Plot with circles is the POE of the $0°$ rotated training target for the filter using the second set of reference targets. Plot with squares is the POE of the $5°$ rotated nontraining target for the same filter using the second set of reference targets. Plots marked with triangles and x symbols are the POE curves of the $0°$ rotated training target and the $10°$ rotated nontraining target, respectively, for the filter using the first set of reference targets. The range of the standard deviations of the additive white Gaussian noise is $[0.1, 0.7]$ with step size 0.1. Background noise is nonoverlapping colored noise that is zero mean and has a standard deviation of 0.1 and a bandwidth of 10.

noise. Background noise is nonoverlapping colored noise that is zero mean and has a standard deviation of 0.1 and a bandwidth of 10. The curve with circles is the POE curve of the correlation outputs for the $0°$ rotated training target. The curve with squares is the POE curve of the correlation outputs for the $5°$ rotated nontraining target. The two curves marked with triangles and x symbols are the POE curves of the correlation outputs for a $0°$ rotated training target and a $10°$ rotated nontrainin target, respectively. The first two curves (marked with circles and squares) are obtained when the filter is constructed with the second set of reference targets; the last two curves are obtained for the filter using the first set of reference targets. The POE of the training target is higher than that of the nontraining target, and the POE of the filter with the training dataset of small increments in rotation angle is higher than that of the filter with the training dataset of larger increments in rotation angle.

Receiver Operating Characteristics Curves

Detection performance of the filter is also evaluated by plotting probability of detection (P_d) versus probability of false alarm (P_{fa}) (see Figs. 6a–7d). Each point on the curve corresponds to a value of (P_d, P_{fa}) for a given threshold between the probability density function (pdf) of the noise and pdf of the noise plus the signal. As the threshold increases, P_{fa} and P_d decrease and vice versa. This type of performance evaluator is called the receiver operating characteristics (ROC). The ROC curve is a standard metric used to compare processor performance and has been used extensively to specify performance of digital communication, radar, and sonar systems. A good receiver/detector structure will yield an ROC curve in the upper left corner of the linearly scaled graph. We conduct 100 Monte Carlo simulations to obtain each ROC curve. For each case, the standard deviation of the additive white Gaussian noise ranges from 0.6 to 1.2 in increments of 0.1. For the purpose of showing the detils of the curves, only the upper left corners of the curves are displayed. Figure 6a shows ROC curves of a $0°$ rotated training target for the filter using the first set of reference targets. When the noise has a standard deviation of less than 0.7, detection is perfect. Even with noise of standard deviation 1.2, the probability of detection is 0.96 and the probability of false alarm is 0.1. Figure 6b shows the ROC curves of a $10°$ rotated nontraining target for the same filter used in Fig. 6a. Although the detection capability is slightly degraded, it is still excellent. When the standared deviation of the noise is 1.2, the probability of detection is 0.95 and the probability of false alarm is 0.1. This amounts to 1% degradation compared to the training target detection case. Figures 6c and 6d are the ROC curves of the $0°$ rotated training target and $5°$ rotated nontraining target, respectively, for the filter using the second set of reference targets. The ROC curves in these cases show better detecting performance for the filter than in the cases of Figs. 6a and 6b. When the noise of standard deviation is 1.2, the probability of detection is more than 0.99 for both cases and the probability of false alarm is 0.1. From these results, we see that the derived filter detects not only the training target but also the nontraining target under very noisy conditions. Figures 7a–7d are ROC curves of training targets and nontraining targets with additive noise as well as nonoverlapping background colored noise. Background noise is nonoverlapping colored noise that is zero mean with a standard deviation of 0.1 and a bandwidth of 10. Figure 7a shows ROC curves of a $0°$ rotated training target for the filter using the first set of reference targets. Compared with the reuslts in Fig. 6a, the detection performance is slightly worse than in the case of ROC curves without the nonoverlapping background noise that is similar to the target. Even with noise of standard deviation 1.2, the probability of detection is 0.885 and the probability of false alarm is 0.1. This result is slightly worse than in the case without the background noise, with a detection probability of 0.96. Figure 7b shows the ROC curves of a $10°$ rotated nontraining target for the same filter used for Fig. 7a. Although the detection capability is slightly degraded, it is still good. When the standard deviation of the noise is 1.2, the probability of detection is 0.87 and the probability of false alarm is 0.1. This amounts to 1.5% degradation compared to the training target detection case. Figure 7c and 7d are the ROC curves of the $0°$ rotated training target and $5°$ rotated nontraining target, respectively, for the filter using the second set of reference targets. The ROC curves in these cases show better detecting performance for the filter than in the cases of Figs. 7a and 7b. When the noise of standard deviation is 1.2, the probability of detection is 0.97 for the training target detection case (Fig. 7c) and 0.95 for the nontraining target detection case (Fig. 7d) and the probability of false alarm is 0.1 (in both cases). From these results, we see that the derived filter detects the training target as well as the nontraining target under very noisy conditions, including nonoverlapping background colored noise and additive overlapping white Gaussian noise.

Output Signal-to-Noise Ratio

Output SNR (signal-to-noise ratio) is a metric used to measure the noise robustness of the filters. The output SNR of the filter is defined as the ratio of the expected value of the output at the target location to its variance:

$$\text{SNR} = \frac{|E[y(0,0)]|^2}{\text{Var}(y(0,0))} \tag{26}$$

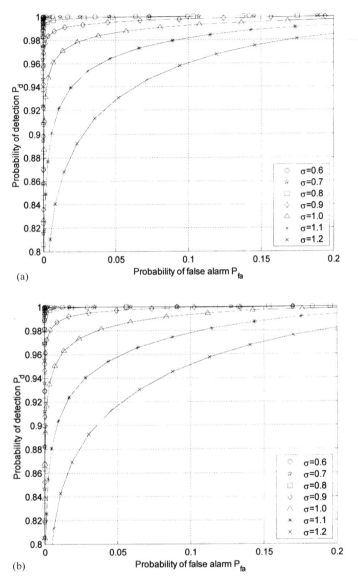

Figure 6 ROC curves of the correlation output. The standard deviations of the additive white Gaussian noise range from 0.6 to 1.2, increasing by 0.1. Only the upper left corners of the curves are depicted to show the details. (a) ROC curves of the 0° rotated training target for the filter using the first set of reference targets. (b) ROC curves of the 10° rotated nontraining target with the same filter as in (a). Only the upper left corners of the curves are depicted to show the details.

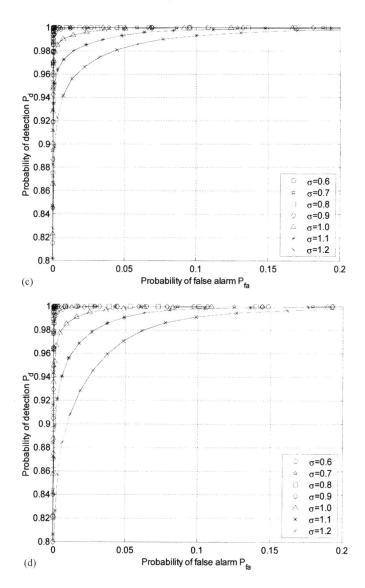

Figure 6 (*Continued*). (c) ROC curves of the 0° rotated training target with a filter using the second set of reference targets. Only the upper left corners of the curves are depicted to show the details. (d) ROC curves of the 5° rotated nontraining target with the same filter as in (c). Only the upper left corners of the curves are depicted to show the details.

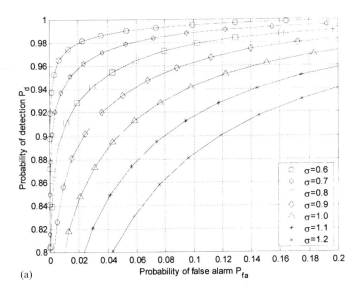

(a)

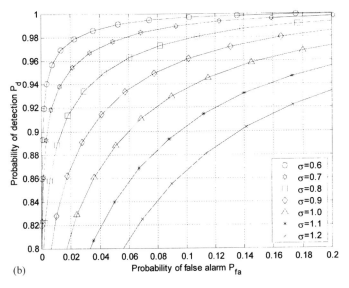

(b)

Figure 7 ROC curves of the correlation output. The standard deviations of the additive white Gaussian noise range from 0.6 to 1.2, increasing by 0.1. Background noise is nonoverlapping colored noise that is zero mean with a standard deviation of 0.1 and a bandwidth of 10. Only the upper left corners of the curves are depicted to show the details. (a) ROC curves of the $0°$ rotated training target for the filter using the first set of reference targets with the nonoverlapping background colored noise. (b) ROC curves of the $10°$ rotated nontraining target with the same filter as in (a) with the same nonoverlapping background colored noise. Only the upper left corners of the curves are depicted to show the details.

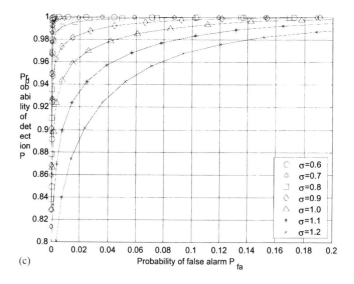

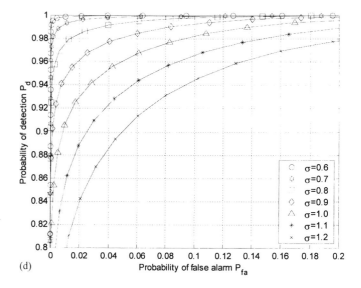

Figure 7 (*Continued*). (c) ROC curves of the $0°$ rotated training target with a filter using the second set of reference targets with the same nonoverlappping background colored noise. Only the upper left corners of the curves are depicted to show the details. (d) ROC curves of the $5°$ rotated nontraining target with the same filter as in (c) with the same nonoverlapping background colored noise. Only the upper left corners of the curves are depicted to show the details.

where Var(·) denotes the variance. We conducted 100 Monte Carlo simulations to obtain the output SNR. For the simulations, the standard deviation of the additive white Gaussian noise ranges from 0.2 to 1.2 in increments of 0.1. Figure 8 shows output SNR curves of training target and nontraining target for the filters using the first set and second set of reference targets. The curve with circles is the output SNR of the correlation outputs for the 0° rotated training target. The curve with squares is the SNR curve of the correlation outputs for the 5° rotated nontraining target. As the standard deviation of the additive noise increases above the 0.2 level, the difference between these outputs SNRs decreases. The output SNRs for both cases with the noise level set at $\sigma = 1.2$ are 14 (11.5 dB). The two curves marked with triangles and x symbols are the SNR curves of the correlation outputs for a 0° rotated training target and a 10° rotated nontraining target, respectively. The output SNRs for these cases with the noise level set at $\sigma = 1.2$ are 20 (13 dB). The first two curves (marked with circles and squares) are obtained when the

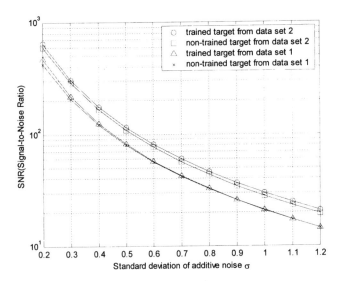

Figure 8 Output SNR curves of the correlation output. The range of the standard deviations of the additive white Gaussian noise is [0.2, 1.2] with step size of 0.1. The plot with circles is the SNR of the 0° rotated training target for the filter using the second set of reference targets. The plot with squares is the SNR of the 5° rotated nontraining target for the same filter using the second set of reference targets. Plots marked with triangles and x symbols are the SNR curves of the 0° rotated training target and the 10° rotated nontraining target, respectively, for the filter using the first set of reference targets.

filter is constructed with the second set of reference targets, and the last two curves are for the filter using the first set of reference targets. The output SNR of the training target is higher than that of the nontraining target, and the output SNR of the filter with the training dataset of small increments in rotation angle is higher than that of the filter with the training dataset of larger increments in rotation angle. Figure 9 shows output SNR curves of training target and nontraining target for the filters using the first set and second set of reference targets with overlapping additive white Gaussian noise and background noise. Background noise is nonoverlapping colored noise that is zero mean with a standard deviation of 0.1 and a bandwidth of 10. The curve with circles is the output SNR of the correlation outputs for the $0°$ rotated training target. The curve with squares is the SNR curve of the correlation outputs for the $5°$ rotated nontraining target. The output SNRs for both cases with the noise of $\sigma = 1.2$ are 13 (11 dB). The two curves

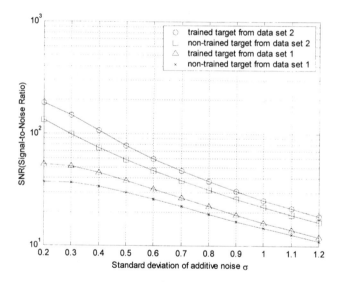

Figure 9 Output SNR curves of the correlation output. The range of the standard deviations of the additive white Gaussian noise is $[0.2, 1.2]$ with step size of 0.1. Background noise is nonoverlapping colored noise that is zero mean with a standard deviation of 0.1 and a bandwidth of 10. The plot with circles is the SNR of the $0°$ rotated training target for the filter using the second set of reference targets. The plot with squares is the SNR of the $5°$ rotated nontraining target for the same filter using the second set of reference targets. Plots marked with triangles and x symbols are the SNR curves of the $0°$ rotated training target and the $10°$ rotated nontraining target, respectively, for the filter using the first set of reference targets

marked with triangles and x symbols are the SNR curves of the correlation outputs for a $0°$ rotated training target and a $10°$ rotated nontraining target, respectively. The output SNRs for these cases with the noise of $\sigma = 1.2$ are 16 and 19 (12 dB, 13 dB) respectively. The first two curves (marked with circles and squares) are obtained when the filter is constructed with the second set of reference targets, and the last two curves are for the filter using the first set of reference targets. The SNR of the training target is 2 dB higher than that of the nontraining target with the background overlaping noise when the standard deviation of the additive noise is low. The output SNR of the filter with the training dataset of small increments in rotation angle is 6 dB higher than that of the filter with the training dataset of larger increments in rotation angle when the standard deviation of the additive noise is low.

11.4 CONCLUSION

We have designed and derived a nonlinear distortion-tolerant filter by considering discriminating capability, noise robustness, and distortion tolerance. The filter has been obtained by minimizing the output energy due to the input scene and the input noise. We have tested the performance of the derived filter using computer simulations. For the images used in the tests, the computer simulations show that this filter is capable of detecting the distorted reference targets (true class training images and true class nontraining images) in a very noisy situation. The input noise includes additive overlapping noise on the input scene and nonoverlapping background noise around the target. To measure the statistical performnce of the filter, we have performed Monte Carlo runs. First, to show the discriminating capability of the derived optimum filter, we performed simulations to obtain POE with a noise level up to $\sigma = 0.7$. To show the detection capabilities of the filter for the training target and the nontraining target, we have performed simulations to obtain the ROC curves with an additive white Gaussian noise up to $\sigma = 1.2$. The resulting ROC curves show excellent capabilitites for the detection of the targets including the nontraining targets. With nonoverlaping background colored noise, the detecting capability is degraded. However, we still obtain excellent curves when the relatively low standard deviation of additive noise is used. We also performed the simulations to obtain the output SNR for different cases with an additive white Gaussian noise up to standard deviation $\sigma = 1.2$.

According to the results of the simulations, the optimum nonlinear distortion-tolerant composite filter was capable of detecting the distorted images (in-plate-rotated images) with the overlapping additive noise and nonoverlapping background noise.

ACKNOWLEDGMENTS

We wish to thank Mr. L. Pan and Dr. N. Towghi for their helpful suggestions.

REFERENCES

1. JW Goodman. Introduction to Fourier Optics. 2nd ed. New York: McGraw-Hill, 1996.
2. A VanderLugt. Optical Signal Processing. New York: Wiley, 1992.
3. A McAulay. Optical Computer Architectures: The Application of Optical Concepts to Next Generation Computers. New York: Wiley, 1991.
4. B Javidi, J L Horner. Real Time Optical Information Processing. New York: Academic, 1994.
5. JL Turin. An introduction to matched filters. IRE Trans Inform Theory 6:311–329, 1960.
6. A VanderLugt. Signal detection by complex filters. IEEE Trans Inform Theory IT-10:139–145, 1964.
7. D Flannery, J Horner. Fourier optical signal processor. Proc IEEE 77:1511–1527, 1989.
8. N Towghi, B Javidi, J Li. Generalized optimum receiver for pattern recognition with multiplicative, additive, and non-overlapping background noise. Opt Soc Am A 15:1557–1565, 1998.
9. Ph Refregier. Bayesian theory for target location in noise with unknown spectral density. J Opt Soc Am A 16:276–283, 1999.
10. RP Lippmann. An introduction to computing neural nets. IEEE ASSP Mag 4–22, April 1987.
11. B Javidi. Nonlinear joint power spectrum based optical correlation. Appl Opt 28:2358–2367, 1989.
12. JL Horner, PD Gianino. Phase-only matched filtering. Appl Opt 23:812–816, 1984.
13. B Javidi, Ph Refregier, P Willett. Optimum receiver design for pattern recognition with nonoverlapping target and scene noise. Opt Lett 18:1160–1162, 1994.
14. D Psaltis, Y Qiao, H Li. Optical face recognition system. Proc of SPIE 1773:59–62, 1993.
15. Ph Refregier, V Laude, B Javidi. Nonlinear joint-transform correlation: an optimal solution for adaptive image discrimination and input noise robustness. Opt Lett 19:405–407, 1994.
16. D Casasent, D Psaltis. Position, rotation and scale invariant optical correlation. Appl Opt 15:1795–1799, 1976.
17. A Mahalanobis. Review of correlation filters and their application for scene matching. In: Optoelectronic Devices and Systems for Processing, Critical Reviews of Optical Science Technology Vol 65, Bellingham, WA: SPIE Press, 1996, pp 240–260.

18. B Javidi, J Wang. Optimum distortion-invariant filter for detecting a noisy distorted target in nonoverlapping background noise. J Opt Soc Am A 12:276–283, 1995.
19. B Javidi, D Painchaud. Distortion-invariant pattern recognition with Fourier-plane nonlinear filters. Appl Opt 35:318–331, 1996.
20. C Hester, D Casasent. Multivariant technique for multiclass pattern recognition. Appl Opt 19:1758–1761, 1980.
21. J Caulfield, W Maloney. Improved discrimination in optical character recognition. Appl Opt 8:2354–2356, 1969.
22. Y Hsu, H Arsenault. Optical pattern recognition using circular harmonic expansion. Appl Opt 21:4016–4019, 1982.
23. Ph Refregier, F Goudail. Decision theoretical approach to nonlinear joint transform correlation. J Opt Soc Am A 15:61–67, 1998.
24. B Javidi, F Parchekani, G Zhang. Minimum-mean-square-error filters for detecting a noisy target in background noise. Appl Opt 35:6964–6975, 1996.
25. JL Horner. Metrics for assessing pattern recognition performance. Appl Opt 31:165–166, 1992.
26. Ph Refregier, J Figue. Optimum trade-off filters for pattern recognition and their comparison with Wiener approach. Opt Comput Process 1:245–265, 1991.
27. B Javidi. Smart Imaging Systems. Bellingham, WA: SPIE Press, 2001.
28. B Javidi, J Wang. Limitations of the classic definition of the signal-to-noise ratio in matched filter based optical pattern recognition. J Appl Opt 31, November, 1992.
29. SM Kay. Fundamentals of Statistical Signal Processing. Volume II: Detection Theory. Englewood Cliffs, NJ: Prentice-Hall, 1998.
30. EKP Chong, SH Zak. An Introduction to Optimization. New York: Wiley, 1996.

12

l_p-Norm Optimum Distortion-Tolerant Filter for Image Recognition

Luting Pan and Bahram Javidi
University of Connecticut, Storrs, Connecticut

12.1 INTRODUCTION

Matched filter [1,2] and many of its variations [3–11] have been proposed for image recognition. In image recognition, it is often desired to design filters or algorithms that are invariant to deterministic distortions of the target and are robust to input noise. This means that the filters need to recognize a target that is contained in the noisy input scene, and this target may be a distorted version of the reference target. For instance, the target may be a rotated or a scaled version of the reference target placed in background noise. These filters are referred to as distortion tolerant filters. The most common approaches to design these filters are based on training the filter with a set of distorted versions of the reference targets [12–16]. What distinguishes such filters from one another is the metric or method used to optimize the filter.

In this chapter, we present the l_p-norm optimum distortion-tolerant filter, which allows greater freedom in adjusting the noise tolerance and discrimination capability. This filter is based on the l_p-norm metric, where $1 \leq p \leq 2$. It is trained with a set of distorted versions of true targets and is designed to generate the same filter output peak for all of the training images. The filter design method is described in Sections 12.2 and 12.3. The derivation of the filter requires solving a constraint minimization problem that minimizes the l_p-norm metric of the filter output. This metric is defined to be the weighted sum of the filter output due to the input scene and the noise. The requirement to generate a constant fixed filter output peak at

349

the location of the training targets introduces the constraints. Solving the above optimization problem for general values of p with multiple constraints poses mathematical difficulties that are not present when $p = 2$. However, the derived filter has an additional parameter p, which can be adjusted to control the robustness of the filter and its discrimination capabilities, which a conventional l_2-norm filter does not have.

A brief review of the l_p-norm filter is given in Section 12.2. The l_p-norm optimum distortion-tolerant filter is developed in Section 12.3. Section 12.4 provides computer simulations to evaluate the performance of the filter. Section 12.5 summarizes the chapter.

12.2 REVIEW OF THE l_p-NORM FILTER

In Ref. 17, a family of linear and nonlinear filters called the l_p-norm optimum filters was designed to detect a known target with known scale and orientation in the presence of input noise. To introduce our technique, we provide a brief review of the development of the l_p-norm filter [17].

In this section, we use one-dimensional notation for simplicity. The l_p-norm filter is designed to detect a known target $r(i)$ in the input scene $s(i)$ which is corrupted by the presence of input noise $n(i)$ $[s(i) = r(i) + n(i)]$. The filter $h(i)$ is designed to generate a sharp output peak at the location of the target while minimizing the pth power of the l_p-norm of the output due to input scene $s(i)$ and the pth power of the l_p-norm's mean of the output due to the additive input noise $n(i)$. The l_p-norm of a sequence of J scalars, $c = \{c_j\}_{j=0}^{J-1}$, is given by

$$\|c\|_p = \left(\sum_{j=0}^{J-1} |c_j|^p \right)^{1/p} \tag{1}$$

The filter $h(i)$ is designed so that the output due to the target r at the origin is

$$\sum_{j=1}^{J} h^*(j)r(j) = C = C(0) \tag{2a}$$

where the superscript asterisk denotes conjugation and $C = C(0)$ is a positive constant. The filter must also minimize the l_p-norm of the output due to the noise and the background:

$$S = a \sum_{j=0}^{J-1} E\left[\left| \sum_{l=0}^{J-1} h(j-1)n(l) \right|^p \right] + b \sum_{j=0}^{J-1} \left| \sum_{l=0}^{J-1} h(j-l)s(l) \right|^p \tag{2b}$$

where $E[\cdot]$ denotes the expected value and $|\cdot|$ denotes the absolute value.

Note that Eq. (2b) is the weighted sum of the pth power of the l_p-norm rather than the weighted sum of the l_p-norms. This will simplify the analytical solution of the minimization problem of the weighted sum of the pth power of l_p-norms. Also, controlling the size of the pth norm amounts to controlling the size of the pth power of the norm.

The weights a and b are suitably chosen positive quantities. If the emphasis is on noise robustness, a is the larger of the two. If the emphasis is on discrimination, b should be the dominant quantity. The rest of this section uses $a = b = 1$.

The minimization problem given by Eqs. (2a) and (2b) is defined in the spatial domain. In Ref. 17, the l_p-norm minimization problem was transferred into the Fourier domain. Let $S(k)$, $R(k)$, $N(k)$, and $H(k)$ denote the Fourier transfer of $s(i)$, $r(i)$, $n(i)$, and $h(i)$, respectively. Equations (2a) and (2b) can be stated in the Fourier domain as follows [17]:

Minimize

$$\frac{1}{J} \sum_{k=0}^{J-1} |H(k)|^q [\hat{\sigma}_q + |S(k)|^q] \tag{3a}$$

subject to

$$\sum_{k=0}^{J-1} H^*(k)R(k) = JC(0) \tag{3b}$$

where $1 \leq p \leq 2$, $q \equiv p/(p-1)$, and $\hat{\sigma}_q = E|N(j)|^q$. The details of this transformation are in Appendix A of Ref. 17.

By solving Eqs. (3a) and (3b) using the Lagrange multiplier method, for $1 < p \leq 2$, we obtain the l_p-norm optimum filter for a single reference target, as a constant multiple of

$$H_q^\sigma(k) = \left(\frac{R(k)}{\hat{\sigma}_q + |S(k)|^q} \right)^{1/(q-1)} \exp(i\Phi_{R(k)}) \tag{4}$$

where $\hat{\sigma}_q = E|N(j)|^q$ and $\Phi_{R(k)}$ is the phase of $R(k)$; that is, $R(k) = |R(k)| \exp[j\Phi_{R(k)}]$. We should point out that Eq. (4) requires the values of $\hat{\sigma}_q = E|N(j)|^q$. With few exceptions, this quantity may be difficult to compute analytically. Therefore, in Ref. 17, $\hat{\sigma}_q$ is determined by using the lower-bound estimate, $E|N(k)|^q \geq [\sigma\sqrt{J}]^q$, where σ is the input noise standard deviation. This estimate holds for $q \geq 2$. We can then derive an approximation of a filter equation for the case $q = \infty$ or $p = 1$:

$$H_\infty^\sigma(k) = \left(\frac{1}{\max[\sqrt{J}\sigma, |S(k)|]} \right) \exp(i\Phi_{R(k)}) \tag{5}$$

The reader can find some lower- and upper-bound estimates for various types of noise processes in Ref. 17.

12.3 DERIVATION OF THE l_p-NORM DISTORTION-TOLERANT FILTER

We now extend the development of the l_p-norm distortion tolerant optimum filter [18] by considering possible multiple true training targets contained in the input scene. Here, we will be using two-dimensional notation. Let $s(x, y)$ denote the value of the input scene at location (x, y), where $x, y = 0, 1, 2, \ldots, J - 1$. Let $n(x, y)$ denote the additive noise [assumed to be white and wide-sense stationary (WSS), with standard deviation σ]. For $i = 1, \ldots, I$, where I is the number of training true class targets, let the ith training target, $r_i(x, y)$, describe a possible distortion of the true class targets at (x, y). We model our noisy input scene by the following equation:

$$s(x, y) = n(x, y) + \sum_{i=1}^{I} a_i r_i(x, y) \tag{6}$$

In Eq. (6), a_i is a binary random variable which takes on values of 0 and 1. It indicates whether or not the target r_i is present in the input scene. For analytical simplicity, the probability that a_i takes on values of 0 or 1 are the same, for $i = 1, 2, 3, \ldots, I$; that is, $p(a_i = 1) = 1/I$ and $p(a_i = 0) = 1 - 1/I$, where $p(\cdot)$ denotes the probability.

Let $h(x, y)$ denote the impulse response of the optimum distortion-tolerant filter and let $H(w, v)$ denote its Fourier transform, where w and v are the Fourier-domain coordinates. The filter is designed such that the output at the origin due to ith training true class target $r_i(x, y)$, located at the origin, is C_i; that is,

$$\sum_{x=0}^{J-1} \sum_{y=0}^{J-1} h^*(x, y) r_i(x, y) = C_i(0, 0) = C_i \tag{7}$$

where $i = 1, \ldots, I$. In our design, we want the filter to generate equal correlation peak intensities for every training target; that is, all C_i are equal to a constant C. To achieve discrimination capability, we minimize the pth power of l_p-norm of the output due to the input scene:

$$E_s = \|h(x, y) \otimes s(x, y)\|_p^p = \sum_{y=0}^{J-1} \sum_{x=0}^{J-1} \left| \sum_{k=0}^{J-1} \sum_{l=0}^{J-1} h(x - 1, y - k) s(l, k) \right|^p \tag{8}$$

where $\otimes$ denotes the convolution operation. To achieve noise robustness, we minimize the expected pth power l_p-norm of the output due to the input noise, $n(x, y)$; that is, we wish to minimize

$$E_n = \|E[h(x, y) \otimes n(x, y)]\|_p^p = \sum_{x=0}^{J-1} \sum_{l=0}^{J-1} E \left| \sum_{k=0}^{J-1} \sum_{l=0}^{J-1} h(x - l, y - k)n(l, k) \right|^p$$

(9)

Thus, we need to minimize the weighted summation of Eqs. (8) and (9);

$$S = a\|E[h(x, y) \otimes n(x, y)]\|_p^p + b\|h(x, y) \otimes x(s, y)\|_p^p \tag{10}$$

subject to the constraints given by Eq. (7). Note that Eq. (10) is similar to Eq. (2b). Again, we are minimizing the weighted sum of the pth power of the l_p-norm instead of the weighted sum of the l_p-norm. Here, a and b are positive weights. To repeat, the larger value controls whether the filter is more discriminant or more noise tolerant, respectively. In our derivation of the filter, we will assume that $a = b = 1$.

To simplify the minimization problem given by Eq. (10), we perform the optimization in the Fourier domain. We refer the reader to Appendix A of Ref. 17 for the mathematical proof of the transformation to the Fourier domain. Let $S(w, v)$ and $N(w, v)$ denote the Fourier transforms of $s(x, y)$ and $n(x, y)$, respectively. Let $R_i(w, v)$ denote the Fourier transform of the ith training target at $r_i(x, y)$; then, the problem can be stated in the Fourier domain as

Minimize

$$\frac{1}{J^2} \sum_{w=0}^{J-1} \sum_{v=0}^{J-1} |H(w, v)|^q [\hat{\sigma}_q + |S(w, v)|^q] \tag{11}$$

subject to the constraint

$$\sum_{w=0}^{J-1} \sum_{v=0}^{J-1} H^*(w, v)R_i(w, v) = J^2 C_i(0, 0), \qquad i = 1, \ldots, I \tag{12}$$

where $1 < p \le 2$, $q = p/(p - 1)$, and $\hat{\sigma}_q = E|(N(w, v)|^q$. We will ignore J^2 in Eq. (11) because it is a constant.

Due to multiple constraints [the I constraints given by Eq. (12)], the minimization problem cannot be solved by the method that was used in Ref. 17. We first break $H(w, v)$ into its real and imaginary components and denote the real and imaginary part of $H(w, v)$ by $a(w, v)$ and $b(w, v)$, respectively. The real and imaginary part of $R_i(w, v)$ are denoted by $c_i(w, v)$ and $d_i(w, v)$, respectively. Thus, we restate Eq. (12) as follows:

$$\sum_{w=0}^{J-1}\sum_{v=0}^{J-1}\{[a(w, v) - jb(w, v)] \times [c_i(w, v) + jd_i(w, v)]\}$$

$$= \sum_{w=0}^{J-1}\sum_{v=0}^{J-1}\{[a(w, v)c_i(w, v) + b(w, v)d_i(w, v)] + j[a(w, v)d_i(w, v)$$

$$- b(w, v)c_i(w, v)]\}$$

$$= J^2 C_i(0, 0) \tag{13}$$

Because J^2 and $C_i(0, 0)$ are real, the above equation can be stated in terms of real variables only as follows:

$$\sum_{w=0}^{J-1}\sum_{v=0}^{J-1}[a(w, v)c_i(w, v) + b(w, v)d_i(w, v)] = J^2 C_i(0, 0) = J^2 C_i \tag{14a}$$

$$\sum_{w=0}^{J-1}\sum_{v=0}^{J-1}[a(w, v)d_i(w, v) - b(w, v)c_i(w, v)] = 0 \tag{14b}$$

where we are looking for the solutions to the unknowns $a(w, v)$ and $b(w, v)$. We use the Lagrange multiplier method to obtain $a(w, v)$ and $b(w, v)$ by solving Eq. (11) subject to the constraints given by Eqs. (14a) and (14b).

More precisely, let

$$\Psi = \sum_{w=0}^{J-1}\sum_{v=0}^{J-1} |H(w, v)|^q[\hat{\sigma}_q + |S(w, v)|^q]$$

$$+ \sum_{i=1}^{I} \lambda_{1i}\left(J^2 C_i - \sum_{w=0}^{J-1}\sum_{v=0}^{J-1} a(w, v)c_i(w, v) - \sum_{w=0}^{J-1}\sum_{v=0}^{J-1} b(w, v)d_i(w, v)\right)$$

$$+ \sum_{i=1}^{I} \lambda_{2i}\left(0 - \sum_{w=0}^{J-1}\sum_{v=0}^{J-1} a(w, v)d_i(w, v) + \sum_{w=0}^{J-1}\sum_{v=0}^{J-1} b(w, v)c_i(w, v)\right)$$

$$\tag{15}$$

where $|H(w, v)| = [a^2(w, v) + b^2(w, v)]^{1/2}$ and λ_{1i} and λ_{2i} are unknown variables (Lagrange multipliers).

We must find $a(w, v)$, $b(w, v)$, λ_{1i}, and λ_{2i} such that for each i and each (w, v), the following equations are satisfied:

$$\frac{\partial\Psi}{\partial a(w, v)} = 0, \qquad \frac{\partial\Psi}{\partial b(w, v)} = 0$$

$$\frac{\partial\Psi}{\partial\lambda_{1i}(w, v)} = 0, \qquad \frac{\partial\Psi}{\partial\lambda_{2i}(w, v)} = 0 \tag{16}$$

A simple calculation confirms that the Hessian of Ψ with respect to $a(w, v)$ and $b(w, v)$ is symmetric and invertible regardless of the values of $a(w, v), b(w, v), \lambda_{1i},$ and λ_{2i}. Thus, it is positive definite and, consequently, Ψ has a unique minimum.

The rest of this section is devoted to solving the above set of equations. To streamline the presentation, we will introduce further notation. In addition, some of the calculations are carried out in the appendices.

Let $D(w, v) = \hat{\sigma}_q + |S(w, v)|^q$,

$$G_{w,v}^a = \frac{\sum\limits_{i=1}^{I} \lambda_{1i} c_i(w, v) + \sum\limits_{i=1}^{I} \lambda_{2i} d_i(w, v)}{qD(w, v)} \tag{17a}$$

and

$$G_{w,v}^b = \frac{\sum\limits_{i=1}^{I} \lambda_{1i} d_i(w, v) - \sum\limits_{i=1}^{I} \lambda_{2i} c_i(w, v)}{qD(w, v)} \tag{17b}$$

Now,

$$\frac{\partial \Psi}{\partial a(w, v)} = q[a^2(w, v) + b^2(w, v)]^{(q-2)/2} a(w, v)D(w, v)$$

$$- \sum_{i=1}^{I} \lambda_{1i} c_i(w, v) - \sum_{i=1}^{I} \lambda_{2i} d_i(w, v)$$

$$= 0 \tag{18a}$$

$$\frac{\partial \Psi}{\partial b(w, v)} = q[a^2(w, v) + b^2(w, v)]^{(q-2)/2} b(w, v)D(w, v)$$

$$- \sum_{i=1}^{I} \lambda_{1i} d_i(w, v) + \sum_{i=1}^{I} \lambda_{2i} c_i(w, v)$$

$$= 0 \tag{18b}$$

In Appendix A, we show that

$$a(w, v) = \frac{G_{w,v}^a}{[(G_{w,v}^a)^2 + (G_{w,v}^b)^2]^{(q-2)/2(q-1)}} \tag{19a}$$

$$b(w, v) = \frac{G_{w,v}^b}{[(G_{w,v}^a)^2 + (G_{w,v}^b)^2]^{(q-2)/2(q-1)}} \tag{19b}$$

However, both $G_{w,v}^a$ and $G_{w,v}^b$ depend on the unknowns λ_{1i} and λ_{2i}. We now concentrate on obtaining λ_{1i} and λ_{2i} in terms of known quantities.

Replacing $a(w, v)$ and $b(w, v)$ in Eqs. (14a) and (14b) with Eqs. (19a) and (19b), we obtain

$$
\sum_{w=0}^{J-1} \sum_{v=0}^{J-1} \left\{ \frac{G_{w,v}^a}{[(G_{w,v}^a)^2 + (G_{w,v}^b)^2]^{(q-2)/2(q-1)}} \, c_i(w, v) \right.
$$
$$
\left. + \frac{G_{w,v}^b}{[(G_{w,v}^a)^2 + (G_{w,v}^b)^2]^{(q-2)/2(q-1)}} \, d_i(w, v) \right\}
$$
$$
= J^2 C_i \tag{20a}
$$

$$
\sum_{w=0}^{J-1} \sum_{v=0}^{J-1} \left\{ \frac{G_{w,v}^a}{[(G_{w,v}^a)^2 + (G_{w,v}^b)^2]^{(q-2)/2(q-1)}} \, d_i(w, v) \right.
$$
$$
\left. - \frac{G_{w,v}^b}{[(G_{w,v}^a)^2 + (G_{w,v}^b)^2]^{(q-2)/2(q-1)}} \, c_i(w, v) \right\} = 0 \tag{20b}
$$

Solving λ_{1i} and λ_{2i} directly using Eqs. (19a) and (19b) does not seem possible because both the numerator and the denominator depend on unknowns λ_{1i} and λ_{2i}. Therefore, we first derive an expression for the denominators of Eqs. (20a) and (20b). We denote the denominator by $L_{w,v}^q$:

$$
L_{w,v}^q = [(G_{w,v}^a)^2 + (G_{w,v}^b)^2]^{(q-2)/2(q-1)} \tag{21}
$$

In Appendix B, an expression for $L_{w,v}^q$ is obtained in terms of known quantities. More precisely, in Appendix B it is shown that

$$
L_{w,v}^q = [(G_{w,v}^q)^2 + (G_{w,v}^q)^2]^{(q-2)/2q-2} = [O_1^2(w, v) + O_2^2(w, v)]^{(q-2)/2} \tag{22}
$$

where $O_1(w, v)$ and $O_2(w, v)$ depend only on the known quantities, namely the training targets. The actual expression for $O_1(w, v)$, and $O_2(w, v)$ require considerable notational setup and are therefore relegated to Appendix B.

With values of the denominator $L_{w,v}^q$ [given by Eq. (22) and Appendix B] in hand, we can solve Eqs. (20a) and (20b) for the unknowns λ_{1i} and λ_{2i}. More precisely, by substituting the values of $L_{w,v}^q$ given by Eq. (22) into Eqs. (20a) and (20b) and by replacing $G_{w,v}^a$ and $G_{w,v}^b$ in the numerator with the values given by Eqs. (17a) and (17b), we obtain

$$
\sum_{i=1}^{I} \lambda_{1i} \sum_{w=0}^{J-1} \sum_{v=0}^{J-1} \frac{c_i(w, v)c_p(w, v) + d_i(w, v)d_p(w, v)}{[O_1^2(w, v) + O_2^2(w, v)]^{(q-2)/2} qD(w, v)}
$$
$$
+ \sum_{i=1}^{I} \lambda_{2i} \sum_{w=0}^{J-1} \sum_{v=0}^{J-1} \frac{d_i(w, v)c_p(w, v) - c_i(w, v)d_p(w, v)}{[O_1^2(w, v) + O_2^2(w, v)]^{(q-2)/2} qD(w, v)}
$$
$$
= J^2 C_p \tag{23a}
$$

$$\sum_{i=1}^{I} \lambda_{1i} \sum_{w=0}^{J-1} \sum_{v=0}^{J-1} \frac{c_i(w,v)d_p(w,v) + d_i(w,v)c_p(w,v)}{[O_1^2(w,v) + O_2^2(w,v)]^{(q-2)/2} qD(w,v)}$$

$$+ \sum_{i=1}^{I} \lambda_{2i} \sum_{w=0}^{J-1} \sum_{v=0}^{J-1} \frac{d_i(w,v)d_p(w,v) - c_i(w,v)c_p(w,v)}{[O_1^2(w,v) + O_2^2(w,v)]^{(q-2)/2} qD(w,v)}$$

$$= 0 \tag{23b}$$

Now, the Lagrange multipliers λ_{1i} and λ_{2i} can be solved for to complete the derivation of $a(w,v)$ and $b(w,v)$. In Appendix C, we solve Eqs. (23a) and (23b) for λ_{1i} and λ_{2i}. Thus, the solution for the minimization problem given by Eq. (11) subject to constraints given by Eq. (12) is a constant multiple of

$$H(w,v) = a(w,v) + jb(w,v)$$

$$= \left[\sum_{i=1}^{I} \lambda_{1i}c_i(w,v) + \sum_{i=1}^{I} \lambda_{2i}d_i(w,v) \right.$$

$$\left. + j\left(\sum_{i=1}^{I} \lambda_{1i}d_i(w,v) - \sum_{i=1}^{I} \lambda_{2i}c_i(w,v) \right) \right]$$

$$\{[O_1^2(w,v) + O_2^2(w,v)]^{(q-2)/2} qD(w,v)\}^{-1}$$

$$= \frac{\sum_{i=1}^{I} [(\lambda_{1i} - \lambda_{2i})R_i(w,v)]}{[O_1^2(w,v) + O_2^2(w,v)]^{(q-2)/2} qD(w,v)} \tag{24}$$

where $R_i(w,v) = c_i(w,v) + jd_i(w,v)[1 < p \leq 2, q = p/(p-1)]$, $D(w,v) = \hat{\sigma}_q + |S(w,v)|^q$, and $\hat{\sigma}_q = E|N(w,v)|^q$. λ_{1i} and λ_{2i} are solved for in Appendix C, and the values of $O_1(w,v)$ and $O_2(w,v)$ are solved for in Appendix B.

12.4 PERFORMANCE OF THE l_p-NORM DISTORTION-TOLERANT FILTERS

In this section, we perform some computer simultations to analyze the performance of the l_p-norm distortion-tolerant filter. The target distortion that we consider in the simulation is out-of-plane rotation. The input noise is computer generated. We add overlapping additive noise [$n(i)$ in Eq. (6)] to the entire scene, including the target, and nonoverlapping noise that is added only to the background of the target [7]. Here, nonoverlapping noise means that the noise is placed in the background around the targets. An overlapping noise distorts the entire input scene, including any targets in the scene. Different input noise statistics are used. Because this chapter only proves the concept of the l_p-norm distortion-tolerant filter and presents its derivation, we do not go further in detail to derive or to prove the optimum value of p toward any realistic noise. We merely demonstrate how a better performance can be achieved through varying the value of p by our computer simulation. Thus, we choose the noises available on our computer simulation tool, and they may not represent any realistic noise. The optimum p value for a specific noise will need more extended research in the future.

First, we define and describe the receiver operating characteristics curve (ROC). It is used to compare the relative performance of the l_p-norm distortion-tolerant filter for different values of p. The curve plots the probability of detection (P_d) versus the probability of false alarm (P_{fa}) for a range of threshold values and given input noise.

In general, a large amount of simulated data is required to evaluate P_d and P_{fa}. In our simulation, 200 samples were used to approximate the probability density function [19]. Because the sampling pool is relatively small, we use Parzen's estimator [19,20] to obtain a smooth probability density function. This estimator falls in the general category of a nonparametric estimation method. The smoothing kernel is chosen to be a Gaussian function with zero mean and unity variance [17]. The method is based on the idea that the density function is smooth. As far as we know, there are no definitive rules for choosing the kernel, and the choice of a Gaussian is generally considered reasonable.

We use a set of training targets to generate the distortion-tolerant filter. This set of true training targets has 10 out-of-plane rotated distortions, shown in Fig. 1. The training set covers the range of out-of-plane rotations from 0° to 90° with increments of 10°. Zero degrees represents the front view of the car.

In the following simulations, we add noise with different statistics into the input scene. The l_p-norm optimum filter in Eq. (24) with different values of p is used to detect the true target.

In the simulations reported in Figs. 2, 4, 6, and 8, we use training target 1 of Fig. 1, which has an out-of-plane rotation of 0°. Additive noise with differrent statistics is used in each figure. In Figs. 3, 5, 7, and 9, we have three targets in the input scene. The output peak marked by No. I is due to a true class training target with 0° of rotation. The output peak marked by No. II is due to a nontraining true class target with 15° of rotation. The output peak marked by No. III is due to a false object. The nontraining true target with 15° of rotation is chosen to represent the worst distortion between 10° and 20°. Also, we corrupted the input scene by adding noise with different statistics.

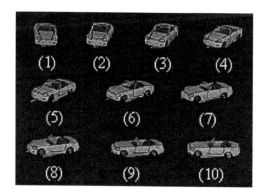

Figure 1 A set of training true targets with different out-of-plane rotational distortions from $0°$ to $90°$ with increments of $10°$ (i.e., $0°, 10°, \ldots, 80°$). Zero degrees (target 1) represents the front view of the car.

Realistic input-scene noise is modeled as color noise [20]. We obtain computer-generated color noise by passing the random white noise through a two-dimensional Gaussian filter function with a Fourier plane bandwidth equal to the desired color-noise power spectrum bandwidth BW. In this experiment, color noise has a chi-square distribution with one degree of freedom. In Fig. 2, the input noise is nonoverlapping background zero-mean color noise, with standard deviation of 0.7. The Gaussian transfer function used to create the color noise has a bandwidth equal to one-fourth of the size of the input scene. In this simulation, $p = 1.2$ produces the best ROC.

We add overlapping color noise to the entire scene and nonoverlapping color noise to the background of the target shown in Fig. 3a. The standard deviations of the overlapping and nonoverlapping noises are equal to 0.2 and 0.4, respectively. The noisy scene is shown in Fig. 3b. In Figs. 3c, 3d, and 3e, we present the filter output produced by $p = 1.8$, $p = 1.5$, and $p = 1.3$, respectively. We observe that $p = 1.3$ has the best result.

Figure 4 presents the ROC curves when the input noise has Poisson distribution with parameter 0.4. It is computer processed to have zero mean. Based on the result, we observe that $p = 1.6$ produces the best performance in these simulations.

In Fig. 5, overlapping noise with standard deviation of 0.3 is added to the entire input scene. Nonoverlapping noise having standard deviation of 0.4 is placed in the background around the targets. Both types of noise are Poisson distributed. The resulting noisy scene is shown in Fig. 5b. Both sources of noise are adjusted to have zero mean. In Figs. 5c, 5d, and 5e, we present the filter output produced by the l_p-norm distrotion-tolerant filter

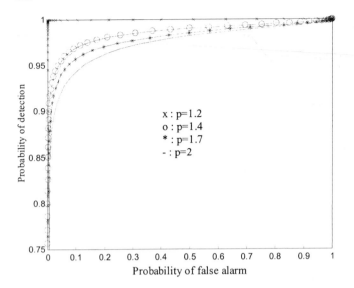

Figure 2   Performance of the l_p-norm distortion-invariant filter with different value of p when the input scene has overlapping color noise added to the entire scene. The noise has a chi-square distribution with one degree of freedom. The standard deviation (σ) of the color noise is 0.7. The results of the l_p-norm filters with $p = 1.2, p = 1.4, p = 1.7$, and $p = 2$ are plotted and indicated by x, o, *, and –, respectively.

for values of $p = 1.8, p = 1.6$, and $p = 1.4$, respectively. It appears that $p = 1.6$ produces the best result in these simulations.

In Fig. 6, the input noise has a Rayleigh distribution. The standard deviation (σ) for both the overlapping and nonoverlapping noise is equal to 0.4, $p = 1.8$ produces the best performance in these simulations.

Figure 7 illustrates the effect of computer-generated overlapping noise with standard deviation of 0.3 and nonoverlapping noise with standard deviation of 0.4. Both sources of noise have Rayleigh distribution, adjusted to have zero mean. The resulting noisy scene is shown in Fig. 7b. In Figs. 7c, 7d, and 7e, we present the filter output for values of $p = 2, p = 1.8$, and $p = 1.6$, respectively. It appears that $p = 1.8$ produces the best result in these simulations.

In Fig. 8, the input noise is overlapping additive white Gaussian noise, with a standard deviation of 1. We compare the relative performance of the filter for different values of p. It is clear that $p = 2$ produces the best performance, as expected.

Figure 9a shows three images in the input scene: one training true class target marked by I, one nontraining true class target marked by II, and one

(*text continues on p. 367*)

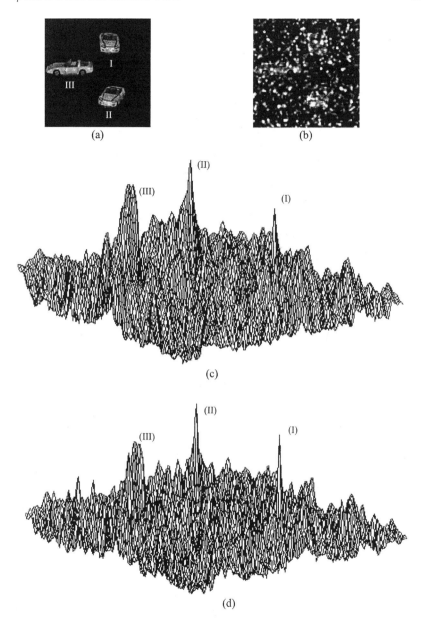

Figure 3 (a) Three objects in the input scene: one true class training target (I), one nontraining true class target (II), and one false object (III). (b) Noisy input scene. The input noise has the same statistics as in Fig. 2. Parts (c), (d), and (e) present the filter output produced by $p = 1.8$, $p = 1.5$, and $p = 1.3$, respectively.

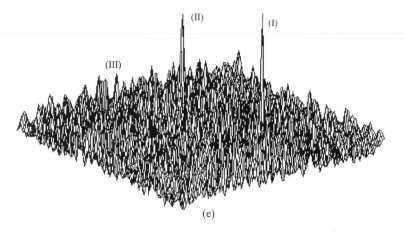

(e)

Figure 3 (*Continued*).

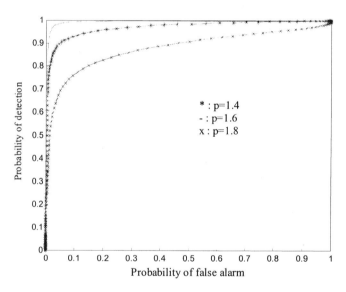

Figure 4 Performance of the l_p-norm distortion-invariant filter with different values of p when the input scene has additive overlapping noise and nonoverlapping (background) noise with Poisson distribution. The standard deviation (σ) of the overlapping and nonoverlapping noise are both equal to 0.4. The results of the l_p-norm filters with $p = 1.4, p = 1.6$, and $p = 1.8$ are plotted and indicated by $*$, $-$, and x, respectively.

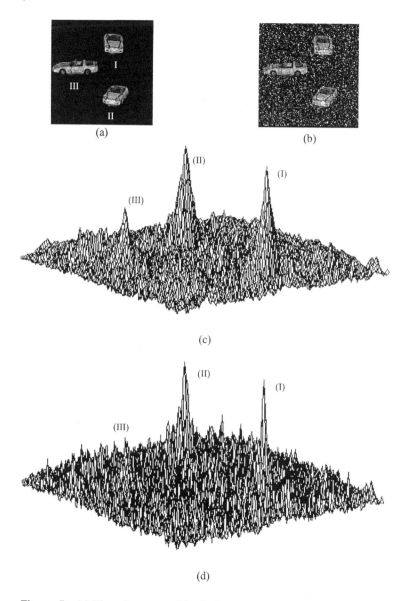

Figure 5 (a) Three images used in the input scene: one true class training target (I), one nontraining true class target (II), and one false object (III). (b) Noisy input scene. The input noise has the same statistics as in Fig. 4. Parts (c), (d), and (e) present the filter output produced by $p = 1.8, p = 1.6$, and $p = 1.4$, respectively.

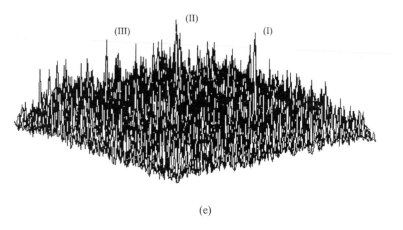

(e)

Figure 5   (*Continued*).

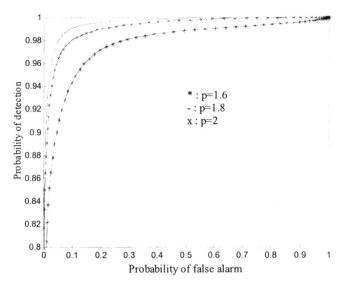

Figure 6   Performance of the l_p-norm distortion-invariant filter with different values of p when the input scene has additive overlapping noise and nonoverlapping noise with Rayleigh distribution. The standard deviation (σ) of the overlapping noise and nonoverlapping noise are both equal to 0.4. The ROC curves of the l_p-norm filters with $p = 1.6, p = 1.8$, and $p = 2$ are plotted and indicated by $*$, $-$, and x, respectively.

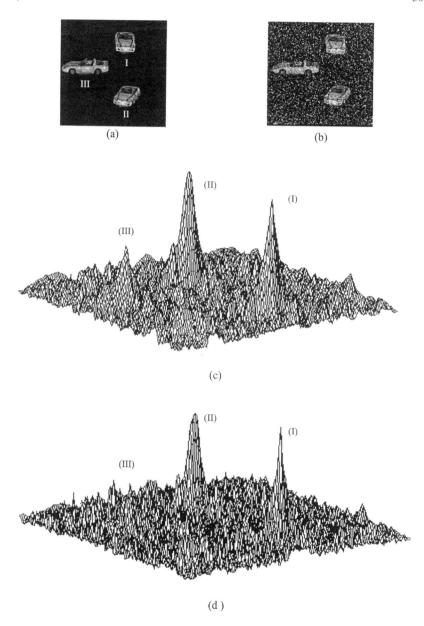

Figure 7 (a) Three images in the input scene: one true class training target (I), one nontraining true class target (II), and one false object (III). (b) Noisy input scene. The input noise has the same statistics as in Fig. 6. Parts (c), (d), and (e) present the filter output produced by $p = 2$, $p = 1.8$, and $p = 1.6$, respectively.

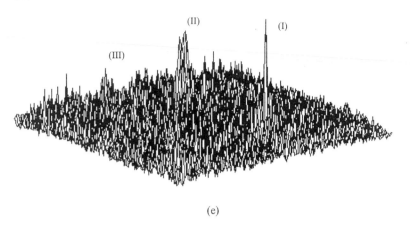

(e)

Figure 7   (*Continued*).

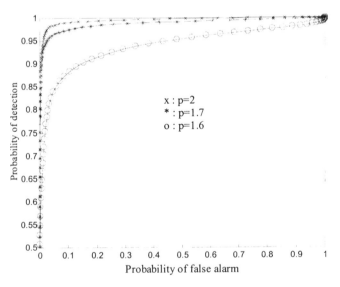

Figure 8   Relative performance of the l_p-norm distortion-tolerant filter for different values of p when the input scene has additive overlapping noise that has Gaussian distribution. The standard deviation (σ) of the overlapping noise is 1. The results of the l_p-norm filters for $p = 1.6, p = 1.7$, and $p = 2$ are plotted and indicated by o, *, and x, respectively.

false object marked by III. We then add to the input scene overlapping noise with standard deviation of 0.5 and nonoverlapping background noise with standard deviation of 0.5 (SNR $= -10\,$dB). The resulting scene is shown in Fig. 9b. In Figs. 9c, 9d, and 9e, we present the filter output produced by the filter for values of $p = 1.9, p = 1.7$, and $p = 1.5$, respectively. We see that $p = 1.9$ has the lowest noise floor in these simulations.

12.5 CONCLUSIONS

We obtained and solved a constraint minimization problem in the Fourier domain to develop a family of distortion-tolerant filters. They are derived by minimizing the pth powers of both the l_p-norm filter of output due to the input scene and the filter output due to the input noise. Thus, a family of nonlinear composite filters is designed that is l_p-norm optimum in terms of noise tolerance and discrimination capability, where p ranges from 1 to 2.

We used receiver operator characteristic (ROC) curves to test the performance of the l_p-norm distortion tolerant filters. Our simulation results showed that for different types of input noise, the performance of the filter can be improved by choosing the appropriate index p. Thus, l_p-norm filters allow greater freedom in adjusting the noise tolerance and discrimination capability.

We have not provided an analytical procedure to determine the best value of p which optimizes the output for a specific noise. We merely demonstrate the potential of using the l_p-norm distortion-tolerant filter to design a system that produces an optimum output under different noise statistics. Further research is required in this direction. However, based on a limited number of simulations, we have noted that, under certain input noise, the performance of the l_p-norm filter improves as p is varied. Limited computer simulations indicate that the filter performance improves when p is not equal to 2 for non-Gaussian color input noise (which is generally the case for background clutter).

ACKNOWLEDGMENTS

We are grateful to S. Kishk for his assistance in finalizing this chapter. We thank N. Towghi for helpful discussions. We thank S. Hong and T. Naughton for reading the manuscript.

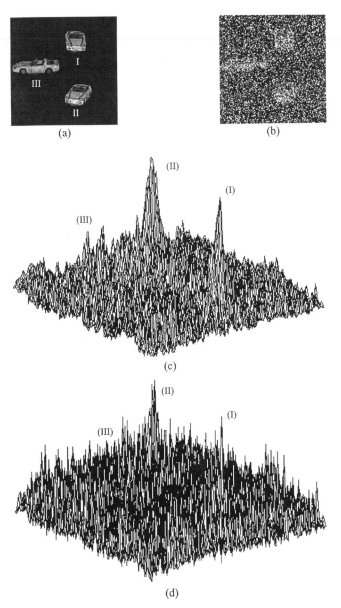

Figure 9   (a) Three images in the input scene: one true class training target (I), one nontraining true class target (II), and one false object (III). (b) An overlapping and a nonoverlapping zero-mean Gaussian noise both with standard deviations equal to 0.5, is added to the scene (SNR= −10 dB). In parts (c), (d), and (e) we present the filter output produced by $p = 1.9, p = 1.7$, and $p = 1.5$, respectively

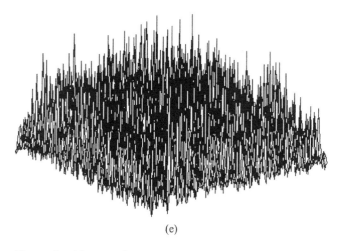

(e)

Figure 9 (*Continued*).

APPENDICES

Appendix A

In this appendix, we find expressions for $a(w, v)$ and $b(w, v)$ in terms of the known values and the Lagrange multipliers λ_{1i} and λ_{2i} as given by Eqs. (17) and (18).

Let $D(w, v) = \hat{\sigma}_q + |S(w, v)|^q$. Note that Eqs. (18a) and (18b) can be written as

$$[a(w, v)^2 + b(w, v)^2]^{(q-2)/2} a(w, v) = \frac{\displaystyle\sum_{i=1}^{I} \lambda_{1i} c_i(w, v) + \sum_{i=1}^{I} \lambda_{2i}(w, v)}{qD(w, v)}$$

(A1a)

$$[a^2(w, v) + b^2(w, v)]^{(q-2)/2} b(w, v) = \frac{\displaystyle\sum_{i=1}^{I} \lambda_{1i} d_i(w, v) - \sum_{i=1}^{I} \lambda_{2i} c_i(w, v)}{qD(w, v)}$$

(A1b)

Recall that, by definition,

$$G_{w,v}^a = \frac{\displaystyle\sum_{i=1}^{I} \lambda_{1i} c_i(w, v) + \sum_{i=1}^{I} \lambda_{2i} d_i(w, v)}{qD(w, v)} \tag{A2a}$$

$$G_{w,v}^b = \frac{\displaystyle\sum_{i=1}^{I} \lambda_{1i} d_i(w, v) - \sum_{i=1}^{I} \lambda_{2i} c_i(w, v)}{qD(w, v)} \tag{A2b}$$

Dividing Eqs. (A1a) by (A1b), and squaring the result, we obtain

$$\left[\frac{a(w, v)}{b(w, v)}\right]^2 = \left[\frac{G_{w,v}^a}{G_{w,v}^b}\right]^2 \tag{A3}$$

Substituting Eq. (A3) into Eqs. (A1a) and (A1b) and solving the equations for $a(w, v)$ and $b(w, v)$, we obtain

$$a(w, v) = \frac{G_{w,v}^a}{[(G_{w,v}^a)^2 + (G_{w,v}^b)^2]^{(q-2)/2(q-1)}} \tag{A4a}$$

$$b(w, v) = \frac{G_{w,v}^b}{[(G_{w,v}^a)^2 + (G_{w,v}^b)^2]^{(q-2)/2(q-1)}} \tag{A4b}$$

Appendix B

In this appendix we will find an expression for $L_{w,v}^q$ in terms of known quantities, where

$$D(w, v) = \hat{\sigma}_q + |S(w, v)|^q, \qquad L_{w,v}^q = [(G_{w,v}^a)^2 + (G_{w,v}^b)^2]^{(q-2)/2(q-1)}$$

$$G_{w,v}^a = \frac{\displaystyle\sum_{i=1}^{I} \lambda_{1i} c_i(w, v) + \sum_{i=1}^{I} \lambda_{2i} d_i(w, v)}{qD(w, v)} \tag{B1a}$$

$$G_{w,v}^b = \frac{\displaystyle\sum_{i=1}^{I} \lambda_{1i} d_i(w, v) - \sum_{i=1}^{I} \lambda_{2i} c_i(w, v)}{qD(w, v)} \tag{B1b}$$

Let $E(w, v) = G_{w,v}^a / L_{w,v}^q$ and $F(w, v) = G_{w,v}^b / L_{w,c}^q$, then,

$$E^2(w, v) + F^2(w, v) = \frac{(G_{w,v}^a)^2 + (G_{w,v}^b)^2}{[(G_{w,v}^a)^2 + (G_{w,v}^b)^2]^{(q-2)/(q-1)}} = (L_{w,v}^q)^{2/(q-2)} \tag{B2}$$

Thus, in order to find an expression for $L_{w,v}^q$ in terms of known quantities, it suffices to find $E^2(w, v) + F^2(w, v)$.

Define the following vectors:

$$\mathbf{V1} = [E(0, 0)\ E(0, 1)\ \cdots\ E(0, J-1)\ E(1, 0)\ E(1, 1)\ \cdots\ E(1, J-1)\ \cdots$$
$$E(J-1, 0)\ E(J-1, 1)\ \cdots\ E(J-1, J-1)]$$
$$\mathbf{V2} = [F(0, 0)\ F(0, 1)\ \cdots\ F(0, J-1)\ F(1, 0)\ F(1, 1)\ \cdots$$
$$F(1, J-1)\ \cdots\ F(J-1, 0)\ F(J-1, 1)\ \cdots\ F(J-1, J-1)]$$

Let $\mathbf{V} = [\mathbf{V1}\ \mathbf{V2}]^T$, where the superscript T denotes the transpose. Therefore, $\mathbf{V}$ is a $2J^2$-dimensional vector ($2J^2 \times 1$ matrix).

For the ith training target, we define the following J^2-dimensional vectors ($J^2 \times 1$ matrix):

$$\mathbf{c}_i = [c_i(0, 0)\ c_i(0, 1)\ \cdots\ c_i(0, J-1)\ c_i(1, 0)\ \cdots\ c_i(1, j-1)\ \cdots$$
$$c_i(J-1, 0)\ \cdots\ c_i(J-1, J-1)]^T$$
$$\mathbf{d}_i = [d_i(0, 0)\ d_i(0, 1)\ \cdots\ d_i(0, J-1)\ d_i(1, 0)\ \cdots\ d_i(1, j-1)\ \cdots$$
$$d_i(J-1, 0)\ \cdots\ d_i(J-1, J-1)]^T$$

where $i = 1, 2, \ldots, I$. Note that $c_i(w, v)$ and $d_i(w, v)$ are the real and imaginary components of $R_i(w, v)$, respectively. Let

$$\mathbf{W}_1 = [\mathbf{c}_1\ \cdots\ \mathbf{c}_I]^T$$
$$\mathbf{W}_2 = [\mathbf{d}_1\ \cdots\ \mathbf{d}_I]^T$$

Therefore, $\mathbf{W}_1$ and $\mathbf{W}_2$ are $I \times J^2$ matrices. Let

$$\mathbf{W} = \begin{bmatrix} \mathbf{W}_1 & \mathbf{W}_2 \\ \mathbf{W}_2 & -\mathbf{W}_1 \end{bmatrix}$$

and

$$\mathbf{P} = [J^2 C_1\ \cdots\ J^2 C_I\ 0\ \cdots\ 0]^T \tag{B3}$$

Then, Eqs. (20a) and (20b) can be written as

$$\mathbf{WV} = \mathbf{P} \tag{B4}$$

where $\mathbf{V} = [\mathbf{V1}\ \mathbf{V2}]^T$ is the unknown vector defined previously. Note that Eq. (B4) does not have a unique solution because $\mathbf{W}$ is a singular matrix. Thus, we opt to choose the minimum energy solution. To that end, note that

$$\mathbf{W}^T\mathbf{WV} = \mathbf{W}^T\mathbf{P} \tag{B5}$$

Note that $\mathbf{W}^T\mathbf{W}$ is a non-negative definite matrix and it depends on the training targets; therefore, for all practical purposes, we may assume that it is invertible. Because, otherwise, by slightly changing the value of one of the entries, the matrix can be changed to a nonsingular matrix. Solving Eq. (B5) for $\mathbf{V}$, we obtain

$$\mathbf{V} = \begin{bmatrix} \mathbf{V_1} \\ \mathbf{V_2} \end{bmatrix} = (\mathbf{W}^T\,\mathbf{W})^{-1}\mathbf{W}^T\mathbf{P} \equiv \begin{bmatrix} \mathbf{O_1} \\ \mathbf{O_2} \end{bmatrix} \tag{B6}$$

where

$$\mathbf{O_1} = [O_1(0,0)\ O_1(0,1)\ \cdots\ (O_1(0, J-1)\ O_1(1,0)\ \cdots\ O_1(1, J-1)\ \cdots$$
$$O_1(J-1,0)\ \cdots\ O_1(J-1, J-1)]^T$$

and

$$\mathbf{O_2} = [O_2(0,0)\ O_2(0,1)\ \cdots\ (O_2(0, J-1)\ O_2(1,0)\ \cdots\ O_2(1, J-1)\ \cdots$$
$$O_2(J-1,0)\ \cdots\ O_2(J-1, J-1)]^T$$

that is,

$$V_1(w,v) = E(w,v) = O_1(w,v) \quad \text{and} \quad V_2(w,v) = F(w,v) = O_2(w,v) \tag{B7}$$

Combining Eqs. (B2) and (B7), we obtain

$$L_{w,v}^q = [(G_{w,v}^q)^2 + (G_{w,v}^q)^2]^{(q-2)/(2q-2)} = [O_1^2(w,v) + O_2^2(w,v)]^{(q-2)/2} \tag{B8}$$

Appendix C

In this appendix we solve Eqs. (23a) and (23b) for the Lagrange multipliers λ_{1i} and λ_{2i} in terms of the known quantities. Recall that Lagrange multiplier for the ith constraint equations (14a) and (14b) are λ_{1i} and λ_{2i}. Let

$$\lambda_1 = [\lambda_{11}\ \lambda_{12}\ \cdots\ \lambda_{1I}]^T \quad \text{and} \quad \lambda_2 = [\lambda_{21}\ \lambda_{22}\ \cdots\ \lambda_{2I}]^T \tag{C1}$$

where, as earlier, the superscript T denotes transpose. As in the main text, $\mathbf{R}_i$ denotes the Fourier transform of the ith training target. $\mathbf{c}_i$ and $\mathbf{d}_i$ are the real and imaginary parts of $\mathbf{R}_i$. Note that $\mathbf{R}_i$, $\mathbf{c}_i$, and $\mathbf{d}_i$ are $J \times J$ matrices and $R_i(w,v)$, $c_i(w,v)$, and $d_i(w,v)$ denote their values at (w,v). Let C_i be the correlation output peak for the ith training target; then,

$$\mathbf{C} = [C_1\ C_2\ \cdots\ C_I]^T \tag{C2}$$

Let $\mathbf{A}$ and $\mathbf{B}$ be $I \times I$ matrices defined by

$$A_{i,j} = \sum_{w=0}^{J-1} \sum_{v=0}^{J-1} \left[\frac{c_i(w, v)c_j(w, v) + d_i(w, v)d_j(w, v)}{[O_1^2(w, v) + O_2^2(w, v)]^{(q-2)/2}qD(w, v)} \right] \tag{C3a}$$

$$B_{i,j} = \sum_{w=0}^{J-1} \sum_{v=0}^{J-1} \left[\frac{d_i(w, v)c_j(w, v) + c_i(w, v)d_j(w, v)}{[O_1^2(w, v) + O_2^2(w, v)]^{(q-2)/2}qD(w, v)} \right] \tag{C3b}$$

where $O_1(w, v)$ and $O_2(w, v)$ are given by Eq. (B6) and $D(w, v) = \hat{\sigma}_q + |S(w, v)|^q$.

Thus, Eqs. (23a) and (23b) can be rewritten as

$$\lambda_1^T \mathbf{A} + \lambda_2^T \mathbf{B} = J^2 \mathbf{C}^T \tag{C4a}$$

$$-\lambda_1^T \mathbf{B} + \lambda_2^T \mathbf{A} = 0 \tag{C4b}$$

Note that because $\mathbf{A}$ and $\mathbf{B}$ are symmetric, non-negative definite, we may assume $\mathbf{A}^{-1}$ and $\mathbf{B}^{-1}$ exist.

Equations (C4a) and (C4b) form a linear system of equations, which we can solve for λ_1 and λ_2 and obtain

$$\lambda_1^T = J^2 \mathbf{C}^T (\mathbf{A} + \mathbf{B}\mathbf{A}^{-1}\mathbf{B})^{-1} \tag{C5a}$$

$$\lambda_2^T = J^2 \mathbf{C}^T (\mathbf{A} + \mathbf{B}\mathbf{A}^{-1}\mathbf{B})^{-1}\mathbf{B}\mathbf{A}^{-1} \tag{C5b}$$

REFERENCES

1. JL Turin. An introduction to matched filters. IRE Trans Inform Theory 6:311–329, 1960.
2. A VanderLugt. Signal detection by complex spatial filtering. IEEE Trans Inform Theory IT-10:139–145, 1964.
3. JL Horner, PD Gianino. Phase-only matched filtering. Appl Opt 23:812–816, 1984.
4. J Caulfield, WT Maloney. Improved discrimination in optical character recognition. Appl Opt 8:2354–2356, 1969.
5. D Casasent, D Psaltis. Position, rotation and scale invariant optical correlation. Appl Opt 15:1795–1799, 1976.
6. P Réfrégier. Filter design for optimal pattern recognition: multicriteria optimization approach. Opt Lett 15:854–856, 1990.
7. B Javidi, J Wang. Limitation of the classic definition of SNR in matched filter based pattern recognition. Appl Opt 31, 1992.
8. B Javidi. Smart Imaging Systems. Bellingham, WA: SPIE Press, 2001.
9. FM Dickey, LA Romero. Normalized correlation for pattern recognition. Opt Lett 16:1186–1188, 1991.

10. P Refregier, V Laude, B Javidi. Nonlinear joint transform correlation: an optimum solution for adaptive image discrimination and input noise robustness. Opt Lett 19:405–407, 1994.
11. HL Van Trees. Detection, Estimation and Modulation Theory. New York: Wiley, 1968.
12. CF Hester, D Casasent. Multivariant technique for multiclass pattern recognition. Appl Opt 19:1758–1761, 1980.
13. D Casasent. Unified synthetic discriminant function computational formulation. Appl Opt 23:1620–1627, 1984.
14. A Mahalanobis. Review of correlation filters and their application for scene matching. In: B Javidi, K Johnson, eds. Optoelectronic Devices and Systems for Processing, Critical Reviews of Optical Science Technology. Vol. 65. Bellingham, WA: SPIE Press, 1996, pp 240–260.
15. YN Hsu, HH Arsenault. Optical pattern recognition using circular harmonic expansion. Appl Opt 21:4016–4019, 1982.
16. B Javidi, J Wang. Optimum distortion-invariant filter for detecting a noisy distorted target in nonoverlapping background noise. J Opt Soc Am A 12:2604–2614, 1995.
17. N Towghi, B Javidi. l_p-norm optimum filters for image recognition. Part I. Algorithms. J Opt Soc Am A 16(8):1928–1935, 1999.
18. L Pan. l_p-Norm distortion tolerant filter for detecting distorted targets in a scene. MS thesis, University of Connecticut, Storrs, CT, 2000.
19. CW Therrien, Decision Estimation and Classification, New York: Wiley, 1989.
20. TY Yong, TW Calvert, Classification, Estimation and Pattern Recognition, Amsterdam: Elsevier, 1974.

13
Image-Based Face Recognition: Issues and Methods

Wen-Yi Zhao
Sarnoff Corporation, Princeton, New Jersey

Rama Chellappa
University of Maryland, College Park, Maryland

13.1 INTRODUCTION

13.1.1 Background

As one of the most successful applications of image analysis and understanding, face recognition has recently gained significant attention, especially during the past several years. This is evidenced by the emergence of specific face recognition conferences such as AFGR [1] and AVBA [2] and systematic empirical evaluation of face recognition techniques (FRT), including the FERET [3,4] and XM2VTS protocols [5]. There are at least two reasons for such a trend: The first is the wide range of commercial and law enforcement applications and the second is the availability of feasible technologies after 35 years of research.

The strong demand for user-friendly systems which can secure our assets and protect our privacy without losing our identity in a sea of numbers is obvious. At present, one needs a PIN to get cash from an ATM, a password for a computer, a dozen others to access the Internet, and so on. Although extremely reliable methods of biometric personal identification exist (e.g., fingerprint analysis and retinal or iris scans), these methods have yet to gain acceptance by the general population. A personal identification system based on analysis of frontal or profile images of the face is nonintrusive and therefore user-friendly. Moreover, personal identity can often be ascertained without the client's assistance. In addition, the need for applying FRT has been boosted by recent advances in multimedia processing along with others such as IP (Internet Protocol) technologies.

In summary, there exist tremendous opportunities and great challenges for FRT. The challenge facing FRT is to perform well under severe conditions. For example, a personal verification system might need to process a low-quality face image which might be acquired using an inexpensive PC camera and transferred over IP; or the image capture happens in an uncontrolled environment with bad lighting and so forth. On the other hand, the opportunity lies in the fact that multimedia is almost ubiquitous and face objects are among the most important multimedia contents. For example, you may want to search for the video clips from home video archives where your baby shows a cute pose. Multimedia applications based on face objects include content-based applications, human–machine interactive applications, security related applications, and so forth. For example, a database software capable of searching for face objects or a particular face object is very useful. Another example is a smart video conference system that is able to automatically track objects and enhance their appearances.

13.1.2 Face Recognition Technology

A general statement of face recognition problem can be formulated as follows. Given still or video images of a scene, identify or verify one or more persons in the scene using a stored database of faces. Available collateral information such as race, age, gender, facial expression, and speech may be used in narrowing the search (enhacing recognition). The solution of the problem involves segmentation of faces (face detection) from cluttered scenes, feature extraciton from the face region, recognition, or verification. In identification problems, the input to the system is an unknown face, and the system reports back the decided identity from a database of known individuals, whereas in verification problems, the system needs to confirm or reject the claimed identity of the input face.

Various applications of FRT range from static, controlled format photographs to uncontrolled video images, posing a wide range of different technical challenges and requiring an equally wide range of techiques from image processing, analysis, understanding, and pattern recognition. One can broadly classify the challenges and techniques into two groups: static and dynamic/video matching. Even within each group, significant differences exist, depending on the specific application. The differences are in terms of image quality, amount of background clutter (posing challenges to segmentation algorithms), the availability of a well-defined matching criterion, and the nature, type, and amount of input from a user. A rich repository of research literature exists after 35 years of research. Particularly, the last 5-year experienced the most active research activities and rapid advances. For an up-to-date critical survey of still- and video-based face recognition research, please see Ref. 6.

13.1.3 Chapter Organization

In this chapter, we present efficient techniques for processing and recognizing face images. We assume an image-based baseline system because image-based approaches are possibly the most promising and practical ones. However, the two-dimensional (2D) images/patterns of three-dimensional (3D) face objects can change dramatically due to lighting and viewing variations. Hence, the illumination and pose problems present significant obstacles for wide applications of this type of approach. To overcome these issues, we propose using a generic 3D model to enhance existing image-based systems. More specifically, we use a 3D model to synthesize the so-called *prototype image* from a given image acquired under different lighting and viewing conditions. This enhancement enables the existing systems to handle both illumination and pose problems specific to face recognition under the following assumption: just *one* image per face object is available.

In Section 13.2, we first review and categorize existing methods proposed to address the pose problem and the illumination problem. We then propose using a generic 3D face model to enhance existing systems in Section 13.3. Instead of being a full 3D approach which directly uses accurate 3D information which is not easy to obtain in practice, this approach synthesizes a 2D *prototype image* from a given 2D image acquired under different lighting and viewing conditions with the aid of a 3D mode. The prototype image is defined as the frontal view of an object under frontal lighting. For the purpose of completeness, a brief introduction to one particular baseline subspace LDA (linear discriminant analysis) system [7,8] has been included. In Section 13.4, we feed prototype images into the subspace LDA system to perform recognition. Finally, we conclude our chapter in Section 13.5.

13.2 EXISTING FACE RECOGNITION TECHNIQUES

Automatic face recognition consists of subtasks in a sequential manner: face detection, face segmentation/normalization, and face recognition/verification. Many methods of face recognition have been proposed [6]. Basically, they can be divided into holistic template-matching-based systems, geometrical local-feature-based schemes, and hybrid schemes. Even though schemes of all these types have been successfully applied to the task of face recognition, they do have certain advantages and disadvantages. Thus, an appropriate scheme should be chosen based on the specific requirements of a given task. For example, the Elastic Bunch Graph Matching (EBGM)-based system [9] has very good performance in general. However, it requires large size image (e.g., 128×128). This severely restricts its possible application to video-based surveillance, where the image size of the face area is very small. On the other hand, the subspace LDA system [7] works well with both large and small images (e.g., 96×84 or 24×21). It also has the best performance for verification tasks according to the most recent FERET test [4].

13.2.1   Two Problems in Face Recognition

Despite the successes of many systems [7,9,10] based on the FERET test, many issues remain to be addressed. Among those issues, the following two are prominent for most systems: (1) the illumination problem and (2) the pose problem.

The illumination problem is illustrated in Fig. 1, where the same face appears differently due to the change in lighting. More specifically, the changes induced by illumination could be larger than the differences between individuals, causing systems based on comparing images to misclassify the identity of the input image. This has been experimentally observed in Ref. 11 with a dataset of 25 individuals. We can also carry out some analysis. For example, the popular eigensubspace projections

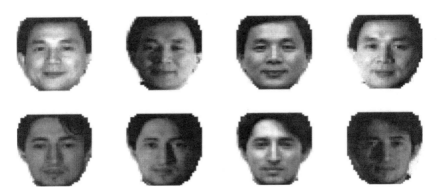

Figure 1   The illumination problem.

used in many systems as features have been analyzed under illumination variation in Ref. 8. The conclusions suggest that significant illumination changes cause dramatic changes in the projection coefficient vectors and, hence, can seriously degrade the performance of subspace-based methods.

For comparison purposes, we plot the variations of projection coefficient vectors due to pure differences in class label (Fig. 2a) along with the variations due to pure illumination change of the same class (Fig. 2b).

The pose problem is illustrated in Fig. 3, where the same face appears differently due to changes in viewing condition. Moreover, when illumination variation also appears in the face images, the task of face recognition becomes even more difficult (Fig. 3). In Ref 8, an analysis and classification of various pose problems are performed using a reflectance model with varying albedo. Using such a model, the difficulty of the pose problem can be assessed and the efficacy of existing methods can be evaluated systematically. For example, the pose problem has been divided into three categories: (1) the simple case with small rotation angles, (2) the most commonly addressed case, when there is a set of training image pairs (frontal and rotated images), and (3) the most difficult case, when training image pairs are not available and illumination variations are present.

Difficulties due to illumination and pose variations have been documented in many evaluations of face recognition systems [3,8,11,12]. An even more difficult case is the combined problem of pose and illumination variations. Unfortunately, this happens when face images are acquired in uncontrolled environments—for instance, in surveillance video clips. In the following, we examine the two problems in turn and review some existing approaches to these problems. More importantly, we point out the pros and cons of these methods so an appropriate approach can be applied to the specific task.

13.2.2 Solving the Illumination Problem

As a fundamental problem in image-understanding literature, the illumination problem is generally quite difficult and has been receiving consistent attention. For face recognition, many good approaches have been proposed utilizing the domain knowledge (i.e., all faces belong to one face class). These approaches can be broadly divided into four types [8]: (1) heuristic methods, including discarding the leading principal components, (2) image comparison methods, for which various image representations and distance measures are applied, (3) class-based methods, for which multiple images of one face under a fixed pose but different lighting conditions are available, and (4) model-based approaches, for which 3D models are employed.

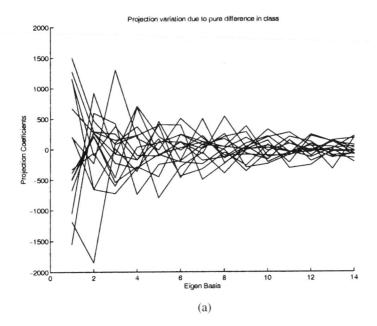

(a)

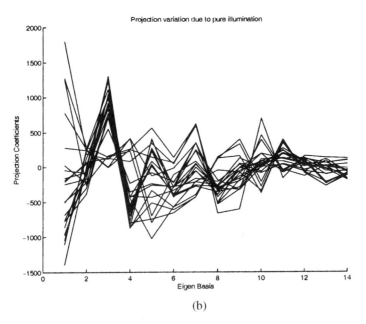

(b)

Figure 2   Change of projection vectors due to (a) class variation and (b) illumination change.

Figure 3   The pose (and illumination) problem. Top row: pure pose problem; bottom row: coupled pose and illumination problem.

Heuristic Approaches

To handle the illumination problem, researchers have proposed various methods. Within the eigensubspace domain, it has been suggested that by discarding the three most significant principal components, variations due to lighting can be reduced. Also, it has been experimentally verified in Ref. 12 that discarding the first few principal components seems to work reasonably well for images under variable lighting. However, in order to maintain system performance for normally lighted images to improve performance for images acquired under varying illumination, we must assume that the first three principal components capture the variations only due to lighting. In Ref. 13, a heuristic method based on face symmetry is proposed to enhance system performance under different lighting.

Image Comparison Approaches

In Ref. 11, statistical approaches based on image comparison have been evaluated. The reviewed methods use different image representations and distance measures. The image representations used are edge maps, derivatives of the gray level, images filtered with 2D Gabor-like functions, and a representation that combines a log function of the intensity with these representations. The different distance measures used are pointwise distance, regional distance, affine-GL (gray level) distance, local affine-GL distance, and LOG pointwise distance. For more details about these methods and the evaluation database, please refer to Ref. 11. One important conclusion drawn is that these representations are not sufficient by themselves to overcome the image variations. More recently, a new image comparison method proposed by Jacobs et al. [14] uses a measure robust to illumination change. Their method is based on the observation that the difference between two images of the same object is smaller than the difference between images of different objects. However, this measure is not strictly illumination invariant because the measure changes for a pair of images of the same object when the illumination changes.

Class-Based Approaches

With assumptions of Lambertian surfaces, no shadowing and three aligned images/faces acquired under different lighting conditions, a 3D linear illumination subspace for a person has been constructed in Refs. 15–17 for a fixed viewpoint. Thus, under ideal assumptions, recognition based on the 3D linear illumination subspace is illumination invariant. More recently, an illumination cone has been proposed as an effective method to handle illumination variations, including shadowing and multiple lighting sources [15,18]. This method is an extension of the 3D linear subspace method [16,17], and, hence, needs three aligned training images acquired under different lightings. One drawback to using this method is that we need more than three aligned images per person. More recently, a new method based on a quotient image has been introduced [19]. The advantage of this approach over existing similar approaches is that it only uses a small set of sample images. This method assumes the same shape and different textures for faces of different individuals. An interesting energy function to be minimized is then formulated. Using this formulation, better results are rendered than using connectionist approaches.

Model-Based Approaches

In their article [20], Atick et al. suggest using principal component analysis (PCA) as a tool for solving the parametric shape-from-shading problem (i.e., obtain the eigenhead approximation of a real 3D head after training on about 300 laser-scanned range data of real human heads. Although the ill-posed shape-from-shading problem is transformed into a parametric problem, they still assume constant albedo. This assumption does not hold for most real face images and we believe that this is one of the major reasons why most SFS algorithms fail on real face images.

13.2.3 Solving the Pose Problem

Researches have proposed various methods to handle the rotation problem. Basically, they can be divided into three classes [21]: (1) multiple-view-based methods, when multiple views per person are available, (2) hybrid methods, when multiple training images are available during training but only one database image per person is available during recognition, and (3) single image/shaped-based methods, when no training is conducted. We have Refs. 22–25 in the first type, and Refs. 26–29 in the second type. Up to now, the second type of approach is the most popular one. The third approach does not seem to have received much attention.

Multiple-View-Based Approaches

Among the first class of approaches, one of the earliest is by Beymer [22], in which a template-based correlation matching scheme is proposed. In this work, pose estimation and face recognition are coupled in an iterative loop. For each hypothesized pose, the input image is aligned to database images corresponding to a selected pose. The alignment is first carried out via a 2D affine transformation based on three key feature points (two eyes and nose), and then optical flow is used to refine the alignment of each template. After this step, the correlation scores of all pairs of matching templates are used to perform recognition. The main restrictions of this method are (1) many images of different views per person are needed in the database, (2) no lighting variations (pure texture mapping) or facial expressions are allowed, and, finally, (3) the computational cost is high because it is an iterative searching approach. More recently, an illumination-based image synthesis method [25] has been proposed as a potential method for robust face recognition handling both pose and illumination problems. This method is based on the well-known approach of *an illumination cone* [15] and can handle illumination variation quite well. To handle variations due to rotation, it needs to completely resolve the GBR (generalized bas relief) ambiguity when reconstructing the 3D shape.

Hybrid Approaches

Numerous algorithms of the second type have been proposed and are, by far, the most popular ones. Possible reasons for this are as follows: (1) it is probably the most successful and practical method up to now and (2) it utilizes prior class information. We review three representative methods here: The first one is the *linear-class*-based method [26], the second one is the graph-matching-based method [30], and the third is the view-based eigenface approach [31]. The image synthesis method in Ref. 26 is based on the assumption of linear 3D object classes and the extension of linearity to images which are 2D projections of the 3D objects. It extends the linear shape model from a representation based on feature points to full images of objects. To implement their method, a correspondence between images of the input object and a reference object is established using optical flow.

Also, correspondence between the reference image and other example images having the same pose are computed. Finally, the correspondence field for the input image is linearly decomposed into the correspondence fields for the examples. Compared to the parallel deformation scheme in Ref. 29, this method reduces the need to compute the correspondence between images of different poses. This method is extended in Ref. 27 to carry an additive error term for better synthesis. In Ref. 30, a robust face recognition scheme based on Elastic Bunch Graphic Matching (EBGM) is proposed. The authors basically assume a planar surface patch in each key feature point (landmark) and learn the transformation of "jets" under face rotation. They demonstrate substantial improvements in face recognition under rotation. Also, their method is fully automatic, including face localization, landmark detection, and, finally, a flexible graph-matching scheme. The drawback of this method is the requirement of accurate landmark localization, which is not an easy task, especially when illumination variations are present. The popular eigenface approach to face recognition has been extended to view-based eigenface method in order to achieve pose-invariant recognition [31]. This method explicitly codes the pose information by constructing an individual eigenface for each pose. Despite their popularity, these methods have some common drawbacks: (1) They need many example images to cover all possible views and (2) the illumination problem is separated from the pose problem.

Single Image/Shape-Based Approaches

Finally, there is the third class of approach which includes low-level feature-based methods, invariant-feature-based methods, and the 3D-model-based method. In Ref. 32, a Gabor wavelet-based feature extraction method is proposed for face recognition and is robust to small-angle rotation. There are many articles on invariant features in the computer vision literature. To our knowledge, serious application of this technology to face recognition has not yet been explored. However, it is worthwhile to point out that some recent work on invariant methods based on images Ref. 33 may shed some light in this direction. For synthesizing face images under different appearances/lightings/expressions, 3D facial models have been explored in the computer graphics, computer vision, and model-based coding communities. In these methods, face shape is usually represented either by a polygonal model or a mesh model that simulates tissue. However, due to its complexity and computational cost, any serious attempt to apply this technology to face recognition has not yet been made, except in Ref. 34.

13.3   3D MODEL ENHANCED FACE RECOGNITION

Based on the assumption of one image per class available, solving the coupled illumination and pose problem is not an easy task. Most previous approaches are either based on different assumptions or treat the two problems separately; hence, it is necessary that we search for methods which can solve both problems simultaneously and efficiently. For example, the 3D-model-based synthesis approaches used in computer graphics and coding communities are usually too complicated and expensive. Instead, we propose using a 3D model to enhance existing 2D approaches. To overcome the constant albedo issue [20] in modeling face objects, a varying albedo reflectance model is proposed [8]. Using this technique, we can convert any input images into prototype images, which are later fed into existing systems. In the recognition experiments carried out later, we choose a particular baseline system based on its simplicity and efficiency: the subspace LDA system.

13.3.1   The Subspace LDA System

The subspace LDA system is shown in Fig. 4 [7]. It was proposed with the motivation of trying to solve the *generalization/overfitting* problem when performing face recognition on a large face dataset but with *very few* training face images available per class. Like existing methods, this method

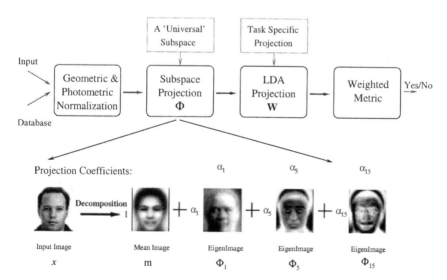

Figure 4   The subspace LDA face recognition system.

consists of two steps: First, the face image is projected into a *face subspace* via PCA, where the subspace dimension is carefully chosen, and then the PCA projection vectors are projected into the LDA to construct a linear classifier in the subspace. Unlike other methods, the authors argue that the dimension of the face subspace is fixed (for a given training set) regardless of the image size as long as the image size surpasses the subspace dimensionality. The property of relative invariance of the subspace dimension enables the system to work with smaller face images without sacrificing performance. This claim is supported by experiments using normalized face images of different sizes to obtain different face subspaces [8]. The choice of such a fixed subspace dimension is mainly based on the characteristics of the eigenvectors instead of the eigenvalues. Such a choice of the subspace dimension enables the system to generate class-separable features via LDA from the *full* subspace representation. Hence, the *generalization/overfitting* problem can be addressed to some extent. In addition, a weighted distance metric guided by the LDA eigenvalues was employed to improve the performance of the subspace LDA method. The improved performance of generalized recognition was demonstrated on FERET datasets [7] and the MPEG-7 content set in a proposal on the robust face descriptor to MPEG-7 [35]. In Ref. 7, experiments were conducted to compare algorithm performances. The authors used a subset of the FERET development dataset for training and the FERET development dataset and other datasets for testings. The results show that subspace LDA (subLDA) is the best compared to PCA (with different eigenvectors: 15, 300, and 1000) and pure LDA algorithms. The detailed ranking of the algorithms is as follows: subLDA300(85.2%) > subLDA1000(80.8%) > LDA(67.0%) > PCA1000(58.3%) > PCA300 (57.4%) > subLDA15(50.4%) > PCA15(47.8%). The numbers in the parentheses are correct top-match scores.

The authors also reported a sensitivity test of the subspace LDA system [7]. They took one original face image and then electronically modified the image by creating occlusions, applying Gaussian blur, randomizing the pixel location, and adding an artificial background. Figure 5 shows electronically modified face images which were correctly identified.

13.3.2 A Varying Albedo Illumination Model for Face

In dealing with 2D–3D transformations, a physical illumination model is needed. There are many illumination models available which can be broadly categorized into diffuse reflectance models and specular models [36]. Among these models, the Lambertian model is the simplest and most popular one for diffuse reflectance and has been used extensively, especially in shape-from-shading (SFS) literature. With the assumption

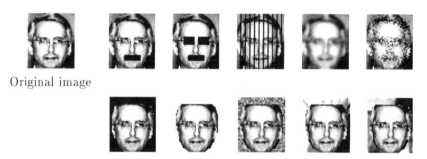

Original image

Figure 5 Electronically modified images which have been correctly identified.

of Lambertian surface reflection and a single, distant light source, we have the following standard equations:

$$I = \rho \cos \gamma$$

or

$$I = \rho \frac{1 + pP_s + qQ_s}{\sqrt{1 + p^2 + q^2}\sqrt{1 + P_s^2 + Q_s^2}} \tag{1}$$

where γ is the angle between the outward normal to the surface $\mathbf{n} = (p, q, 1)$ and the negative illumination vector $-\mathbf{L} = (P_s, Q_s, 1)$ which represents the direction opposite to the distant light source, and ρ is the albedo. The surface orientation can also be represented using two angles; slant and tilt. Similarly, the light source can be represented by illuminant direction slant and tilt. The illuminant direction *slant* α is the angle between the negative $\mathbf{L}$ and the positive z axis: $\alpha \in [0°, 180°]$; and the illuminant direction *tilt* τ is the angle between the negative $\mathbf{L}$ and the x-z plane: $\tau \in [-180°, 180°]$. To relate these angle terms to P_s and Q_s, we have $P_s = \tan\alpha \cos\tau$ and $Q_s = \tan\alpha \sin\tau$.

Because we allow for arbitrary albedo, both (p, q) and ρ are functions of locations (x, y). However, we impose symmetry constraint for front-view face objects as follows (with an easily-understood coordinate system):

$$p[x, y] = -p[-x, y]$$
$$q[x, y] = q[-x, y] \tag{2}$$

and

$$\rho[x, y] = \rho[-x, y] \tag{3}$$

To show that the varying albedo Lambertian model is a good model, we compare the image synthesis results obtained using constant albedo and varying albedo assumptions. In Fig. 6, image synthesis results are compared one by one; that is, a pair of images (in the same column) are synthesized exactly the same way except that one is using a constant albedo model and the other is using a varying albedo model. To obtain a realistic albedo, we use a real face image and a generic 3D face model. To align this 3D model to the input image, we normalize both of them to the same size, with two eye pairs kept in the same fixed positions. Because the input image and model are not from the same object, we can see that some parts of the synthesized images are not perfect (e.g., around the nose region).

13.3.3 The Self-Ratio Image r_I

The concept of *self-ratio image* was initially introduced in Ref. 8 to address the additional parameter (albedo) issue. The idea of using two aligned images to construct a ratio has been explored by many researchers [14,37]. However, it was extended to a single image in Ref. 8. Based on this concept, a new SFS scheme has been developed. We can also use it to help us to obtain the prototype images from given images.

Let us substitute (3, 2) into the equations for $I[x, y]$ and $I[-x, y]$ and then add them, giving

$$I[x, y] + I[-x, y] = 2\rho \frac{1 + qQ_s}{\sqrt{1 + p^2 + q^2}\sqrt{1 + P_s^2 + Q_s^2}} \qquad (4)$$

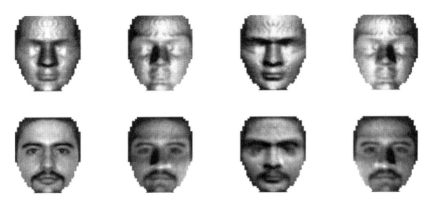

Figure 6 Image synthesis comparison under various lighting conditions. First row: constant albedo Lambertian model; second row: varying albedo Lambertian model.

Similarly, we have

$$I[x, y] - I[-x, y] = 2\rho \frac{pP_s}{\sqrt{1 + p^2 + q^2}\sqrt{1 + P_s^2 + Q_s^2}} \tag{5}$$

To simplify the notation, let us define

$$I_+[x, y] = \frac{I[x, y] + I[-x, y]}{2}$$

and

$$I_-[x, y] = \frac{I[x, y] - I[-x, y]}{2}$$

Then, the *self-ratio image* r_I can be defined as

$$r_I[x, y] = \frac{I_-[x, y]}{I_+[x, y]} = \frac{pP_s}{1 + qQ_s} \tag{6}$$

Solving for shape information using Eq. (6) combined with Eq. (1) is called symmetric SFS [8]. The main result of symmetric SFS is the following theorem [8]:

Theorem 1   *If the symmetric depth z is a $\mathscr{C}^2$ surface and the symmetric albedo field is piecewise constant, then both the solution for shape (p, q) and albedo ρ are unique except in some special conditions. Significantly, the unique global solution consists of unique local solutions at each point simultaneously obtained using the intensity information at that point and the surrounding local region under the assumption of a $\mathscr{C}^2$ surface.*

When symmetric objects are rotated, we cannot directly apply the symmetric SFS theorem/algorithm. However, we can generate a virtual front view from the given image. Also, the virtual front-view image can be obtained using the following relation between the rotated (in the x-z plane about the y axis) image $I^\theta[x', y']$ and the original image $I[x, y]$ [8]:

$$I^\theta[x', y'] = 1_{z,\theta} I[x, y](\cos\theta - p[x, y]\sin\theta)$$
$$\times \frac{\tan(\theta + \theta_0)P_s + [q\cos(\theta_0)/\cos(\theta + \theta_0)]Q_s + 1}{pP_s + qQ_s + 1} \tag{7}$$

where $\tan\theta_0 = p[x, y]$ and $1_{z,\theta}$ is the indicator function indicating possible occlusion determined by the shape and rotation angle. After the virtual front view is obtained, the symmetric SFS algorithm can then be applied. Of course, this is useful only if we can solve the correspondence problem: $[x, y] \rightarrow [x', y']$. In the following subsection, we propose using a 3D model to obtain an approximate solution to this problem.

13.3.4 Using a Generic 3D Face Shape

In theory, we can apply symmetric SFS to recover the complete 3D shape of a symmetric object. Also, we have proposed simple algorithms which work well on objects with complex/face shape and piecewise albedo field. In Fig. 7, we plot the input and reconstructed face images, with partial derivatives side by side for the piecewise constant albedo case. However we still have difficulties in recovering the case when both the shape and albedo are arbitrary. Moreover, there are many practical issues that prevent us from applying it for face recognition: (1) The unique solution might be very sensitive to possible violations of the assumptions, such as the $\mathscr{C}^2$ surface (possibly caused by digitizing the surface), (2) the solution might be sensitive to noise in measurement I (and, hence, r_I), and (3) the solution might be sensitive to possible violations of the single light source assumption. To be more practical for face recognition, we propose using a simple 3D model to bypass this 2D-to-3D process. This technique has been successfully applied in Ref. 38 to address the pure illumination problem with pose fixed. This technique can also be extended for rotated face images [21].

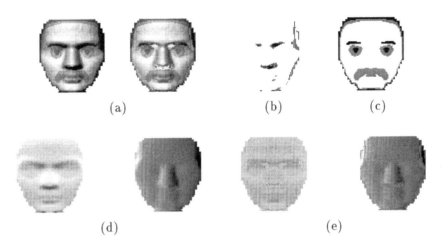

(a)   (b)   (c)

(d)   (e)

Figure 7 Face shape and image reconstruction results using symmetric SFS: the piecewise constant albedo case. The plots in the left-half of the first row are input and reconstructed images (a). Plot (b) represents the shadow map in the input images which explains the holes in the reconstructed images. Plot (c) is the recovered albedo filed. The plots in the second row are the true partial derivatives (d) and recovered partial derivatives (e).

Front-View Case

Let us write the image equation for the prototype image I_p with $\alpha = 0$:

$$I_p[x, y] = \rho \, \frac{1}{\sqrt{1 + p^2 + q^2}} \tag{8}$$

Comparing Eqs. (4) and (8), we obtain

$$I_p[x, y] = \frac{K}{2(1 + qQ_s)} \left(I[x, y] + I[-x, y] \right) \tag{9}$$

where K is a constant equal to $\sqrt{1 + P_s^2 + Q_s^2}$. This simple equation directly relates the prototype image I_p to $I[x, y] + I[x, -y]$, which is already available. It is worthwhile to point out that this *direct computation* of I_p from I offers the following advantages over the two-step procedure which uses SFS to recover 3D information and then synthesizes new 2D images:

- There is no need to recover the varying albedo $\rho[x, y]$.
- There is no need to recover the *full* shape gradients (p, q).

The only parameter that needs to be recovered is the partial shape information q and we approximate this value with the partial derivative of a 3D face model. To guarantee good synthesis quality, we use the self-ratio image equation (6) as a consistency-checking tool.

Rotated Case

Combing Eqs. (7) and (9), we have a direct relation between a rotated (in the x-z plane about the y axis) image $I^\theta[x', y']$ and the prototype image I_p [21]:

$$I^\theta[x', y'] = 1_{z, \theta} I_p[x, y](\cos\theta - p[x, y]\sin\theta) \, \frac{1}{\sqrt{1 + P_s^2 + Q_s^2}}$$
$$\times \left(\tan(\theta + \theta_0)P_s + \frac{q\cos(\theta_0)}{\cos(\theta + \theta_0)} Q_s + 1 \right), \tag{10}$$

To actually apply the above techniques for face recognition, we need the illumination direction of the light source and the face pose. In this way, we do not need to synthesize *all* views and illuminations in our database in order to recognize input images under various viewing and illumination conditions. Instead, we can synthesize the prototype view defined in the database from an input image acquired under different views and illumination directions.

13.3.5 Light Source and Pose Estimation

Frontal-View Case

Many source-from-shading algorithms are available, but we found that none of them work well for both the tilt and slant angles [8]. Instead, we propose a new model-based symmetric source-from-shading algorithm [21]. Basically, we can formulate a minimization problem as

$$(\alpha^*, \tau^*) = \arg_{\alpha,\tau} \min(r_{I_{M_F}}(\alpha, \tau)) - r_I)^2 \tag{11}$$

where r_I is the self-ratio image, and $r_{I_{M_F}}$ is the self-ratio image generated from the 3D face model M_F given hypothesized α and τ. One advantage of using a 3D face model is that we can take into account both attached-shadow and cast-shadow effects, which are not utilized in the traditional statistics-based methods. Compared to other model-based methods [20], this method produces better results because it adopts a better model. To compare, other methods can also be formulated as a minimization problem:

$$(\alpha^*, \tau^*) = \arg_{\alpha,\tau} \min(I_M(\alpha, \tau)) - I)^2 \tag{12}$$

where I is the input image and I_M is the image generated from a 3D generic shape M based on Lambertian model (1) with constant albedo given hypothesized α and τ. For a simple comparison of these two model-based methods, we ran both of these algorithms on real face images. In Fig. 8, we plot one face image along with the error-versus-slant curve for each method.

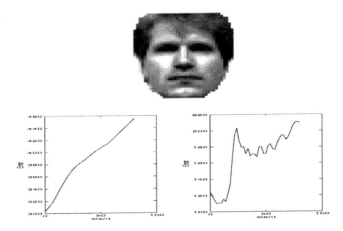

Figure 8 Comparison of model-based source-from-shading algorithms. The correct slant value was recovered using algorithm (11) (right figure), whereas it was missed using algorithm (12) (middle figure).

As can be seen, the correct (subjective judgment) value of slant (8°) has been recovered by the symmetric method (11). However, it is missed using method (12). This new symmetric source-from-shading method has been successfully applied to more than 150 real face images as the preprocessing step prior to illumination–normalization for face recognition [38].

Rotated Case

Most existing face pose estimation algorithms use some prior class knowledge; that is, all face object are similar. Instead of using 2D example images, we propose using a simple 3D model to estimate the pose. Further, to incorporate the estimation of the light source, we formulate the following problem:

$$(\theta^*, \alpha^*, \tau^*) = \arg_{\theta, \alpha, \tau} \min[I_{M_F}^R(\theta, \alpha, \tau) - I^R]^2 \tag{13}$$

However, such formulations ignore the reality of a varying albedo. To better address this problem, the *self-ratio* image is used. However, in order to apply this method to rotated images of symmetric objects, we need additional processing. Using Eq. (7), we can formulate a new estimation problem:

$$(\theta^*, \alpha^*, \tau^*) = \arg_{\theta, \alpha, \tau} \min[r_{I_{M_F}}(\alpha, \tau) - r_{I^F}(\theta, \alpha, \tau)]^2 \tag{14}$$

where $r_{I^F(\theta, \alpha, \tau)}$ is the self-ratio image for the virtual frontal view generated from the original image I_R via image warping and texture mapping using Eq. (7).

13.4 EXPERIMENTS

13.4.1 Shadow and Implementation Issues

One important issue we have not discussed in detail is the attached-shadow and cast-shadow problem. By definition, attached-shadow points are those where the image intensities are set to zero because $(1 + pP_s + qQ_s) \leq 0$. A cast shadow is the shadow cast by the object itself. It has been shown in Ref. 8 that the shadow points can still be utilized in both source estimation and image rendering. For example, in the case of source estimation, one advantage of using a 3D face model is that we can take into account both attached-shadow and cast-shadow effects, which are not utilized in the traditional statistics-based methods. However, these points contribute significantly and correctly to the computation of slant and tilt angles. Hence, the model-based method can produce a more accurate estimate if the 3D face model is a good approximation to the real 3D face shape.

In addition to these shadow points, we need to single out the "bad" points, or outliers in statistical terms, for stable source estimation and prototype image rendering. This is because we need to compute the self-ratio image which may be sensitive to image noise. Let us denote the set of all "bad" points by $\mathscr{B}$; at these points, the values cannot be used. From a robust statistics point of view, these "bad" points are outliers. Hence, our policy for handling these outliers is to reject them and mark their locations. We then use values computed at good points to interpolate/extrapolate at the marked bad points. Many interpolation methods are available such as nearest-neighbor interpolation, polynomial interpolation, spline interpolation, and so forth. Because we may have an irregular structure of good points, we use triangle-based methods. For example, to detect these bad points in the process of computing the prototype images, we employ the *consistency check*

$$\mathscr{B} = \left\{ I[x, y] \middle| \left| r_I - \frac{p_{M_F} P_s}{1 + q_{M_F} Q_s} \right| > \epsilon \right\} \tag{15}$$

13.4.2 Solving the Illumination Problem

Rendering Prototype Images

The faces we used in our experiments are from the FERET, Yale, and Weizmann databases [3,12,11]. The Yale database contains 15 persons, including 4 images obtained under different illuminations. The Weizmann database contains 24 persons, also including 4 images obtained under different illuminations. We have applied our light-source estimation and direct prototype image rendering method to more than 15 face images from the Yale and Weizmann databases. Although the purpose of rendering prototype images is to improve the recognition performance, we would like to visualize the quality of the rendered images and compare them to the images obtained using a local SFS algorithm [39]. These results (Fig. 9) clearly indicate the superior quality of the prototype images rendered by our approach. More rendered prototype images using only the direct computation are plotted in Fig. 10.

Enhancing Face Recognition

In this experiment, we demonstrate that the generalized/predictive recognition rate of subspace LDA can be greatly enhanced. We conducted two independent experiments on the Yale and Weizmann databases. For the Yale database, we have a testing database composed of a gallery set con-

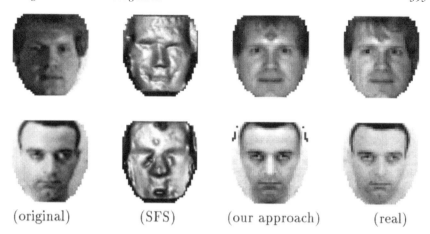

(original)   (SFS)   (our approach)   (real)

Figure 9 Image-rendering comparison.

taining 486 images from several face databases, including 15 (1 image per class) from the Yale database, and a probe set containing 60 images also from the Yale database. For the Weizmann database, we have a testing database composed of a gallery set containing 495 images from several face databases, including 24 (image per class) from the Weizmann database, and a probe set containing 96 images from the same database. Figure 11 shows the significant improvement in performance using the prototype images in both databases.

13.4.3 Solving the Pose Problem

Currently, light source and pose estimation (14) is not conducted. Instead, very rough pose information is given manually. The light source is also assumed to be frontal, although, in fact, it may be not. Basically, we have only experimented on model-based pure image warping and plan to implement full SFS-based view synthesis in the near future. However, as we have shown in Ref. 21, this is a good approach for eigensubspace-based method even when the Lambertian model and frontal lighting are assumed.

The database we used her is drawn from FERET and Stirling databases [7]. To compare, the quality of this dataset is lower than the one reported in Ref. 26 (which is also used in several subsequent works on face recognition such as Ref. 19) and the size of normalized images we are using is much smaller than those in Ref. 26 and others. There are 108 pairs of face images: front view and quarter-profile view, all normalized to the size of 48 × 42 with respect to the two eyes. The poses of these faces are

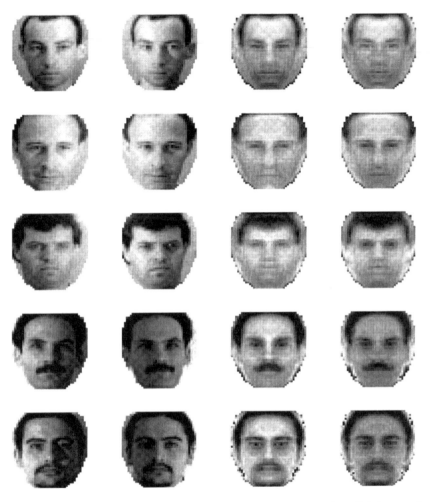

Figure 10   Rendering the prototype image. The images with various lightings are in the first two columns, and the corresponding prototype images are shown in the last two columns, respectively.

not quite consistent, and we only apply a unique rotation angle picked manually to all images.

Visualization of Image Synthesis

As we mentioned earlier, the poses of these faces are not consistent and only one unique rotation angle is chosen for all the images. Hence, some of the

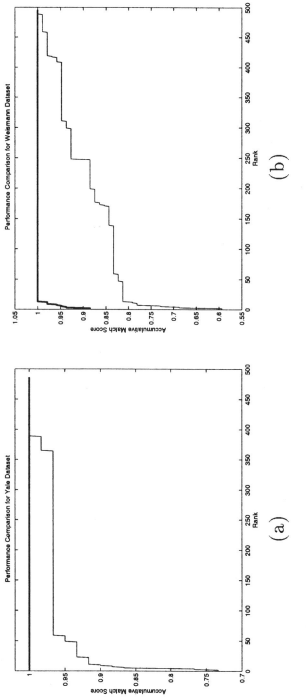

Figure 11 Enhancing the subspace LDA. The thin lines represent the cumulative scores when applying the existing subspace LDA to the original images, and the thick lines represent the scores when applying it to the prototype images. The curves in (a) are for the Yale face database, and the curves in (b) are for the Weizmann database. Note that no retraining on PCA or LDA is performed.

synthesis results are good (the first three columns in Fig. 12) if the actual rotation angle agrees with the preset value, and some are bad (the last three columns in Fig. 12).

Comparison of Recognition Results

To test and compare the efficacy of various methods for robust face recognition, we have tried two subspace LDA methods: (I) subspace LDA [7] on the original images and (II) subspace LDA on the synthesized frontal images.

As mentioned earlier, the database we used have 108 pairs of images, of which only about 42 pairs are good images in terms of the correctness of rotation angles we manually picked (refer to Fig. 12). We use all of the frontal views as the database and all of the rotated images as the testing images. We report the recognition performances of subspace LDA on the original images and on the virtual frontal images in Table 1. Some conclusions can be drawn here: Using the virtual frontal views, the performance of subspace LDA, which does not have the generalization problem and does not need retraining

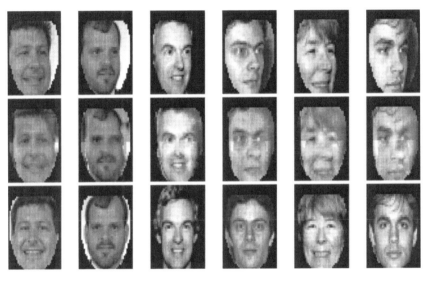

Figure 12 Some images used in the database. The first row are the rotated views, the second row are the synthesized frontal views, and the third view are the real frontal views. The first three columns are good image pairs, and the last three columns are bad pairs

Table 1 Performance Comparison of
Subspace LDA on Original Images (I)
and on Virtual Images (II).

	Method	
	I	II
Score (%)	39.8/61.9	46.3/66.7

The scores on the right are for the 42 good
images, and the scores on the left are for all
108 images.

of the subspace and the LDA classifier, can be improved, and the extent of
the improvement depends on the quality of the virtual views.

13.5 CONCLUSION

We have proposed simple and efficient techniques for processing and recognizing face objects. The characteristics of these techniques are very suitable
for many applications. We first identified two key issues in the face recognition literature: the illumination and pose problems. We then examined existing methods of handling these two problems extensively. To handle pose
and illumination problems in a uniform framework, we proposed a reflectance model with varying albedo for the 3D face and introduced a new
concept: the self-ratio image. Finally, we proposed using a 3D model to
synthesize the prototype image from a given image under any lighting and
viewing conditions. This technique alone can be used to synthesize new
images (i.e., enhancing appearance). Adding this technique into existing
subspace LDA systems, we basically propose an enhanced system. In the
future, we plan to improve our method by deforming the 3D model to fit
individuals better or using multiple 3D face models, as in Ref. 20.

REFERENCES

1. Proceedings of the International Conferences on Automatic Face and Gesture
   Recognition, 1995, 1996, 1998, 2000.
2. Proceedings of the International Conferences on Audio- and Video-Based
   Person Authentication, 1997 and 1999.
3. PJ Phillips, H Moon, P Rauss, SA Rizvi. The FERET evaluation methodlogy
   for face-recognition algorithms. Proceedings of the Conference on Computer
   Vision and Pattern Recognition, 1997, pp 137–143.

4. SA Rizvi, PJ Phillips, H Moon. The FERET verification testing protcol for face recognition algorithms. Proceedings of the International Conference on Automatic Face and Gesture Recognition, 1998.

5. K Messer, J Matas, J Kittler, J Luettin, G Maire. XM2VTSDB: The extended M2VTS database. Proceedings of the International Conference on Audio- and Video-Based Person Authentication, 1999, pp 72–77.

6. W Zhao, R Chellappa, A Rosenfeld, P J Phillips. Face recognition: A literature survey. Technical Report CS-TR-4167, Center For Automation Research, University of Maryland, 2000.

7. W Zhao, R Chellappa, PJ Phillips. Subspace linear discriminant analysis for face recognition. Technical Report CAR-TR-914, Center for Automation Research, University of Maryland, 1999.

8. W Zhao. Robust image based 3D face recognition. PhD thesis, University of Maryland, 1999.

9. K Okada, J Steffans, T Maurer, H Hong, E Elagin, H Neven, C v d Malsburg. The Bochum/USC face recognition system and how it fared in the FERET phase III test. In: H Wechsler, P J Phillips, V Bruce, F F Souli, T S Huang, eds. Face Recognition: From Theory to Applications. New York: Springer-Verlag, 1998, pp 186–205.

10. B Moghaddam, A Pentland. Probabilistic visual learning for object representation. IEEE Trans PAMI PAMI-19(7):696–710, 1997.

11. Y Adini, Y Moses, S Ullman. Face recognition: The problem of compensating for changes in illumination direction. IEEE Trans PAMI PAMI-19:721–732, 1997.

12. PN Belhumeur, JP Hespanha, DJ Kriegman. Eigenfaces vs. Fisherfaces: Recognition using class specific linear projection. IEEE Trans PAMI PAMI-19(7):711–720, 1997.

13. W Zhao. Improving the robustness of face recognition. Proceedings of the International Conference on Audio- and Video-based Person Authentication, 1999, pp 78–83.

14. DW Jacobs, PN Belhumeur, R Basri. Comparing images under variable illumination. Proceedings of the Conference on Computer Vision and Pattern Recognition, 1998, pp 610–617.

15. PN Belhumeur, DJ Kriegman. What is the set of images of an object under all possible lighting conditions? Proceedings of the Conference on Computer Vision and Pattern Recognition, San Juan, 1997, pp 52–58.

16. P Hallinan. A low-dimensional representation of human faces for arbitrary lighting conditions. Proceedings of the Conference on Computer Vision and Pattern Recognition, 1994, pp 995–999.

16. Y Adini, Y Moses, S Ullman. Face recognition: the probelm of compensating for changes in illumination direction. IEEE Trans PAMI PAMI-19:721–732, 1997.

17. A Shashua. Geometry and photometry in 3D visual recognition. PdD thesis, MIT, Cambridge, MA, 1994.

18. AS Georghiades, DJ Kriegman, PN Belhumeur. Illumination cones for recognition under variable lighting: Faces. Proceedings of the Conference on Computer Vision and Pattern Recognition, 1998, pp 52–58.
19. T Riklin-Raviv, A Shashua. The quotient image: Class based rerendering and recognition with varying illuminations. Proceedings of the Conference on Computer Vision and Pattern Recognition, 1999, pp 566–571.
20. J Atick, P Griffin, N Redlich. Statistical approach to shape from shading: Reconstruction of three-dimensional face surfaces from single two-dimensional images. Neural Comput 8:1321–1340, 1996.
21. W Zhao, R Chellappa. SFS based view synthesis for robust face recognition. Proceedings of the Conference on Automatic Face and Gesture Recognition, 2000.
22. DJ Beymer. Face recognition under varying pose. Technical Report 1461 MIT Artificial Intelligence Laboratory, 1993.
23. S Akamatsu, T Sasaki, H Fukamachi, N Masui, Y Suenaga. An accurate and robust face identification scheme. Proceedings of the International Conference on Pattern Recognition, The Hague, The Netherlands, 1992, pp 217–220.
24. S Ullman, R Basri. Recognition by linear combinations of models. IEEE Trans PAMI PAMI-13:992–1006, 1991.
25. AS Georghiades, PN Belhumeur, DJ Kriegman. Illumination-based image synthesis: Creating novel images of human faces under differing pose and lighting. Proceedings of the Workshop on Multi-View Modeling and Analysis of Visual Scenes, 1999, pp 47–54.
26. T Vetter, and T Poggio. Linear object classes and image synthesis from a single example image. IEEE Trans PAMI PAMI-19(7):733–742, 1997.
27. E Sali, S Ullman. Recognizing novel 3-D objects under new illumination and viewing position using a small number of example views or even a single view. Proceedings of the Conference on Computer Vision and Pattern Recognition, 1998, pp 153–161.
28. T Maurer, C v d Malsburg. Single-view based recognition of faces rotated in depth. Proceedings of the International Workshop on Automatic Face and Gesture Recognition, 1996, pp 176–181.
29. DJ Beymer, T Poggio. Face recognition from one example view. Proceedings, International Conference on Computer Vision, 1995, pp 500–507.
30. L Wiskott, J-M Fellous, C von der Malsburg. Face recognition by elastic bunch graph matching. IEEE Trans PAMI PAMI 19:775–779, 1997.
31. A Pentland, B Moghaddam, T Starner. View-based and modular eigenspaces for face recognition. Proceedings of the Conference on Computer Vision and Pattern Recognition, 1994.
32. BS Manjunath, R Chellappa, C v d Balsburg. A feature based approach to face recognition. Proceedings of the Conference on Computer Vision and Pattern Recogntion, 1992, pp 373–378.
33. R Alferez, YF Wang. Geometric and illumination invariants for object recognition. IEEE Trans PAMI PAMI-21:505–536, 1999.

34. G Gordon. Face recognition based on depth maps and surface curvature. In: Geometric Methods in Computer Vision. SPIE Proceedings Vol. 1570. Bellingham, WA: SPIE Press, 1991, pp 234–247.

35. W Zhao, D Bhat, N Nandhakumar, R Chellappa. A reliable desciptor for face objects in visual content. J Signal Process Image Commun 16:123–136, 2000.

36. SK Nayar, K Ikeuchi, T Kanade. Surface reflection: Physical and geometrical perspectives. IEEE Trans PAMI PAMI-13:611–634, 1991.

37. LB Wolff, E Angelopoulou. 3-D Stereo using photometric ratios. Proceedings of the European Conference on Computer Vision. Berlin: Springer-Verlag, 1994, pp 247–258.

38. W Zhao, R. Chellappa. Illumination-insensitive face recognition using symmetric shape-from-shading. Proceedings of the Conference on Computer Vision and Pattern Recognition, 2000, pp 286–293.

39. PS Tsai, M Shah. A fast linear shape from shading. Proceedings of the Conference on Computer Vision and Pattern Recognition, Urbana-Champaign, IL, 1992, pp 459–465.

40. BKP Horn, M J Brooks. Shape from Shading. Cambridge, MA: MIT Press, 1989.

41. R Chellappa, CL Wislon, S Sirohey. Human and machine recognition of faces, a survey. Proc IEEE 83:705–740, 1995.

14

Image Processing Techniques for Automatic Road Sign Identification and Tracking

Elisabet Pérez
Polytechnic University of Catalunya, Terrassa, Spain

Bahram Javidi
University of Connecticut, Storrs, Connecticut

14.1 INTRODUCTION

In this chapter, the design of an on-board processor which enables recognition of a given road sign from a vehicle in motion is presented. A safety system to be installed in vehicles could be based on this processor in order to automatically detect and identify road signs. Afterward, the recognition system could make an objective decision according to the information detected. One of the greatest difficulties in achieving this goal lies in the number of different distortions that may simultaneously modify the reference sign. Variations in scale, in-plane and out-of-plane rotations, background clutter, partially occluded signs, and variable illumination are some examples of distortions that can affect road signs. To overcome these problems, a number of techniques have been studied in pattern recognition [1–13]. Some of them have been applied to road sign recognition. For instance, an optical correlator for scale-invariant road sign detection was proposed in Refs. 14 and 15. Recently, partial tolerance to in-plane rotations and scale invariance has been obtained by using partially invariant filters in a multiple correlator [16,17].

In general, a given recognition technique is designed to provide satisfactory results when dealing with a particular distortion of the object.

However, the same strategy usually gives poorer results if another type of distortion affects the object. Analysis and comparison of different techniques are carried out in this work. A recognition system simultaneously scale invariant and tolerant to slight tilts or out-of-plane rotations due to different view angles of the acquisition system is obtained by combining various strategies. Tolerance to illumination fluctuations is needed in order to enable a recognition system to work under different illumination or weather conditions. Robustness to cluttered background is also important for a road sign recognition processor that analyzes images captured in real environments.

The proposed distortion-tolerant processor is based on a nonlinear correlator [18], which is described in detail in Section 14.2. Some principles of pattern recognition based on digital image processing are introduced in the same section. Section 14.3 concentrates on some filtering techniques applied to distortion-tolerant systems. In Section 14.4, the influence on the recognition system of a variety of distortions, such as scale variations and in-plane and out-of-plane rotations, is studied. Section 14.5 considers postprocessing of the correlation results to improve road sign recognition results. Section 14.6 presents the application of the proposed distortion-tolerant recognition system to images with real cluttered background and which include road signs affected by several of the aforementioned distortions. A summary is presented in Section 14.7 to conclude the work.

14.2 PATTERN RECOGNITION BY IMAGE PROCESSING

14.2.1 Linear and Nonlinear Correlators

The correlation operator can compute a measure of similarity between two objects. The mathematical expression of correlation between two functions, $s(x, y)$ and $r(x, y)$, is defined by [19,20]

$$c(x, y) = s(x, y) \otimes r(x, y) = \int_{-\infty}^{\infty} \int_{-\infty}^{\infty} s(\xi, \eta) r^*(\xi - x, \eta - y) \, d\xi \, d\eta \quad (1)$$

where the asterisk denotes the complex conjugate and $\otimes$ correlation. If functions $s(x, y)$ and $r(x, y)$ describe a scene to be analyzed and a reference target, respectively, correlation between them is a measure of the overlapping of objects contained in the scene and the reference. In that sense, correlation could be considered as an estimation of their degree of similarity.

Correlation can be also expressed in terms of Fourier transforms. If the hat symbol ˆ is used to denote the Fourier transform of a function, the correlation operation expressed in Eq. (1) can be defined in an equivalent way by [19,20]

$$c(x, y) = \text{TF}^{-1}\{\hat{s}(x, y)\hat{r}^*(x, y)\} \tag{2}$$

where TF^{-1} represents the inverse Fourier transform.

Equation (2) shows that correlation between two functions can be obtained by multiplying their Fourier transforms in the frequency domain and by inverse Fourier transforming this product. Because Fourier transforms and product operations can be achieved optically, correlation between two functions can also be implemented optically. Systems that perform correlation are named corrrelators and they permit real-time processing of a large amount of information using optics. The combination of advantages given by optics along with some properties provided by electronics has made feasible the implementation of powerful hybrid optoelectronic processors for solving different pattern recognition tasks.

Linear corrrelators described by Eq. (1) have many limitations for recognizing objects in background noise. In addition, these correlators are not tolerant to image distortions such as scale, rotation, and illumination fluctuations [21–25]. Applying different spatial filtering techniques prior to multiplying the Fourier transforms in the frequency domain permits one to overcome these problems. For instance, it is possible to recognize an object that presents some distortions such as scale variations or rotations, or even to detect an object when the signal is degraded by noise [1–12].

When a nonlinear operator modifies Fourier transforms of the scene and the reference target, we consider the processor to be a nonlinear correlator [18]. Figure 1 contains a detailed scheme of the operations performed in a nonlinear processor. In this work, nonlinear filtering is being used in the correlator due to its superior performance in comparison with linear filtering techniques in terms of discrimination capability, correlation peak sharpness, and noise robustness [10–11,18]. In a kth-law processor [18], the nonlinear operator applied symmetrically to the scene and to the reference Fourier transforms is defined by

$$g(\hat{f}) = \text{sgn}(\hat{f})|\hat{f}|^k \exp[i\phi_f], \qquad k \le 1 \tag{3}$$

Parameter k controls the strength of the applied nonlinearity. For $k = 1$, a linear filtering technique is obtained, whereas $k = 0$ leads to a binarizing nonlinearity. Intermediate values of k permit one to vary the features of the processor, such as its discrimination capability or its illumination invariance. The precise index k needs to be determined to obtain a good performance of the processor depending on the application.

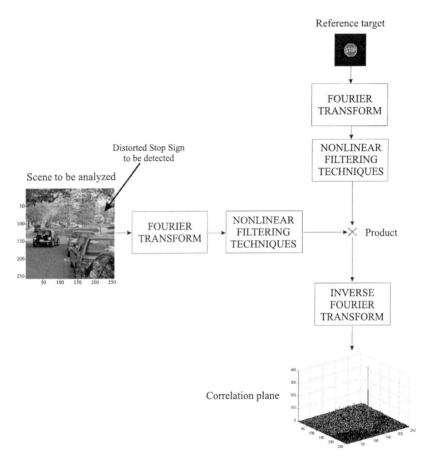

Figure 1   Diagram of a nonlinear processor.

14.2.2   Metrics for Recognition Evaluation

There exists many different metrics to evaluate correlation filter performance. Some of them are described in summarizing works elsewhere [26]. To evaluate correlation results in our experiments, we will use a criterion based on the peak-to-correlation energy (PCE) parameter, which is defined as [26]

$$\mathrm{PCE} = \frac{|c(0, 0)|^2}{\int\int |c(x, y)|^2 \, dx \, dy} \tag{4}$$

This parameter measures the ratio between the intensity value of the output peak at the target location ($|c(0, 0)|^2$) and the total energy of the output plane ($\int \int |c(x, y)|^2 \, dx \, dy$). In general, a high and sharp correlation peak is expected when there is an object in the scene that matches the reference target, thus leading to a high value for the PCE parameter. A better match between an object of the scene and the reference, a closer value to unity for the PCE parameter will be reached. For this reason, PCE parameter seems to be a reliable criterion on which to base the final recognition decision.

A thresholding operation permits one to accept a true target or to reject a false object. The threshold level is sometimes established arbitrarily. However, it can be also determined by means of a learning algorithm. A set of training images, containing true targets and false objects, permits one to measure the probability of error in the recognition process depending on the threshold value. A final threshold level for the recognition procedure is established by considering a null probability of error in the identification of the training objects. In our case, based on the PCE criterion, objects that obtain PCE values above the threshold are considered as true targets. On the contrary, PCE values below the threshold imply the rejection of the object in the recognition process.

Another assumption could be taken into account. Sometimes, objects contained in the analyzed scene are compared, simultaneously or sequentially, to different reference targets. In such a situation, different correlation planes are computed for each scene. The final recognition result for the analyzed scene can be processed in different ways depending on arithmetic or logical operations applied to the correlation outputs. In this work, a winner-take-all model is used. The PCE parameter is computed for all the output planes and the output plane with the maximum PCE value is selected as the final response of the system. Only if the maximum PCE value is above the threshold, established in the learning process, will an object contained in the scene be recognized as similar to the target.

14.3 NONLINEAR FILTERING TECHNIQUES FOR DISTORTION-TOLERANT RECOGNITION

Different approaches to obtain distortion-tolerant recognition systems exist. They have in common the need of storing information of the reference target taking into account different distortions that can affect it.

The most straightforward way to keep the information of the distorted versions of a target is to design a single filter for each type of distortion to be considered. In this case, we talk about a filter bank. To determine if a target, distorted or not, is included in a given scene, it is necessary to correlate the scene with the multiple filters belonging to the bank. This technique could be time-consuming. To avoid a large comput-ing time, composite filters were introduced.

In a general approach, the information included in a composite filter consists of various views of the target under different situations (different rotation angles, scale variations, changes in illumination, etc.). The synthesis of all of the information in a unique composite filter is carried out by taking into account different constraints. The constraint operation used in the synthesis of a composite filter provides desirable features for the correlation output, such as sharp correlations peaks, noise robustness, a low-output noise floor, and so forth.

The principal advantage of a composite filter in front of a bank of filters is the reduction of time in the processing step. Only a single correla-tion can be enough to compare a given image with the whole set of distorted versions of the sought reference. However, composite filters can sometimes lack noise robustness and discrimination capability. The number of images (distorted versions) of the reference included in a composite filter should be limited in order to obtain a successful recognition performance.

In this work, Fourier-plane nonlinear filters [10,11] are used as com-posite nonlinear filters. They are modifications of other well-known distor-tion-invariant filters [1,2]. It has been shown that Fourier-plane nonlinear filters have tolerance to in-plane and out-of-plane rotations, as well as good performance in the presence of different types of noise [10,11]. The kth-law equal-correlation-peak synthetic discriminant function (ECP-SDF) is tested in the following simulation experiments to design a distortion-tolerant road sign recognition system.

14.3.1 kth-Law ECP-SDF

Let $s_1(x, y), s_2(x, y), \ldots, s_N(x, y)$ represent N training images. Let P be the total number of pixels contained in each image. Instead of a matrix to represent an image, we use a vector notation by means of lexicographic ordering. A P-element column vector is obtained for each image by rear-ranging the rows of the matrix. This operation is performed from left to right and from top to bottom. We construct a training data image $\mathbf{S}$ that has the vector s_i as its ith column. Therefore, $\mathbf{S}$ is a $P \times N$ matrix. The lexicographical ordered composite filter $h(x, y)$ (i.e., ECP–SDF) can be expressed as [1]

$$\mathbf{h} = \mathbf{S}(\mathbf{S}^+\mathbf{S})^{-1}\mathbf{c}^* \tag{5}$$

where $\mathbf{S}^+$ is the complex-conjugate transpose of $\mathbf{S}$, and $(\ )^{-1}$ denotes the inverse matrix. Vector $\mathbf{c}$ contains the desired cross-corrleation peak value for each training image, and $\mathbf{c}^*$ is the complex conjugate of $\mathbf{c}$. Equation (5) can be rewritten in the frequency domain as

$$\hat{\mathbf{h}} = \hat{\mathbf{S}}(\hat{\mathbf{S}}^+\hat{\mathbf{S}})^{-1}\mathbf{c}^* \tag{6}$$

where the hat symbol $\hat{\ }$ denotes Fourier transform. The kth-law nonlinear composite filter is obtained by replacing $\hat{\mathbf{S}}$ in Eq. (6) by $\hat{\mathbf{S}}^k$, where nonlinearity is applied on each element of the matrix. The nonlinear operator for the element of rth row and lth column of $\hat{\mathbf{S}}^k$ is defined as

$$|\mathbf{S}_{rl}|^k \exp(j\phi_{\mathbf{S}_{rl}}) \tag{7}$$

and the corresponding kth-law ECP–SDF filter is [10,11]

$$\hat{\mathbf{h}}^k = \hat{\mathbf{S}}^k[(\hat{\mathbf{S}}^k)^+\hat{\mathbf{S}}^k]^{-1}\mathbf{c}^* \tag{8}$$

14.4 ROAD SIGN RECOGNITION SYSTEM TOLERANT TO VARIOUS DISTORTIONS

In this section, we separately analyze tolerance of the proposed recognition system to scale variations and in-plane and out-of-plane rotations of the objects. A scale-invariant road sign recognition system is introduced in Section 14.4.1. A description of the procedure that allows the system's invariance to scale modifications of the object is provided. An accurate analysis could be found in a previous work [27]. The most relevant conclusions of that work are pointed out in Section 14.4.1. Distortions due to both types of rotation, in plane and out of plane, are studied in detail in Sections 14.4.2 and 14.4.3, respectively. Recognition results are also provided.

There are some common steps in the analysis of the aforementioned distortions. Several images have been captured in a real environment. A stop sign is being used as a true target to be detected. Pictures containing a stop sign have been divided into two groups: the set of true target images that train the recognition system, and a different set of nontraining stop signs for testing the system's performance. Another set of images containing a different road sign (false object) is used to train the system and to test its discrimination capability. A training target, a nontraining target, and a false object are shown in Figs. 2a, 2b, and 2c, respectively. Nonlinear filtering for image processing is carried out by centering the training true targets on a zero background. Figure 3 displays an example of a reference target used to build nonlinear filters.

Each image is 128×128 pixels. They are normalized to have a maximum gray scale of unity and then zero padded to 256×256 pixels. The

(a)   (b)   (c)

Figure 2   Samples of images considered as (a) training target, (b) nontraining target, and (c) false object.

normalized images are Fourier transformed and kth-law nonlinearity is applied to them. The nonlinear correlation output is obtained by taking the inverse Fourier transform of the product between the nonlinearly modified spectra of both the input signal and the reference target. As a reference target, we will consider either a single sign (Fig. 3) to synthesize a nonlinear single filter or multiple views of a sign to synthesize a composite nonlinear filter. A nonlinear single filter is obtained by applying the nonlinear operator of Eq. (3) to the Fourier transform of the reference (stop sign shown in Fig. 3). Figure 4 describes the operations carried out to synthesize a composite nonlinear filter. The composite nonlinear filter displayed in this figure is obtained by applying Eq. (8) with $k = 0.1$ to 6 view of the target varying in scale.

Our previous analysis [27] has shown that nonlinearity of $k = 0.1$ improves correlation results in terms of peak sharpness, discrimination

Figure 3   Reference target.

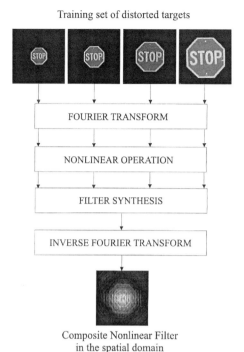

Figure 4   Diagram of the synthesis of a nonlinear composite filter.

capability and partial illumination invariance of the system. These results are in accordance with other results obtained for tolerance to target rotations [10,11]. Thus, value of $k = 0.1$ is selected for all the simulations.

14.4.1 Tolerance to Scale Variations

Scale invariance is required in a road sign recognition system to achieve detection of signs even if the acquisition system is in motion. The wider the range of tolerance to variations in scale, the better the capability of the recognition system to detect objects located at far distances. A nonlinear processor based on a bank of nonlinear single filters makes the detection of road signs varying in scale feasible [27]. In a previous work, performance of a nonlinear filter bank was compared to two types of composite nonlinear filter, the kth-law ECP-SDF and the kth-law minimum average correlation energy (MACE) filters [10,11]. Composite nonlinear filters did not satisfy the requirements of discrimination capability in a scale-invariant system.

However, results obtained by the bank of nonlinear single filters were successful and encouraging.

Images of a stop sign changing in size with a variable-scale increment were used to build the bank of nonlinear single filters. A nonuniform scale increment of the signs is equivalent to considering frames from a captured video sequence at equal time intervals, provided that the vehicle has constant speed; that is, the number of filters is larger for road signs located at far distances than for signs located at closer distances. Varying the increment of scale in the set of single filters improves the system's tolerance to scale distortion, especially for low-resolution objects [27].

As an example of the proposed recognition system's capabilities, we show results from the analysis of a video sequence. The reference target is a stop sign. Also, the registered video sequence contains a stop sign in a noisy and real background. In some frames, the sought sign appears distorted by drops of water due to the rain. In this video sequence, captured stop signs were at far distances from the acquisition system, so that they have low spatial resolution.

Only a few images extracted from the video sequence are shown in Fig. 5. This set of images contains the stop sign varying in scale as it is approaching to the on-board camera. The proposed nonlinear processor using a bank of nonlinear single filters is applied to each scene. Afterward, correlation outputs are obtained and they are displayed next to each corresponding image in Fig. 5. In all the cases, a sharp and high correlation peak located at the same position of the sign in the scene correctly identifies the stop sign.

14.4.2 Tolerance to In-Plane Rotations

In this subsection, we examine the system's performance with respect to in-plane rotation of the objects. Two different methods can provide a recognition system with tolerance to in-plane rotations: synthesis of nonlinear composite filters by using in-plane rotated versions of the reference or rotation of the input signal followed by its correlation with nonrotated versions of the target. In both cases, a digital algorithm to obtain rotated versions of the images is considered.

Synthesis of Composite Nonlinear Filters

First, each training stop sign centered in a zero background is digitally rotated in increments of $3°$ from $-9°$ to $9°$ around its vertical position. Rotated versions of the training sign are used to construct a nonlinear composite filter by applying Eq. (8) with $k = 0.1$. We synthesize a composite

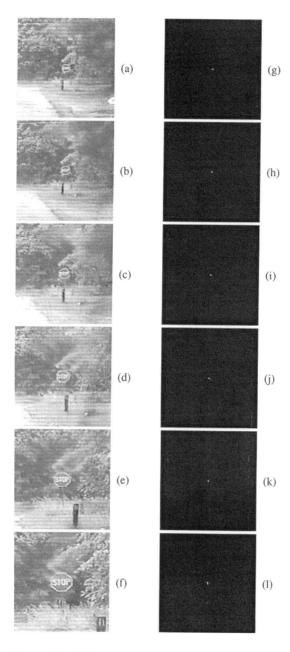

Figure 5 (a)–(f) Images extracted from a video sequence; and (g)–(l) the corresponding correlation outputs obtained after applying a scale-invariant processor.

filter for each training stop sign captured at a different distance from the camera in order to maintain a scale-invariant system.

In the learning process, recognition results are obtained for the entire training set, which is composed from true targets and false targets. The maximum PCE value is considered to classify signs as similar to the reference or to discriminate them from the sought sign. PCE output values above the threshold correspond to objects considered to be similar to the true target, whereas PCE values below the threshold imply the rejection of the object in the recognition process.

Figure 6a plots of the probability of error versus the threshold value in the recognition of training road signs. A nonlinear ECP-SDF filter is used. The solid line in the graph indicates the probability of error in the detection of true target and a dashed line plots the probability of error in the rejection of false targets, depending on the threshold value. A minimum threshold value can be established when the probability of misclassification of false targets reaches the value of zero.

Next, we test the performance of the recognition system by using a set of nontraining stop signs captured with an in-plane rotation angle of 4°. Results for nonlinear ECP-SDF filters are summarized in Fig. 6b. In this graph, the maximum PCE value achieved among the different output planes is plotted for all of the images. A horizontal solid line is plotted at the value of the chosen threshold level. In general, stop signs obtain PCE values above the established threshold level. PCE values for false signs are below the threshold. However, we note that a PCE value obtained for a nontraining stop sign is below the threshold. This implies that a false alarm appears in the recognition process. Furthermore, some of the correlation peaks for other testing images do not coincide with the actual target position in the scene. This is observed in Fig. 6c. This figure plots the position of the maximum correlation peak versus the actual position of the sign in the scene. The position is computed by using the distance of the center of the sign to the origin of the image [pixel (0, 0) located on the left top corner]. In this graph, the incorrect position of some correlation peaks is noted. They correspond also to false alarms.

Rotation of the Input Scene

In this second method, tolerance to in-plane rotations is achieved by rotating the input scene and it is compared to nonlinear single filters of the bank. This bank of filters contains information of the reference varying in scale (Section 14.4.1) to allow a scale-invariant recognition system.

A digital algorithm for rotating the signal is used to obtain in-plane rotated versions of the scene to be analyzed. We rotate the input scene

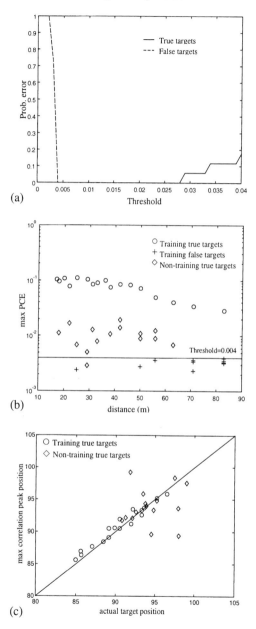

Figure 6 Recognition results for nonlinear ECP–SDF filters tolerant to in-plane rotations: (a) probability of error in the classification of training images; (b) classification of true targets and false targets with respect to the established threshold; (c) correlation peak position versus the actual target position in the scene.

from $-9°$ to $9°$ in increments of $3°$. Rotated versions of the scene are now correlated with filters belonging to the bank. The output of the recognition system is related to the best match between the rotated versions of the input signal and the reference targets. Thus, the output coincides with the maximum PCE value.

Improvement of recognition results can be seen in Fig. 7. In Fig. 7a, we observe that the interval with null probability of error increases, whereas in Figs. 7b and 7c, the successful recognition task is indicated: that is, the training and nontraining stop signs are correctly detected and located at the right position. They are also successfully distinguished from the other road signs used to test the discrimination capability of the system.

From comparison of the obtained results in Sections 14.4.1 and 14.4.2, we conclude that if some tolerance to in-plane rotation is required in the recognition system, better results are achieved by using a single nonlinear filter and rotating the input image, rather than designing nonlinear composite filter for rotation invariance.

14.4.3 Tolerance to Out-of-Plane Rotations

We focus our analysis on the system's tolerance to out-of-plane rotations. Due to the difficulty of generating digital out-of-plane rotated versions of the images, we implement them optically. Thus, stop signs are out-of-plane rotated from $-9°$ to $9°$ in increments of $3°$ during the acquisition process. They are used as training images. These signs, centered in a zero background, are used to construct nonlinear composite filters. A nonlinear composite filter [Eq. (8) with $k = 0.1$] is obtained for each distance between the sign and the acquisition system to maintain scale invariance.

A learning algorithm allows establishing the threshold value for the output of the recognition system. The value of the threshold is determined based on the results of Fig. 8a.

Several nontraining images slightly out-of-plane rotated are captured and used to test the system's tolerance to this type of distortion. Pictures are taken with a view angle of $4°$. A wide range of distances between the road sign and the acquisition camera are also considered to keep scale invariance.

Recognition results, once the established threshold level is applied, are shown in Fig. 8b. Recognition of stop signs is always achieved by a PCE value larger than the threshold level. They are also discriminated from other signs. Correlation peaks corresponding to stop signs are located at the same position as the sign in the scenes (Fig. 8c). Results contained in Figs. 8b and 8c show that the proposed recognition system is able to detect a partially out-of-plane rotated road sign at different distances from the acquisition system. This is due to the bank of composite filters that are being used.

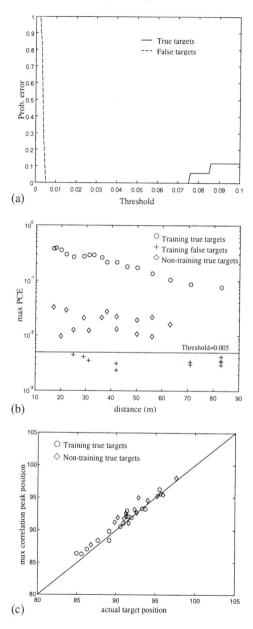

Figure 7 Recognition results for nonlinear single filters and in-plane rotation of the input scene: (a) probability of error in the classification of training images; (b) classification of true targets and false targets with respect to the established threshold value; (c) correlation peak position versus the actual target position in the scene.

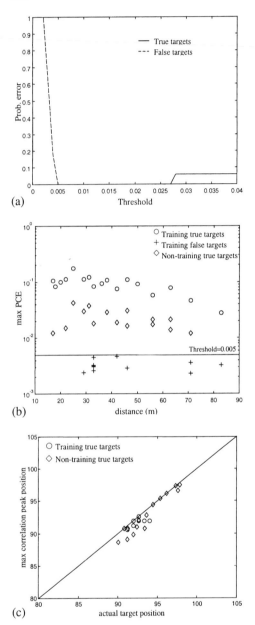

(a)

(b)

(c)

Figure 8   Recognition results for nonlinear ECP–SDF filters tolerant to out-of-plane rotations: (a) probability of error in the classification of training images; (b) classification of true targets and false targets with respect to the established threshold value; (c) correlation peak position versus the actual target position in the scene.

Information of out-of-plane rotation is included in nonlinear composite filters and allows detecting the sign even if it is slightly out-of-plane rotated or if it is captured with a different view angle by the acquisition system.

14.5 POSTPROCESSING OF CORRELATION RESULTS

Improvement of recognition results for the distortion-tolerant system can be achieved by postprocessing the obtained output results [28]. This means that not only is an image (or a frame of a video sequence) taken into account for the final recognition result but also a set of images (or frames) captured at different distances. Following this scheme, it is feasible to reduce the number of false alarms due to objects or background other than the sought sign.

Recognition results comes by considering correlation outputs from four different images. Figure 9 shows the improvement of the recognition results in comparison with the previous method without postprocessing. During the learning process, the null probability of error in the classification of training true targets and false signs is achieved in a wide range of threshold values.

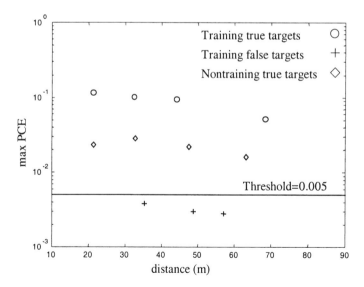

Figure 9   Recognition results by postprocessing of correlation outputs of Fig. 8b. Classification of true targets and false targets are with respect to the established threshold value.

Figure 9 shows PCE values obtained for the entire set of images. Nontraining true targets are always correctly classified as similar to the reference. Differences of PCE values are larger when four images are postprocessed together than when PCE value is obtained from a single image (Fig. 8b).

14.6 ANALYSIS OF REAL IMAGES IN NONOVERLAPPING BACKGROUND

The separate analyses of different distortions have shown an encouraging performance of the recognition system as a distortion-tolerant processor. We now apply the distortion-tolerant system to new captured images. They are selected as samples where it is difficult to recognize the road sign due to the amount of involved distortions. Selected images include stop signs modified by several distortions. They are captured under varying illuminations due to shadows or different weather conditions, and in some cases, the sign to be detected has been vandalized or appears partially occluded. In all the cases, a real cluttered background surrounds stop signs.

The designed recognition system is based on a nonlinear processor that uses a bank of composite nonlinear filters. The bank of filters serves to achieve scale invariance in a wide range of distances from the sign to the acquisition system. Composite nonlinear filters, in particular kth-law ECP-SDF filters, provide tolerance to out-of-plane rotation of targets. Finally, rotation of the input signal allows tolerance to in-plane rotations. A certain degree of tolerance to illumination fluctuations is achieved as a consequence of using a nonlinear processor with parameter k close to zero ($k = 0.1$).

Figure 10a corresponds to an analyzed scene which includes two stop signs to be detected. These signs are located at both sides of the road and they have different illuminations. The stop sign on the left has a low average energy due to a shadow that completely covers it. This sign is partially in-plane and out-of-plane rotated. The stop sign on the right, however, has a nonuniform illumination due to shadows caused by the leaves and it has been vandalized. This sign is tilted, so that tolerance to in-plane rotation is needed to detect it correctly. It is also in-plane and out-of-plane rotated. We observe that the background of the picture is quite cluttered and there are areas with larger energy than the energy of stop signs. Figure 10b shows the output correlation plane where two high intensity peaks appear and coincide with the position of the two true targets. Accuracy in the location of the peaks is easily observed in the 2D representation of the correlation plane (Fig. 10c). We remark that the recognition of both stop signs is achieved

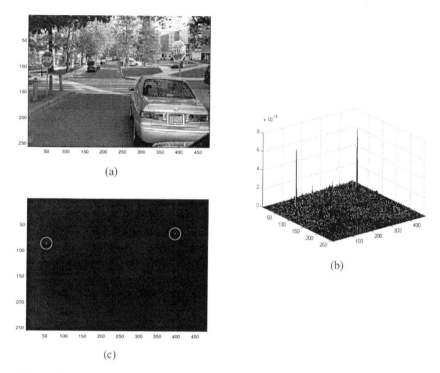

(a)

(b)

(c)

Figure 10   Recognition results for a distortion-tolerant system: (a) input scene; (b) 3D representation of the output plane; (c) 2D representation of the output plane.

under different illumination conditions of the signs. This is mainly due to the nonlinearity applied in the nonlinear process.

A second sample consists of a stop sign strongly faded (Fig. 11a). The sign appears in a cluttered background and with an inverse contrast. However, a high and sharp correlation peak appears in the actual position of the road sign (Figs. 11b and 11c). This implies a satisfactory recognition of the sought sign.

The last example corresponds to the analysis of the image displayed in Fig. 12a. The stop sign that is contained in the scene appears partially occluded by a tree. The detection and location of this sign are also satisfactory, as can be seen from the 3D output graph of Fig. 12b or from the 2D representation in Fig. 12c. A high and sharp peak is obtained in a low-output noise floor.

From the results, we conclude that the proposed recognition system is able to detect and locate road sign in real background images. The detection

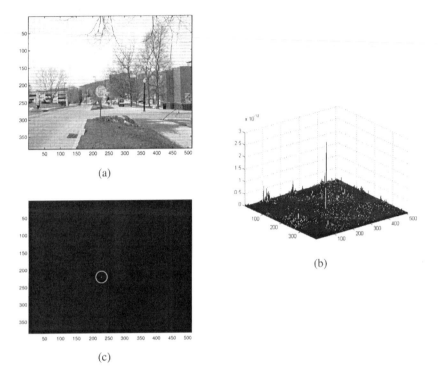

Figure 11 Recognition results for a distortion-tolerant system: (a) input scene; (b) 3D representation of the output plane; (c) 2D representation of the output plane.

is successfully achieved even when the road sign is varying in scale, slightly rotated, illuminated under different conditions, faded, or partially occluded.

14.7 SUMMARY

A road sign recognition system has been proposed based on nonlinear processors. Analysis of different filtering methods allows us to select the best techniques to overcome a variety of distortions. The most frequent distortions when dealing with road sign detection are scale variations, in-plane and out-of-plane rotations, and illumination variations of the targets.

The entire processor performs several correlations between different input scenes and a set of reference targets. Multiple correlation results are then processed to give a single recognition output. A learning process is

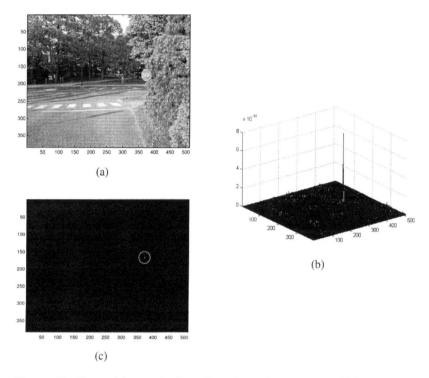

Figure 12 Recognition results for a distortion-tolerant system: (a) input scene; (b) 3D representation of the output plane; (c) 2D representation of the output plane.

carried out to establish a threshold value, which determines whether or not any object contained in an input scene is similar to the target.

Scale invariance is provided to the recognition system by means of a bank of nonlinear filters. A filter bank recognition system shows a better performance than nonlinear composite filters. Images of a true target captured from different distances constitute the set of filters in the bank. A nonuniform increment of variation in scale is established to properly recognize signs located at far distances from the acquisition system. Apart from locating a true sign, this method allows to approximately determine the distance between the acquisition system and the road sign.

In-plane rotation invariance is achieved by rotating the input scene. Recognition results obtained by this method are compared to results obtained for nonlinear composite filters. Composite filters are constructed by using digital rotated versions of the reference target. In-plane rotation of the input scene allows better detection results than composite filters.

Moreover, in the design of composite filters, the maximum number of images included in a composite filter is limited, whereas the range of the input scene rotation can be determined based on the application.

Using nonlinear composite filters rather than using individual filters in the filter bank can satisfy tolerance requirements for out-of-plane rotation of the targets. In particular, kth-law ECP–SDF filters are used.

The entire recognition system has been tested in real static images as well as in a video sequence. Scenes were captured in real environments, with cluttered backgrounds and contained many distortions simultaneously. Recognition results for various images show that the proposed recognition system is able to properly detect a given road sign even if it is varying in scale, slightly tilted, or viewed under different angles. In addition, the system is robust to changes in illumination due to shadows or weather conditions. It is also able to locate a faded or vandalized sign along with partially occluded road signs. Obviously, the processor can be designed for different varieties of road signs in noisy background scenes.

This work has also pointed out that postprocessing of correlation outputs allows one to significantly improve recognition of distorted road signs.

ACKNOWLEDGMENT

E. Pérez wishes to thank the Comisión Interministerial de Ciencia y Tecnología (CICYT), Spain for financial support (project DPI2000-0991).

REFERENCES

1. CF Hester, D Casasent. Multivariant technique for multiclass pattern recognition. Appl Opt 19(11):1758–1761, 1980.
2. A Mahalabonis, BVK Vijaya Kumar, D Casasent. Minimum average correlation energy filters. Appl Opt 26(17):3633-3640, 1987.
3. B Javidi, JL Horner, A Fazlollahi, J Li. Illumination-invariant pattern recognition with a binary nonlinear joint transform correlator using spatial frequency dependent threshold function. Proc SPIE 2026:100–106, 1993.
4. Ph Refregier, B Javidi, G Zhang. Minimum mean-square-error filter for pattern recognition with spatially disjoint signal and scene noise. Opt Lett 18(17):1453–1455, 1993.
5. Ph Refregier, V Laude, B Javidi. Nonlinear joint-transform correlation: An optimal solution for adaptive image discrimination and input noise robustness. Opt Lett 19(6):405–407, 1994.
6. B Javidi, JL Horner, eds. Real-Time Optical Information Processing. New York: Academic Press.

7. B Javidi, J Wang. Optimum distortion-invariant filter for detecting a noisy distorted target in nonoverlapping background noise. J Opt Soc Am A 12(12):2604-2614, 1995.

8. B Javidi, J Wang. Distortion tolerant composite filter to detect a target in nonoverlapping background noise. Opt Lett 20(4):401–403, 1995.

9. E Silvera, T Kotzer, J Shamir. Adaptive pattern recognition with rotation, scale, and shift invariance. Appl Opt 34(11):1891–1900, 1995.

10. B Javidi, D Painchaud. Distortion-invariant pattern recognition with Fourier-plane nonlinear filters. Appl Opt 35(2):318–331, 1996.

11. B Javidi, W Wang, G Zhang. Composite Fourier-plane nonlinear filter for distortion-invariant pattern recognition. Opt Eng 36(10):2690–2696, 1997.

12. B Noharet, R Hey, H Sjoberg. Comparison of the performance of correlation filters on images with real-world non-overlapping noise: influence of the target size, illumination and in-plane rotation distortion. Proc SPIE 3386:212–221, 1998.

13. B Javidi, ed. Smart Imaging Systems. Bellingham, WA: SPIE Press, 2001, Vol. PM91.

14. L Guibert, G Keryer, A Sevel, M Attia, H S MacKenzie, P Pellat-Finet, J L de Bougrenet de la Tocnaye. On-board optical joint transform correlator for real-time road sign recognition. Opt Eng 34(1):135–143, 1995.

15. TD Wilkinson, Y Perillot, RJ Mears, JL de Bougrenet de la Tocnaye. Scale-invariant optical correlators using ferroelectric liquid-crystal spatial light modulators. Appl Opt 34(11):1885–1890, 1995.

16. M Taniguchi, K Matsuoka. Detection of road signs by using a multiple correlator and partial invariant filters. Proc SPIE 3490:178–181, 1998.

17. M Taniguchi, Y Mokuno, K Kintaba, K Matsuoka. Correlation filter design for classification of road sign by multiple optical correlators. Proc SPIE 3804:140–147, 1999.

18. B Javidi. Nonlinear joint power spectrum based optical correlation. Appl Opt 28(12):2358–2367, 1989.

19. JW Goodman. Introduction to Fourier Optics. 2nd ed. New York: McGraw-Hill, 1996.

20. JD Gaskill. Linear Systems, Fourier Transforms, and Optics. New York: Wiley, 1987.

21. J Campos, F Turon, LP Yaroslavsky, MJ Yzuel. Some filters for reliable recognition and localization of objects by optical correlators: a comparison. Int J Opt Comput 2:341–365, 1991.

22. OK Ersoy, M Zeng. Nonlinear matched filtering. J Opt Soc Am A 31(6):636–648, 1989.

23. B Javidi. Generalization of the linear matched concept to nonlinear matched filters. Appl Opt 29(8):1215–1224, 1990.

24. Ph Réfrégier. Filter design for optical pattern recognition: Multicriteria optimization approach. Opt Lett 15(15):854–856, 1990.

25. Ph Réfrégier. Optimal trade-off filters for noise robustness, sharpness of the correlation peak, and Horner Efficiency. Opt Lett 16(11):829–831, 1991.

26. JL Horner. Metrics for assessing pattern-recognition performance. Appl Opt
     31:165–166, 1992.
27. E Pérez, B Javidi. Image processing for intelligent transportation systems:
     Application to road sign recognition. In: Smart Imaging Systems.
     Bellingham, WA: SPIE Press, 2001, Vol. PM91.
28. B Javidi, Q Tang. Optical implementation of neural networks for face recogni-
     tion by the use of nonlinear joint transform correlators. Appl Opt 34(20):3950–
     3962, 1995.

15

Development of Pattern Recognition Tools Based on the Automatic Spatial Frequency Selection Algorithm in View of Actual Applications

Christophe Minetti and Frank Dubois
Université Libre de Bruxelles, Bruxelles, Belgium

15.1 INTRODUCTION

Ever since the first use of the optical correlator for implementing matched spatial filtering [1], many studies have been made to develop new algorithms that increase the performances of the correlation filters. Correlation filters are generally computed on the basis of reference images containing the objects to be recognized or rejected. The set of reference images is chosen in order to cover the different occurrences of the objects in the best way. However, in practical applications, input images may differ strongly from their reference model. In this case, the maximum detected intensity in the correlation plane decreases, which may lead to false detections. In order to have an efficient recognition system robust to input distortions, one must be able to realize filters insensitive to pattern distortions, illumination problems, and additive and substitutive noise appearing in the input images.

When the reference image set can be modeled explicitly by mathematical tools (distortions such as scaling, rotation, additive Gaussian noise, etc.), specific techniques can be used to give invariant correlation outputs [2–10]. Nevertheless, in most of the applications, the distortion model (such as out-of-plane rotations, non-uniform illuminations changes, missing parts of the objects, etc.) cannot be described by explicit mathematical models. In this case, several composite filters based on a set of reference images have been proposed to reduce the sensitivity of the correlation intensity when distortions appear between the reference and the input images [11–16]. The Synthetic Discriminant Functions (SDF) [12] played a central role in those developments: The correlation filter is computed as a linear combination of the reference images to give a priori fixed central correlation amplitudes. In its simplest form, pattern recognition reliability with SDF filters is very sensitive and induces, in the correlation plane, large sidelobe energies which causes false detections. To overcome this sensitivity, several modifications of the SDF algorithm have been introduced [17,18].

Powerful modifications of the SDF algorithm have been proposed by adding an additional constraint to the central correlation constraints that forces the filter to be optimized with respect to a cost function. The first attempt in this was the Minimum Variance Synthetic Discriminant Functions (MVSDF) [19] algorithm, where the fluctuations of the central correlation amplitude are minimized when the input images are corrupted by additional white noise.

Another development based on the SDF algorithm has been proposed by Mahalanobis et al. with the Minimum Average Correlation Energy filters (MACE) [20]. The MACE filters also give predefined responses when correlated with the reference images and are, furthermore, designed to minimize, on average, the correlation energy. This last constraint performs a drastic high-spatial-frequency enhancement that reduces the sidelobes energy substantially and brings about narrow correlation peaks. At the same time, MACE filters are very sensitive to distortions in the input images. To overcome this limitation, Mahalanobis and Casasent combined the MACE algorithm with a training set selection to optimize the choice of the reference images with respect to the considered application [21,22]. Different approaches have been proposed to reduce further the sensitivity of the correlation intensity when imput images differ from the reference set [23–26].

Réfrégier et al. extended the concept of the MVSDF and the MACE algorithms by introducing the Optimal Trade-off Filters (OTF) [26–28], where the filter is forced to minimize a constraint that is a linear combination of constraints like the MACE and the MVSDF constraints. This approach introduces in the filter design a degree of freedom that allows one to match the implementation of specific applications in an efficient way.

Javidi et al. extended successfully the concept of linear matched filters to nonlinear matched filters [29–33] by introducing a nonlinear transfer function in the Fourier plane in order to threshold the linear matched filter function. It has been shown that this new kind of filter could increase the capabilities of correlation filters by reducing the sidelobes energies in the correlation plane and allows easier detections.

In this chapter, we summarize the approach that has been developed by our research group. It is based on the Automatic Spatial Frequency Selection algorithm (ASFS) and our approach is essentially led by implementation of actual applications. In the following, an application of automatic counting of biological cells in division is described. It is a true application with important impact for the medical world. With this applied point of view, it is mandatory to look at global solution with respect to the application constraints. This is the reason why we kept the ASFS approach and that we adapted it to the various constraints that we encountered. This is also the reason why we disregarded very nice approaches of other authors that are very powerful in solving specific problems but that are difficult to adapt to the multiple constraints of the applications. In this chapter, we also wanted to show that we started from academic problems to go toward the implementation of a true application.

We first present the ASFS algorithm that allows one to include, during the filter computation, arbitrary distortions of the reference images. Further improvements of the ASFS algorithm are proposed by cascading two correlation processes with a nonlinear stage at the output of the first correlation stage. It is shown that this correlation architecture increases significantly the performances of the recognition and enlarges the field of applications.

As a solution to substitutive noise problems, we propose two different methods that allows one to reduce the sensitivity of the recognition. The first method includes additional constraints during the filter computation, which reduce the influence of the surrounding background on the central correlation amplitude. The second method is a postcomputation method based on the application of a mask to the correlation filter in order to reduce the spatial extensions of numerical correlation filters. The sensitivity of the correlation to illumination problems is investigated and a method based on the statistical properties of the correlation histograms is proposed. It consists of a postprocess that allows one to become insensitive to illumination changes in the input images.

To conclude this chapter, we present the results obtained with unsupervised learning systems. In this schema, a postsegmentation system is placed after the correlation system, in order to extract detected objects from their background. These segmented objects are used to compute new correlation filters. In this configuration, the system is capable of tracking objects whose shape may vary substantially in time. This work currently in progress, enlarges the field of applications of recognition systems.

15.2 AUTOMATIC SPATIAL FREQUENCY SELECTION ALGORITHM

15.2.1 Introduction

We proposed a method, based on the MACE concept that allows the user to include, in the reference set, arbitrary distortions depending on the application. This process performs an automatic selection of the spatial frequency content of the filter with respect to the possible distortions of the reference images. The main difference between this algorithm and the other ones is that, once the pattern recognition problem is established, the frequency selection process (depending on the problem) is directly integrated into the algorithm. This method is called the Automatic Spatial Frequency Selection algorithm (ASFS) [34].

We present in the following section the theoretical background and give results of numerical simulations of the correlation intensities. Those intensities are compared to the MACE algorithm. To demonstrate the flexibility of the ASFS method, we consider a recognition problem invariant to in-plane rotation.

15.2.2 The ASFS Algorithm as Basic Concept for the Development of Robust Recognition Process

We consider filters and images as discrete function of $n \times n$ spatial variables (s, t). In the following, we will use lowercase letters for functions and variables in the spatial domain and the corresponding uppercase letters in the Fourier domain. Let us consider a multiclass problem with K classes. The reference images x_{kq} are arranged in the K different classes. Inside each class k $(k = 1, \ldots, K)$, each image x_{kq} $[q = 1, \ldots, L(k)]$ has to be recognized in the same way. For the sake of simplicity, we will consider one filter per class. The K correlation filters h_m have to give fixed responses r_{kqm} (which is the central correlation amplitude) when they are correlated with the reference images x_{kq}:

$$r_{kqm} = \sum_{s,t=0}^{n-1} h_m(s, t) x_{kq}(s, t) \tag{1}$$

Using Parseval's theorem, Eq. (1) can be rewritten as

$$r_{kqm} = \sum_{U,y=0}^{n-1} H_m(U, V) X_{kq}^*(U, V) \tag{2}$$

where $H_m(U, V)$ and $X_{kq}(U, V)$ are the discrete Fourier transforms of h_m (s, t) and $x_{kq}(s, t)$, respectively. We can write the filter and the images vectorially by lexicography scanning. Equations (1) and (2) become

$$r_{kqm} = x_{kq} \cdot h_m \tag{3}$$

$$r_{kqm} = X_{kq} \cdot H_m \tag{4}$$

where bold letters are used to indicate vectors.

In practical case of interest, the number of reference images (typically *20* or *30*) is small in comparison to the pixels number of the reference images $x_k(s, t)$ (typically 64×64 or 128×128). Therefore, the system of Eqs. (4) is underdetermined and it is possible to add additional constraints. The goal of those additional constraints is to achieve well-shaped correlation peaks, allowing easy target recognition while keeping a high robustness of the recognition process. Indeed, the system of Eqs. (4) only defines the central correlation amplitude and does not consider the other points of the correlation plane. To involve a model of the possible distortions, we associate a set C_{kq} of distorted versions $y_{kqi}(s, t)$ $(i = 1, \ldots, I_k)$ to each reference image $x_{kq}(s, t)$. The distorted images have to be chosen during the definition phase of the application and have to be identified by the recognition process as the image x_{kq}. The distortions depend on the application concerned. For example, if the filter must be robust to noise, the distorted images y_{kqi} will be noisy versions of the references x_{kq}. If the system must be invariant to inplane rotation, the y_{kqi} will be slightly rotated versions of the x_{kq}. This allows a great flexibility of the possible distortions supported by the algorithm. Instead of the minimization of the average correlation energy (MACE), the filters are constrained to minimize the following expression:

$$F = \sum_{U,V=0}^{n-1} \sum_{kqi} \left| H_m(U, V)\{X_{kq}(U, V) - Y_{kqi}(U, V)\}^* \right|^2 \tag{5}$$

When a reference component $X_{kq}(U, V)$ is close to the $Y_{kqi}(U, V)$ component, the difference in Eq. (5) is small and the minimization constraint of F on the $H_m(U, V)$ amplitude is weak. On the contrary, when the difference in Eq. (5) becomes larger, the impact of the minimization operated on the H_m (U, V) component is increased, leading to a decrease of its amplitude. As the distortion image is a small perturbation of the reference image, the significant difference between $X_{kq}(U, V)$ and $Y_{kqi}(U, V)$ often occurs for high spatial frequencies. Therefore, the minimization of the cost function F will act, in comparison with the MACE, as a low-spatial-frequency filtering process. Due to the modulation of the spatial frequencies operated by the minimization of Eq. (5), F is called the *ASFS cost function*.

It is known that the information carried out by the low spatial frequencies decreases the discrimination of the filters. It is then necessary to remove them before computing the filter. We used a simple frequency selection criteria based on the statistical correlation coefficients between the components of the reference images. We set

$$X_{kq}(U, V) = 0$$

if

$$\frac{1}{2LK} \sum_{k,k',q,q'} \frac{X_{kq}(U, V)x^*_{k'q'}(U, V)}{|K_{kq}(U, V)X_{k'q'}(U, V)|} \geq \eta \tag{6}$$

otherwise $X_{kq}(U, V)$ is kept. η is a parameter adjusting the strength of the filtering. Let us consider the X matrix by ordering side by side the T columns vectors X_{kq}. By keeping the same ordering sequence as for the matrix X, we form the column vector r_m with the r_{kqm} responses. We define de diagonal matrix E as follows:

$$E(Z, Z) = \sum_{kqi} |X_{kq}(Z) - Y_{kqi}(Z)|^2 \tag{7}$$

where $X_{kq}(Z)$ and $Y_{kqi}(Z)$ are the Z components of the vectors X_{kq} and Y_{kqi}, respectively. Using vectorial notation, the ASFS cost function is written

$$F_m = H_m^+ \cdot E \cdot H_m \tag{8}$$

where the superscript plus denotes the transpose conjugate operation. The solution of the filter H_m that verifies the system of Eqs. (4) and that minimizes the ASFS cost function is obtained by using the Lagrange multiplicators [20] and is given by

$$H_m = E^{-1}X(X^+E^{-1}X)^{-1}\vec{r}_m \tag{9}$$

Let us consider two particular cases of interest. If no distortion model is included in the computation of the filter [$y_{kqi}(s, t) = 0$], the cost function F_m becomes

$$F_m = \sum_{kq} \sum_{U,V} |H_m(U, V)X_{kq}(U, V)|^2 \tag{10}$$

which is the energy in the correlation plane. In this particular case, the solution of the filter is reduced to the MACE one [20]. When input images are made of reference images with additive Gaussian white noise of zero mean, the $Y_{kqi}(U, V)$ are expressed by

$$Y_{kqi}(U, V) = X_{kq}(U, V) + N_i(U, V) \tag{11}$$

where $N_i(U, V)$ is the Fourier transform of particular samples of the zero-mean Gaussian white noise in the input plane. The cost function is then expressed by

$$F = \sum_{U,V} \sum_{kqi} |H_m(U, V)N_i^*(UY, V)|^2 \tag{12}$$

When the number of random patterns is large enough, the minimization of Eq. (12) leads to minimizing the variance of the filter responses with respect to the noise in the input plane. This solution is then equivalent to the MVSDF's [19]. These two particular cases show that both the MACE and the MVSDF algorithm are particular cases of the ASFS algorithm and demonstrates the self-adaptive flexibility of the ASFS filters in practical applications.

15.2.3 Evaluation of the ASFS Algorithm on a Test Problem

The ASFS algorithm was first tested on a two class problem where two flat cutting tools have to be recognized regardless of their in-plane rotation (see Fig. 1).

Reference images of 64×64 pixels were taken with an incremental angle of $6°$ for each of the two objects. Because of the symmetries of the objects, 15 images of object 1 [$L(1) = 15$] and 18 images of object 2 [$L(2) = 18$] were involved in the reference set. As the in-plane rotation invariance is required, the distorted images $y_{kqi}(s, t)$ are chosen as intermediate rotations of the reference images. One distorted image $y_{kqi}(s, t)$, is associated to the $x_{kq}(s, t)$ reference image: $y_{kq}(s, t) = x_{kq+1}(s, t)$ (with $x_{kL(k)+1} = x_{kl}$).

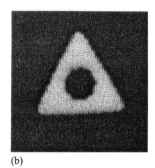

(a)   (b)

Figure 1 Images of the objects of each class: (a) image of object 1; (b) image of object 2.

With the help of the reference and distorted image sets, two correlation filters were computed to recognize each object according to Table 1.

The lower spatial frequencies were removed by using criterion (6) with a value of $\eta = 0.9$. The identification of the filter's response is achieved by measuring the maximum intensity compared to a threshold value. Examples of correlation intensities of response 1 and response 0 with filter h_1 are presented in Fig. 2. The intensities of response 1 are sufficiently high compared to those of response 0 to allow the user to place a decision threshold that recognizes the object without any ambiguity.

To test the robustness of the filters with respect to the different rotations of the input images, a set of input images with an incremental rotation angle of $3°$ was considered (including reference images, distorted images, and images not involved in the reference set). The maximum correlation intensities as a function of the rotation angle are shown in Fig. 3.

We observe in Fig. 3 that the maximum intensities detected closely follow the imposed responses (during the computation of the filter, a central correlation intensity of 127 is imposed for the images that have to produce responses 1), even for images not involved in the reference set. On the same graph, results obtained with the MACE algorithm and computed with the same reference set are provided. The peak intensities of response 1 corresponding to intermediate rotations are affected, in the case of the MACE, by larger losses of intensity that are due to the high sensitivity to input distortions. Although the intensity levels of response 0 for the MACE algorithm are not so much lower. In order to completely characterize the performances of the filters, four performance criteria have been considered: discriminability, peak sharpness, signal-to-noise ratio, and Horner efficiency (for a potential optical implementation [35–36]). A measure of the discriminability is given by the G ratio derived from the Fisher ratio:

$$G = \frac{I_1 - I_0}{SD_1 + SD_0} \qquad (13)$$

Table 1 Responses of the Filter with Respect to the Object Images

	Filter h_1	Filter h_2
Image of object 1	1	0
Image of object 2	0	1

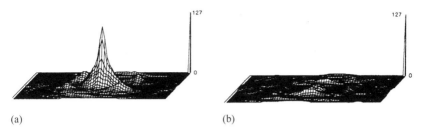

(a)   (b)

Figure 2 Three-dimensional representation of the correlation intensities obtained with filter h_1: (a) example of correlation intensity of response 1; (b) example of correlation intensity of response 0.

where I_1 and I_0 are respectively the mean maximum intensity of response 1 and response 0 correlation intensities and SD_1 and SD_0 are their corresponding standard deviations. The peak sharpness is estimated by using the peak-to-correlation energy (PCE) defined as

$$PCE = \frac{|y_1(0,0)|^2}{\sum_{s,t=0}^{n-1} |y_1(s,t)|^2}$$ (14)

where $y_1(s,t)$ is the correlation amplitude of a response 1.

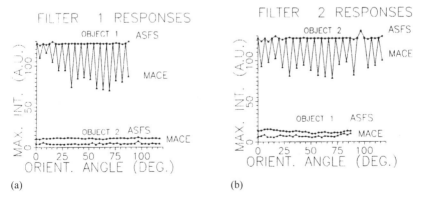

(a)   (b)

Figure 3 Maximum correlation intensities as a function of the rotation angle: (a) maximum correlation intensities obtained with filter h_1 for different rotation angles; (b) maximum correlation intensities obtained with filter h_2 for different rotation angles.

For the estimation of the signal-to-noise ratio (SNR), 30 uniformly distributed random patterns were sequentially added to input images that had to give a response 1 correlation peak. The SNR is then computed in the following way:

$$\text{SNR} = \frac{|E[y_1(0,0)]|^2}{\text{Var}[y_1(0,0)]} \tag{15}$$

where $E[\,]$ denotes the expected value.

To estimate the optical efficiency of the filters, we use the Horner efficiency defined as

$$\text{HE} = \frac{|y_1(0,0)|^2}{\sum\limits_{U,V} |X(U,V)|^2} \tag{16}$$

where the $X(U,V)$ component are kept for the lowest spatial frequencies. The values of these four quantities are reported in Table 2 for the ASFS and the MACE algorithm. The results show a better discrimination with the ASFS algorithm. Indeed, I_1 is increased with the ASFS. Even if I_0 is lower with the MACE, the variance are strongly decreased with the ASFS filters, leading to better discrimination ratios. Better PCE values are obtained with the MACE algorithm. Indeed, MACE filters provide narrow peaks leading to very high peak-to-correlation energy. As it can be seen from the discrimination ratio, the peak is very sensitive to input distortion with the MACE algorithm. The values of the SNR obtained with the ASFS algorithm show a strong resistance of the filters to additive noise in the input images. The low values of the SNR obtained with the MACE filters can be

Table 2 Summary of the Results Obtained in the Correlation Simulation with the ASFS and the MACE Filters

	I_1	I_0	SD_1	SD_0	G	PCE	SNR	HE
Filter 1								
ASFS	126.94	11.099	0.835	0.493	87.22	0.0087	160.59	8.87^{E-4}
MACE	107.3	4.50	21.93	0.762	4.53	0.109	1.148	7.74^{E-6}
Filter 2								
ASFS	127.01	12.998	1.880	1.488	33.85	0.0197	56.50	1.47^{E-3}
MACE	110.2	7.51	18.99	1.295	5.06	0.241	0.339	2.92^{E-5}

I_1 and I_0 are the mean maximum intensities of response 1 and response 0; SD_1 and SD_0 are the corresponding variances. G is the discrimination ratio, PCE is the peak sharpness, SNR is the signal-to-noise ratio, and HE is the Horner efficiency.

expected because the high-spatial-frequency content of the filters make them sensitive to additional noise. The Horner efficiency value is also increased with respect to the MACE filters, expecting better optical implementations.

15.2.4 Remarks on the ASFS Algorithm

Simulation results have demonstrated a better discrimination ratio, SNR, and Horner efficiency. The distortion model included by samples during the computation of the filter gives a great flexibility. The reference images are directly involved in the filter computation, whereas distorted images optimize the spatial frequency weight. Note that the set of distorted images can be a subset of the reference images. Therefore, the ASFS algorithm provides an efficient self-adaptive capability to select the spatial frequencies that are significant for a pattern recognition problem. The results obtained with the linear ASFS correlation system encourages us to use the ASFS methodology as the central concept of our further developments. In the following section, we describe a nonlinear cascaded system that takes benefit from the ASFS algorithm.

15.3 NONLINEAR CASCADED CORRELATION PROCESS WITH THE ASFS ALGORITHM

15.3.1 Introduction

The scope of this section is to show, by theoretical developments and by numerical simulations, that one can obtain improvements of the ASFS algorithm by cascading two correlation processes with a nonlinear transmission stage in the first correlation plane. As it will be shown in the following, the ASFS cost function of the second correlation stage is improved when the intraclass and interclass similarities of the reference images are respectively increased and decreased. The first filter is then calculated to achieve this goal. Because a nonlinear stage is involved in the process, no analytical solution can be found to compute the two filters and it is necessary to iteratively compute them by decreasing a cost function. Theoretical developments are presented in Section 15.3.2. The optimization of the ASFS cost function and the numerical simulations are presented in Section 15.3.3 and 15.3.4. Concluding remarks are reported in Section 15.3.5.

15.3.2 The Nonlinear Cascaded Correlation Architecture

The system we describe and analyze here consists of a first correlation stage, a nonlinear transmission stage, and a second correlation stage (see Fig. 4).

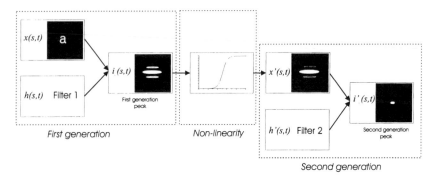

Figure 4 Global scheme of the nonlinear cascaded correlation architecture.

To introduce the notations, let us follow how an input image is processed. The input image $x(s, t)$ is first correlated with the first generation filter $h(s, t)$ to obtain the correlation intensity $i_c(s, t)$. Each point of the correlation product $i_c(s, t)$ is processed by a nonlinear stage to give the input image $i_c'(s, t)$ for the correlation with the second filter:

$$i_c'(s, t) = S(i_c(s, t)) \tag{17}$$

The function $S(\ )$, which defines the nonlinear transmission stage, is the sigmoid activation function commonly used in neural networks:

$$S(x, \theta_0, \theta_i) = \frac{1}{1 + \exp[-(x - \theta_0)/\theta_i]} \tag{18}$$

where the parameters θ_0 and θ_i allow one to adjust the slope of the activation function. By analogy with neural networks, its role is to perform a smooth threshold between the high and the low intensities.

The processed intensity $i_c'(s, t)$ is the input image for the second correlation process. The image $i_c'(s, t)$ is then correlated with the second generation filter $h'(s, t)$ to give the correlation intensity $i_c''(s, t)$, which is compared to a reference threshold value T to determine if an event is recognized in the input image $x(s, t)$. The filters and the parameters of the activation function have to be set to match the application requirements.

As in the case of the one-step ASFS algorithm, we consider a set of reference images $x_{kq}(s, t)$ and distorted images $y_{kqi}(s, t)$. The computations of the two filters are interdependent. Assuming that the first filter $h(s, t)$ has been computed, we consider first the computation of the second filter $h'(s, t)$. After the first correlation stage with the reference and distorted images x_{kq} and y_{kqi} and the nonlinear stage, we obtain the images $x_{kq}'(s, t)$ and $y_{kqi}'(s, t)$.

Those images are used as reference and distorted images for the computation of the second generation filter $h'(s, t)$ according to the ASFS algorithm.

Let us analyze the analytical computation of the second-generation filter h'. As in previous section, uppercase letters denote the Fourier transform of the corresponding quantity in lowercase letters. It is then required that

$$X'^+ H' = r \tag{19}$$

The ASFS algorithm further requires the following quantity to be minimized:

$$F' = H'^+ E' H' \tag{20}$$

where X' and E' are defined, for the second correlation stage, as X and E for the first stage (see the previous section) with the image X'_{kq} and Y'_{kqi}. As the processes between the $x'_{kq}(s, t)$ images and the output of the system is a linear correlation, the second correlation filter is computed as the linear ASFS filter of the previous section. The solution of the second-generation filter H' is given by

$$H' = E'^{-1} X'(X'^+ E'^{-1} X')^{-1}r \tag{21}$$

Let us now interpret the role of the first correlation filter. Its iterative computation is described in the next section. By developing the expression of the cost function F', we obtain, combining Eqs. (20) and (21):

$$F' = r^t(X'^+ E'^{-1} X')^{-1}r \tag{22}$$

Equation (22) is the measure of the algorithm to adapt itself to the ASFS constraint (minimize the difference between reference and distorted images). In a one-stage system, the cost function is completely determined by the reference and distorted images. However, with a two-stage system, the first stage and the nonlinearity can be adapted to obtain reference and distorted images (X'_{kq} and Y'_{kqi}) that further reduce the value of the cost function. Therefore, the first filter H must be chosen to reduce F'.

It is valuable to see how the first correlation stage and the nonlinearity can improve the value of the cost function F'. For this purpose, we define the W' matrix as follows:

$$W' = E'^{-1/2} X' \tag{23}$$

where $E'^{-1/2}$ is the diagonal matrix defined by

$$E'^{-1/2}(Z, Z) = \frac{1}{\left[\displaystyle\sum_{kqi} |X'_{kq}(Z, Z) - Y'_{kqi}(Z, Z)|^2 \right]^{1/2}} \tag{24}$$

Using Eq. (23), one can easily rewrite Eq. (22) as

$$F' = r^t(W'^+ W')^{-1}r \tag{25}$$

Equation (25) shows that the cost function depends on the inverse of the covariance matrix of the secondary reference images (reference images of the second-generation filter) filtered by the matrix $E'^{1/2}$. Because the matrix $(W'^+ W')^{-1}$ is symmetric and as the cost function F' is defined positive, it has T orthogonal eigenvectors $p_j (j = 0, \ldots, T - 1)$ of nonzero positive eigenvalues λ_j such that $(W'^+ W')^{-1}$ may be rewritten as

$$(W'^+ W')^{-1} = \sum_j \lambda_j p_j p_j^t \tag{26}$$

As a result of Eq. (26) and because the response vector r is set a priori, we see that Eq. (25) is minimized if the eigenvector of the smallest eigenvalue is identical to the response vector r.

The first correlation stage and the nonlinear stage have to transform the reference images in such a way that this condition is fulfilled as far as possible. As the interpretation of the covariance matrix is clearer than its inverse one, we note that it can be decomposed in the same way as was Eq. (26), with the eigenvalues λ_j^{-1}.

The cost function F' [Eq. (25)] will be minimal if the vector r is identical to the eigenvector of the largest eigenvalue λ_j^{-1} of matrix $W'^+ W'$. To give a physical interpretation of this concept, consider a two-class pattern recognition problem in which images of the first class have to give a response 1 and images of the second class have to give a response 0. The reference images are sorted in two classes C_0 and C_1:

$$\begin{aligned} C_0 &= \{X_{00}, X_{01}, \ldots, X_{0M}\} \\ C_1 &= \{X_{10}, X_{11}, \ldots, X_{1L}\} \end{aligned} \tag{27}$$

where $M + 1$ and $L + 1$ are the number of reference images in class 1 and class 2, respectively. With respect to the ASFS algorithm, as set of distorted images is associated to the reference image set. As images of class C_0 have to give response 1, the M first components of vector r are set to 1:

$$r^t = (1, \ldots, 1, 0, \ldots, 0) \tag{28}$$

The first correlation and the nonlinear transmission stage yield images X'_{0i} and X'_{i1}. By defining the vectors $W'_{kq} = E'^{-1/2} X'_{kq}$, the matrix $W'^+ W'$ is defined by

$$
W'^{+}W = \begin{pmatrix}
a_{00} & \cdots & a_{0L} & b_{00} & \cdots & b_{0M} \\
\cdots & \cdots & \cdots & \cdots & \cdots & \cdots \\
a_{L0} & \cdots & a_{LL} & b_{L0} & \cdots & b_{LM} \\
b_{00} & \cdots & b_{0L} & c_{00} & \cdots & c_{0M} \\
\cdots & \cdots & \cdots & \cdots & \cdots & \cdots \\
b_{M0} & \cdots & b_{ML} & c_{+M0} & \cdots & c_{MM}
\end{pmatrix}
\tag{29}
$$

where

$$
\begin{aligned}
a_{ij} &= W_{0i}'^{+} W_{0j}' \\
b_{ij} &= W_{0i}'^{+} W_{1j}' \\
c_{ij} &= W_{1i}'^{+} W_{1j}'
\end{aligned}
\tag{30}
$$

a_{ij} and c_{ij} are the intraclass inner products of the W_{ij} vectors and b_{ij} are the interclass inner products of the W_{ij} vectors. By performing the product $W'^{+}W'r = s$, we obtain the following components for the vector s:

$$
s_j = \sum_{l=0}^{L} a_{jl} \qquad \text{for } j < L + 1
\tag{31}
$$

$$
s_j = \sum_{l=0}^{L} b_{j-L-1l} \qquad \text{for } j > L
\tag{32}
$$

The vector r is an eigenvector of the largest eigenvalue if the s_j component defined by Eq. (31) are equal and maximized and if the s_j component defined by Eq. (32) are equal to zero. Indeed, combining Eqs. (25), (29), (31), and (32), we obtain

$$
F'^{-1} = r^{t} \cdot s
\tag{33}
$$

The value of the inverse of the cost function will be maximum if the vectors r and s are parallel. As the scalar product is a measure of the similarities between two images, the above discussion show that the intraclass inner products of response 1 images have to be increased and interclass products have to be decreased. In practice, r will never be exactly the eigenvector of the targets eigenvalue λ_j^{-1}, but matrix $W'^{+} W'$ has to be chosen by iterative computation in such a way that this condition is fulfilled as closely as possible. The first-generation filter and the nonlinearity have to process the reference images in such a way that the different classes are orthogonal and the similarities of the images of response 1 are increased. The next subsection describes the method of computing the first-generation filter H.

15.3.3 Iterative Computation of the First Filter

The Simulated Annealing Algorithm

As the cascaded correlation architecture involves a nonlinearity, it is not possible to find an analytical solution for the first filter h. Therefore, it is necessary to use iterative processes that decreases the cost function F' progressively. We have chosen the simulated annealing algorithm (SAA) [37]. because it is conceptually simple to implement. SAA has already been used for the computation of phase-only filters [38].

We assume that the discrete Fourier transform of filter $H^{(m)}$, resulting of the mth iteration, gives rise to the cost function $F'^{(m)}$. At iteration $m + 1$, we are looking for a new filter $H^{(m+1)}$ that will give a smaller cost function $F'^{(M+1)}$ than $F'^{(m)}$. Filter $H^{(m+1)}$ is obtained from filter $H^{(m)}$ by perturbing each component randomly:

$$H^{(m+1)}(Z) = H^{(m)}(Z) + \Delta H^{(m+1)}(Z) \tag{34}$$

The real and imaginary parts of $\Delta H^{(m+1)}(Z)$ are chosen in an interval $[-V_m, +V_m]$. With the help of $H^{(m+1)}(Z)$, the complete nonlinear two-correlation process is performed and $F'^{(m+1)}$ is evaluated. If $F'^{(m+1)}$ is smaller than $F'^{(m)}$, $H^{(m+1)}$ is kept; otherwise, it is kept conditionally, according to the following probability condition:

$H^{(m+1)}$ *is kept if*

$$p < \exp\left(-\frac{F'^{(m+1)} - F'^{(m)}}{K(T - m)}\right) \tag{35}$$

where p is a random number chosen in the interval $(0, 1)$, K is a constant to adjust the strength of the condition, and T, called the initial temperature, is introduced to increase the strength of the condition during the iterative process. When the iteration number m increases, the exponential term decreases and the condition becomes more restrictive. The probability condition (35) avoids the system to be blocked in a local minimum. Because the second correlation stage is linear, the second filter H' is computed on the basis of the ASFS algorithm, with the images obtained at the output of the first stage, after the nonlinearity.

Problems of Linear Dependencies in the Reference Images

As outlined above, the first and the nonlinear transmission stage must *prepare* input images for the second correlation stage by increasing the intraclass similarities and decreasing interclass similarities. However, this optimization of the similarities can lead to redundancy in the reference image set and singularities in the computation of the second-generation filter H'. Indeed, when the similarities between two reference images of the second correlation stage is close to 1, the vectors are almost collinear and the matrix $(X'^+ E'^{-1} X')$ of the reference images becomes singular. In this case, the cost function defined by

$$F' = r^t \cdot (X'^+ E'^{-1} X')^{-1} \cdot r \tag{36}$$

diverges because at least two columns of the $(X'^+ E' - 1 \ X')$ matrix are nearly identical. To avoid this instability of the cost function, it is necessary to remove, inside the reference image set, the redundancy between the different images. One way to proceed is to compute the angle α between the different vectors x_k. When the angle α is smaller than a reference value ε, the vectors are considered as collinear and one of the two vectors is removed from the reference set (see Fig. 5). This procedure is performed for each couple of vectors $(x_k, x_{k'})$ of the matrix X.

By performing this linear dependency analysis, we pass from a singular matrix $(X'^+ E'^{-1} X)$ of dimension N to a nonsingular matrix of smaller dimension N'. It is an important point because it shows that the constraints applied to the filter of the first correlation stage H are relaxed for the filter of the second correlation stage H'. Note that this analysis of the linear dependencies of the reference images at the input of the second stage does not eliminate the possibility of having a reference image made of a linear combination of the other reference images. However, in practical applications, we never observed this situation.

Choice of the Initial Solution

As simulated annealing algorithm is converging very slowly, it is necessary to start with a solution $H^{(0)}$, which gives a cost function $F'^{(0)}$ that is not too far from a good solution. We have found it convenient to choose, as an initial step, an ASFS filter with the same reference set and the same response vector r. This choice results from the following heuristic approach: Because ASFS filters reduces the influence of the distortions, it is then expected to keep uniform correlation values with the successive reference images inside

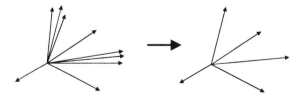

Figure 5   Colinearity between vectors of the reference set.

each class, which should increase the intraclass similarities. On the other hand, response 1 correlation intensities should present a narrow peak and very low sidelobes, whereas response 0 correlation intensities should present very low intensity peaks, which should decrease the interclass similarities. It is important to note that this particular choice of initial filter is an important modification of the SAA algorithm that may change the course of the optimization and that the final filter (and cost function) can correspond only to a local minimum and not to the true optimal solution. However, the numerical simulations show that important improvements of the correlation performances are obtained.

During the iteration process, it is expected to obtain lower cost function values and to be closer to the optimal solution. In this case, the maximum perturbation amplitude $[-V_m, +V_m]$ must decrease in comparison to the first iterations. Indeed, if the filter becomes closer to the optimal one, we observed that smaller perturbation amplitudes increase the convergence speed. For this purpose, we added a decreasing exponential factor to the perturbations $\Delta H_m(Z)$:

$$H^{(m+1)}(Z) = H^{(m)}(Z) + \Delta H^{(m+1)}(Z) \exp\left\{-\frac{m}{C}\right\} \tag{37}$$

where C is a constant to adjust the strength of the decreasing factor.

We now have to determine the choice of the initial maximum perturbation amplitude V_m. Because we use an ASFS filter H^0 as initial filter with a given range of amplitudes, we have to be careful in the choice of V_m. If V_m is too large, the filter $H^{(1)}$ will be completely different from $H^{(0)}$ and will be almost random. In this case, we lose the advantage of choosing an ASFS filter as initial filter. On the contrary, if we choose a value of V_m too small, the convergence will be very slow and time-consuming.

15.3.4   Numerical Simulations of the Correlation Intensities

Evaluation of the Performance on a Two-Class Pattern Recognition Problem

The nonlinear two-step correlation architecture has been tested in a two-class pattern recognition that request invariance to out of plane rotation. In this problem, we want to recognize a car and reject a truck, regardless of their out-of-plane orientation. Examples of images of class 1 and 2 are shown by Fig. 6.

For each class, 10 reference images of 64×64 pixels are taken with an incremental angle of $8°$ around a vertical axis, in order to cover the interval $[0°, 80°]$. A distorted image is associated to each reference image, which correspond to an intermediate rotation of the corresponding reference

(a) Example of image to be
recognized

(b) Example of image to be
rejected

Figure 6 Example of class 1 and 2 images: (a) image to be recognized; (b) image to be rejected.

image. A couple of filter $(H(U, V), H'(U, V))$ has been computed on the basis of the modified SAA described earlier. The different parameters are described in Table 3.

After the learning phase of the system, filters H and H' were used to obtain the numerical correlation intensities. Forty images per class were used as input images, corresponding to incremental angles of $2°$. The maximum intensities detected as a function of the orientation of the input images are shown in Fig. 7, and compared to a one-stage ASFS filter.

Examples of correlation peaks of response 1 and response 0 for a linear one-stage and the (NL)-nonlinear two-step correlation systems are compared in Fig. 8.

Several observations can be made:

The difference between intensity levels of response 1 and response 0 are significantly improved with respect to a one-stage linear ASFS.

Table 3 Parameters of the Numerical Simulations of the Correlation Intensities

Total Iteration nb	K	Initial temperature	V_m
10,000	1/10,000	10,000	$\text{Max}\{H^{(0)}/10\}$
C	ε (maximum similarities)	θ_0 (sigmoid parameter)	θ_i (sigmoid parameter)
2,171	0.95	128	4

The decreasing factor C has been chosen in such a way that the perturbation amplitude at the end of the iterative process is 100 times smaller than in the begin of the iterative process.

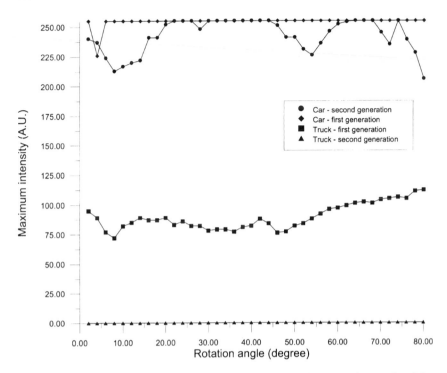

Figure 7 Maximal correlation intensities as a function of the rotation angle of the input images. Results with the nonlinear two-step system are compared to a linear one-stage system.

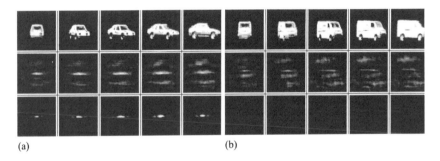

(a)   (b)

Figure 8 Examples of correlation peaks obtained by numerical simulations. The first row of the browser represent in the input images; the second row correspond to the correlation peaks of a one-stage ASFS system; the third row to the nonlinear two-step system. (a) Examples of response 1 correlation peaks; (b) Examples of response 0 correlation peaks.

Even if intensities of response 1 are fluctuating in the case of the NL two-step system, the intensities of response 0 are drastically decreased to almost the zero-intensity level in comparison to the one-stage linear system.

The nonlinear two-step system provides very narrow and stable correlation peaks compared to a one-stage system. Sidelobes energies appearing in the correlation peaks of the one-stage system are completely removed with the NL two-step architecture.

The standard performance measures (identical to those presented in previous section) are shown in Table 4. The mean intensity of response 0 for the nonlinear two-step system are drastically decreased in comparison to the one-stage linear ASFS. It can be observed that fluctuations of response 1 are increased in the case of NL two-step correlation system. Nevertheless, we observe a drastic improvement in the discrimination ratio G. The peak sharpness is also increased by the NL two-step system. For the estimation of the SNR, as described in Section 15.3.3, 30 uniformly distributed random white-noise patterns (of maximum amplitude limited to 255) were sequentially added to a response 1 input image. The SNR is then obtained by using Eq. (15). The maximum amplitude of the noise was 0.7 times the maximum amplitude in the input image (70%). We observe improvements of the SNR for the nonlinear two-stage system. It has been observed that increasing the parameter θ_i could increase the SNR of the nonlinear two-stage system. Indeed, increasing θ_i correspond to a smoother slope of the sigmoid function. In this case, the separation between the low and the high intensities at the output of the first correlation stage is smooth. A loss of correlation intensity (due to the presence of noise in the input image) will be shortly increased or decreased at the output of the nonlinear stage. The correlation intensity will then be reestablished at the output of the second correlation

Table 4   Performance Measurements for the One-Stage and the Nonlinear Two-Step Correlation System

I_1	SD_1	I_0	SD_0	G	PCE	SNR
		One-stage ASFS system				
254.05	2.76	86.89	20.67	7.14	0.05	3663.66
		Nonlinear two-step correlation system				
254.12	5.47	0.00	0.00	46.46	0.11	5634.34

Note: I_1 and I_0 are the mean maximum detected of response 1 and response 0, SD_1 and SD_0 are the corresponding standard deviations, G is the discrimination ratio, PCE is the peak to correlation energy, and SNR is the signal-to-noise ratio.

stage. On the other hand, when θ_i decreases, the separation between high and low intensities appear for a reduced number of input gray levels (*strong nonlinearity*). In this case, a loss of intensity in the first-generation correlation peak is strongly increased or decreased by the nonlinearity, creating a loss of intensity and fluctuations at the output of the second correlation stage. As a consequence, when the noise level increases in the input image, the SNR at the output of the second stage decreases. However, in the following subsection, methods are presented to automatically compensate the loss of intensity at the output of the first correlation stage that appear with illumination changes or presence of noise in the input images.

Evolution of the Cost Function F'

The evolution of the cost function F' as a function of the iteration number is shown in Fig. 9. The first value of the cost function corresponds to the cost function obtained with an ASFS filter $H^{(0)}$ for the first-stage correlation as

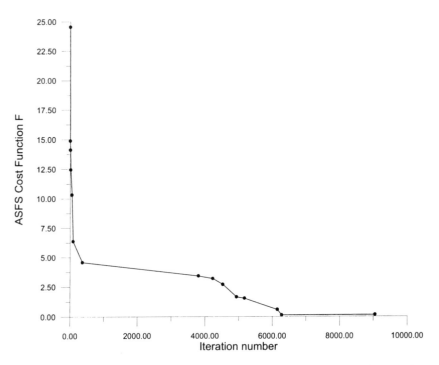

Figure 9   Evolution of the cost function F' as a function of the iteration number.

discussed earlier. We define a measurement of the decreasing of the function as follows:

$$C = \frac{F'^{(0)} - F'^{(k)}}{F'^{(0)}} \tag{38}$$

where k is the number of the last iteration that gives rise to a decreasing of the cost function F'. We obtain a value of C equal to 0.994.

The value of the cost function F' for the first iteration is 24.54 and 0.136502 for the last iteration. These results show that the SAA is suitable for the optimization of the nonlinear two-step correlation system.

15.3.5 Remarks on the Two Nonlinear Two-Step Correlation System

The nonlinear two-step correlation system is composed of two cascaded linear correlation stages with a nonlinear transmission function at the output of the first correlation stage. It is shown by theoretical developments and by numerical simulations that this system increases the performances of the recognition significantly compared to a one-stage linear system. In this configuration, the first correlation and the nonlinearity *prepare* the input images for the second correlation stage by increasing the intraclass similarities and decreasing the interclass similarities. The first correlation stage plays the role of a discriminator, whereas the second correlation stage is a classificator. The first filter H is obtained by a simulated annealing algorithm, whereas the second filter H' is an ASFS filter. It has been outlined that, by a proper choice of the parameters, the standard performance measurements were strongly increased with the nonlinear two-step system. Furthermore, such a system can be used for pattern recognition that cannot be implemented in a linear one-stage correlation system (when reference images are linearly dependent) [39].

15.4 REDUCTION OF THE CORRELATION SENSITIVITY TO NONOVERLAPPING NOISE

15.4.1 Introduction

As outlined in the previous sections, ASFS correlation filters present very high SNRs when additive noise is present in the input images. However, as recently pointed out by several authors [40–49], the performances of the recognition by correlation are very sensitive to substitutive noise (noise present in the background but not inside the object). Indeed, as explained below, the nonoverlapping background acts negatively on the correlation amplitude, decreasing the correlation peak and leading to false detections. In this part, we propose two different methods to reduce the sensitivity of the ASFS algorithm to background noise. However, it has to be noted that those methods are not restricted to this particular kind of filters and can be easily applied to other algorithms. The first method is included directly in the filter computation. The second one is based on a multiplicative mask applied to the filter and is thus performed after the filter computation. The two methods are presented and simulation results are provided.

15.4.2   First Method: The Window Method

Introduction

The window method allows to sensibly decrease the sensitivity of the filters when input images are corrupted by nonoverlapping noise. The method integrates, during the computation phase of the filter, the window functions of all the reference images and imposes additional constraints to the filter. This method increases the number of reference images by a factor of 2, which increases the computation time of the filter. Nevertheless, the method is efficient and can be easily applied to other algorithms found in the literature.

Analysis

Let us consider the set of reference images $x_{kq}(s, t)$ and their corresponding distortions $y_{kqi}(s, t)$. It is requested to recognize the set of reference images regardless of the background $b_{kq}(s, t)$ present around the objects. The input images are described as follows:

$$z_{kq}(s, t) = x_{kq}(s, t)w_{kq}(s, t) + [1 - w_{kq}(s, t)]b_{kq}(s, t) \qquad (39)$$

where $w_{kq}(s, t)$ is the window support function of the image $x_{kq}(s, t)$ defined as

$$w_{kq}(s, t) = 1 \quad \text{when} \quad x_{kq}(s, t) \neq 0$$
$$w_{kq}(s, t) = 0 \quad \text{when} \quad x_{kq}(s, t) = 0$$

It is well known that the information carried out by the lowest spatial frequencies decreases the discrimination of the filter. Those components are removed before the computation of the filter by setting $X_{kq}(0, 0) = 0$ and $Y_{kqi}(0, 0) = 0$. The suppression of the zero order of the Fourier transform of the reference images directly implies a suppression of the filter component $H(0, 0)$.

Let us now point out the negative effect of the background on the central correlation amplitude. We consider a constant background $b(s, t)$. The Fourier transform of $b(s, t)$ is a discrete Dirac function and is different from zero only at the origin $(0, 0)$. If the image $b(s, t)$ is correlated with an ASFS filter (where the zero order has been suppressed), the correlation product $c(s, t)$ is null because the only nonzero frequency of $B(U, V)$ is blocked by the filter $H(U, V)$:

$$c(s, t) = \sum_{s', t'=0}^{N-1} h(s + s', t + t')b(s', t') = 0 \tag{40}$$

The constant background can be decomposed in two separate images: one containing the window support $w_{kq}(s, t)$ of a reference image and the other one containing the nonoverlapping background $b_{kq}(s, t)$ corresponding to the window support function $b_{kq}(s, t) = 1 - w_{kq}(s, t)$:

$$b(s, t) = w_{kq}(s, t) + b_{kq}(s, t) \tag{41}$$

By combining Eqs. (40) and (41), we obtain

$$c(0, 0) = \sum_{s, t=0}^{n-1} h(s, t)b(s, t) = \sum_{s, t=0}^{N-1} h(s, t)\{w_{kq}(s, t) + b_{kq}(s, t)\} = 0 \tag{42}$$

and

$$\sum_{s, t=0}^{N-1} h(s, t)w_{kq}(s, t) = -\sum_{s, t=0}^{N-1} h(s, t)b_{kq}(s, t) \tag{43}$$

Equation (43) shows that the contribution of the nonoverlapping background to the central correlation amplitude is just the opposite of the contribution of the support window functions. As the support window functions have strong correlations with the reference images, this contribution is important, demonstrating the negative impact of the spatially disjoint background on the central correlation amplitude. In order to remove this negative contribution, we impose additional constraints to the filter:

$$\sum_{s, t=0}^{N-1} h(s, t)w_{kq}(s, t) = 0 \tag{44}$$

In this case, as the contribution of the substitutive background is exactly the opposite of the contribution of the window function, the spatially disjoint background will have no effect on the response of the filter. As surrounding nonuniform realistic background clutters are often made of low spatial frequencies, it is expected that, in this case, constraints (44) will also improve the results. However, it has been shown [48] that suppressing other Fourier components (by applying a ring filtering with an increasing diameter) could decrease the sensitivity of ASFS filters to realistic nonoverlapping backgrounds.

Because constraints (44) are applied for each kq, the filter includes a set of reference images with twice the number of reference and distorted images. It is important to note that it is a general method that is not restricted to the ASFS algorithm. The filter is constrained to give a response 1 for the reference images and a response 0 for their corresponding window functions. Because the $(0, 0)$ spatial frequency of all the reference images is suppressed, the window functions $w_{kq}(s, t)$ are filtered and only the external contours give significant contributions. The filter will keep only significant information present inside the objects. It results that this method is not appropriate for binary objects. Specific methods have been proposed in the case of binary objects [49].

We observe that the performances can further be improved by an appropriate choice of the distorted images:

$$y_{kqi}(s, t) = x'_{kq}(s, t) + b_{kqi}(s, t) \tag{45}$$

where $b_{kqi}(s, t)$ is a spatially disjoint background clutter associated with the image $y_{kqi}(s, t)$ and $x'_{kq}(s, t)$ is a distorted version of the reference image $x_{kq}(s, t)$. For each reference image, we associate a distorted version x'_{kq} and a different background b_{kqi}. It has to be emphasized that the introduction of the surrounding background clutter or noise in the distorted images does not require any assumption on the nonoverlapping noise model. It is made in a natural way with actual situations.

Results of the Numerical Simulations

According to the previous discussion, an ASFS filter that reduces the sensitivity to nonoverlapping noise has been computed for a two-class pattern recognition problem. It consists, as in previous section, to recognize a car and to reject a truck independently of their out-of-plane orientation. As in the previous section, 10 reference images per class are involved in the filter computation with an incremental angle of $8°$. For each reference image, we associate a distorted image corresponding to an incremental angle of $4°$ with respect to the corresponding reference image. A nonoverlapping background clutter is associated to each distorted image. The background clutters are different for each distorted image and represent general landscape view. Examples of distorted images are shown in Fig. 10. For each reference and distorted image $x_{kq}(s, t)$ and $x'_{kq}(s, t)$, respectively, the window function $w_{kq}(s, t)$ and $w'_{kq}(s, t)$ is computed and added to the reference image set.

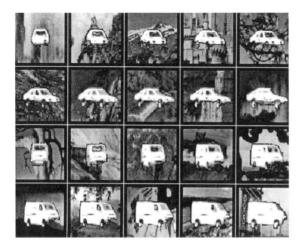

Figure 10 Examples of distorted images involved in the computation of the filter. The two first lines correspond to class 1 images (to be recognized); the two last lines correspond to class 2 images (to be rejected).

The filter was tested on 80 images (40 images per class), corresponding to an incremental angle of 2° and covering the interval [0°, 80°]. This set takes into account the reference images, the distorted images, and intermediate orientations that were not involved in the filter computation. In order to test the robustness of the filter to nonoverlapping textures, we took a different spatially disjoint background clutter for each input image. The major part of those background clutter are a priori unknown for the filter (not involved in the distorted image set). Furthermore, we also used random uniformly distributed noise ($\mu = 127$, $\sigma^2 = 73$ and maximal amplitude equal to 255) as nonoverlapping noise in order to test the robustness of the algorithm. The results of the numerical correlation intensities are summarized in Table 5 and in Fig. 11 and compared to a normal ASFS (without additional constraints).

The performance of the recognition is increased when the additional constraints (44) are included. Indeed, the G ratio is appreciably increased for the optimized ASFS filter. It can be seen in Fig. 11 that the correlation intensities of response 1 and response 0 of a normal ASFS are not sufficiently separated to set a recognition threshold. On the contrary, intensity levels of response 1 and response 0 with the optimized ASFS are well separated and allow to fix a threshold level without ambiguity. The discrimination ratio G presents higher values for the optimized filter and is strongly improved in the case of background clutters (which is the most realistic practical situation). On the other hand, a normal ASFS presents better values of the G ratio when input images are not corrupted by nonoverlapping noise, which indicates

Table 5 Summary of the Numerical Correlation Intensities

Filter 1	I_1	SD_1	I_0	SD_0	PCE	G	SNR
Without substitutive noise	129.98	15.97	34.67	4.41	1.168×10^{-2}	4.6778	
With a background clutter	108.59	15.95	17.99	6.01	3.225×10^{-2}	4.1269	6.85×10^2
With a substitutive noise	119.87	20.04	19.25	4.24	1.938×10^{-2}	4.1442	6.25×10^4

Filter 2	I_1	SD_1	T_0	SD_0	PCE	G	SNR
Without substitutive noise	128.15	9.96	35.92	6.81	1.25×10^{-2}	5.4994	
With a background clutter	64.87	19.61	23.15	7.11	1.003×10^{-2}	1.5617	0.86×10^2
With a substitutive noise	50.30	8.49	15.78	2.47	1.443×10^{-2}	3.1474	5.70×10^2

Note: Filter 1 is an ASFS filter optimized for the nonoverlapping noise. Filter 2 is a normal ASFS filter. I_1 is the mean maximum intensity of response 1; I_0 is the mean maximum intensity of response 0. SD_1 and SD_0 are the corresponding standard deviations. PCE is the peak sharpness, G is the discrimination ratio, and SNR is the signal-to-noise ratio.

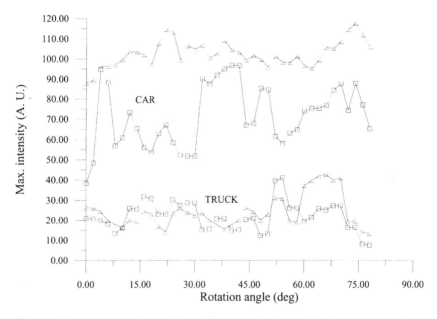

Figure 11 Maximum detected intensities as a function of the input image orientation with spatially disjoint background. The squares correspond to a normal ASFS and the triangles correspond to an optimized ASFS filter for spatially disjoint noise.

that the additional constraints are strong and decreases shortly the discrimination power of ASFS filters when input images are presented on a black scene. The PCE and SNR values are also increased in the case of filter 1.

Remarks

We have shown that the ASFS algorithm can be modified to handle non-overlapping background noise or textures. This is achieved by imposing response 0 for the window functions of the reference images and by filtering the zero order of the reference images. This method requires a larger number of reference images (twice the number) and is limited for applications with nonbinary images. Those limits lead us to find another way to handle the substitutive noise problem. This new method is presented in the next subsection.

15.4.3 Second Method: The Mask Method

Introduction

As outlined in the window method, the substitutive noise contributes negatively to the central correlation amplitude. The concept of the mask method is to limit the spatial extension of the filter to the average spatial extension of the reference objects by multiplying the filter $h(s, t)$ by a mask $m(s, t)$. In this case, the influence of the spatially disjoint background is strongly reduced. Indeed, the numerical filter $h(s, t)$ results from an optimization algorithm in the Fourier space regardless to the actual extension of the objects in the input images. As a consequence, the filter $h(s, t)$ presents nonzero values even when all the corresponding components $x_{kq}(s, t)$ are null.

Note that although these components result from the optimization of the ASFS cost function, they do not have actual significance. Therefore, it seems natural to limit them, because they disturb the filter performances when the input targets have surrounding backgrounds.

Analysis

Let us consider a set of reference images $x_k(s, t)$ and a random noise $n(s, t)$ of mean μ and standard deviation σ^2 that is modeling the background clutter. We have

$$E\{n(s, t)\} = \mu$$
$$E\{n^2(s, t)\} = \sigma^2$$

(46)

where $E\{\alpha\}$ defines the expected value of the random variable α.

The images $x'_k(s, t)$ surrounding by the spatially disjoint noise $n(s, t)$ are described as

$$x'_k(s, t) = x_k(s, t)w_k(s, t) + [1 - w_k(s, t)]n(s, t) \qquad (47)$$

where, as earlier, $w_k(s, t)$ is the support window function of the reference image $x_k(s, t)$. When the input image $x'_k(s, t)$ is correlated with a filter $h(s, t)$, the central correlation amplitude $c(0, 0)$ becomes

$$
\begin{aligned}
c(0, 0) &= (h \otimes x'_k)(0, 0) \\
&= \sum_{s,t=0}^{N-1} x_k(s, t)h(s, t) + \sum_{s,t=0}^{n-1} [1 - w_k(s, t)]n(s, t)h(s, t)
\end{aligned}
\qquad (48)
$$

The first term of Eq. (48) is the contribution of the reference image. The second one is the contribution of the nonoverlapping background. As shown in the previous section, this term gives contributions because the filter components have nonzero numerical values outside of the object to be recognized. To reduce the influence of those insignificant components, we should consider a filter $h(s, t)$ whose spatial extension is comparable to the ones of the reference images $x_k(s, t)$. For this purpose, we multiply the filter $h(s, t)$ by a mask $m(s, t)$ computed on the basis of the reference images. It is requested that the multiplicative mask $m(s, t)$ must have the smallest possible impact on the correlation peak height. Therefore, we define a cost function f to be minimized as follows:

$$f = \sum_{k=1}^{k} \sum_{s,t=0}^{N-1} \{m(s, t)h(s, t)x'_k(s, t) - h(s, t)x_k(s, t)\}^2 \qquad (49)$$

where K is the total number of reference images involved in the filter computation. The expected value of f is computed as follows:

$$E\{f\} = \sum_{k=1}^{K} E\{h^2 x_k^2 + h^2 m^2 x'^2_k - 2h^2 x_k m x'_k\}$$

$$E\{f\} = \sum_{k=1}^{K} h^2 x_k^2 + E\{h^2 m^2 x'^2_k\} - 2h^2 x_k m E\{x'_k\} \qquad (50)$$

$$E\{f\} = \sum_{k=1}^{K} h^2 x_k^2 + h^2 m^2 E\{x'^2_k\} - 2h^2 x_k m E\{x'_k\}$$

where the (s, t) dependency has been removed to keep simple notations.

The expected values of x_k' and $x_k'^2$ are expressed as

$$E\{x_k'\} = E\{x_k w_k + (1 - w_k)b\} = x_k w_k + (1 - w_k)\mu$$
$$E\{x_k'^2\} = x_k^2 w_k^2 + (1 - w_k)^2 \sigma^2 + 2E\{x_k w_k n - x_k^2 w_k^2 n\} \qquad (51)$$
$$E\{x_k'^2\} = x_k^2 w_k^2 + (1 - w_k)^2 \sigma^2$$

By combining Eqs. (50) and (51), we obtain

$$E\{f\} = \sum_{k=1}^{K} h^2 x_k^2 + h^2 m^2 [x_k^2 w_k + (1 - w_k)\sigma^2 \qquad (52)$$
$$- 2h^2 x_k m [x_k w_k + (1 - w_k)\mu]$$

By annulling the derivative of $E\{f\}$ with respect to $m(s, t)$, we obtain the solution of $m(s, t)$:

$$m(s, t) = \frac{\displaystyle\sum_{k=1}^{K} [x_k w_k + (1 - w_k)\mu]x_k}{\displaystyle\sum_{k=1}^{K} [x_k^2 w_k + (1 - w_k)\sigma^2]} \qquad (53)$$

which reduces to

$$m(s, t) = \frac{\displaystyle\sum_{k=1}^{K} x_k^2}{\displaystyle\sum_{k=1}^{K} [x_k^2 w_k + (1 - w_k)\sigma^2]} \qquad (54)$$

Because the reference images components appear in the numerator of Eq. (54), $m(s, t)$ is null when all the components $x_k(s, t)$ are null. Consequently, the mask $m(s, t)$ will play the role of a support function limited to the zone where the targets are situated. When the $x_k(s, t)$ are not null, we realize a bounded function whose importance depends on the number of pixels of x_k (s, t) different from zero.

The mask $m(s, t)$ defines, in average, the regions of interest of the reference images (the zone where the reference images are not null). For each pixel (s, t), the value of $m(s, t)$ is a weighted linear combination of the nonzero components of the reference images $x_k(s, t)$. If, for a value of (s, t), all of the components $x_k(s, t)$ are not null, the value of $m(s, t)$ for that particular point will be maximum: $m(s, t) = 1$. On the contrary, if, for a point (s, t), some images $x_k(s, t)$ have zero components, the value of $m(s, t)$ will decrease. The larger σ^2 is, the stronger the decrease will be. Increasing the value of σ^2 will narrow the mask around the zones where all of the reference images are not null.

Results of the Numerical Correlation Intensities

According to the previous discussion, an ASFS filter $h(s, t)$ and a mask $m(s, t)$ (with a value of $\sigma^2 = 200$) have been computed. The filter has been designed to recognize a car (set 1) and to reject a truck (set 2), independently of their out-of-plane rotation. Twenty images per class have been selected, corresponding to an incremental angle of $4°$. The filter has been tested on 40 images per class, corresponding to an incremental angle of $2°$. For the numerical simulations of the correlation intensities, we used, as in the previous section, two kinds of nonoverlapping background: a set of background clutter representing general landscape views and a set of uniformly distributed random noise ($\mu = 127, \sigma^2 = 73$ with a maximal amplitude limited to 255).

The results of the simulated correlation intensities are reported in Table 6 and in Figs. 12 and 13 with and without the application of the mask.

The use of the mask increases the correlation intensities of response 1 (object to be recognized) while keeping the comparable response 0 correlation intensities. It is observed that the discrimination ratio is increased by the application of the mask for both of the input image sets. SNR values are also strongly improved.

Remarks

We have shown that reducing the spatial extension to the mean extension of the reference images could increase the recognition performances of the correlation filter when the spatially disjoint background is present in the input image. This is performed by the application of a mask to the filter.

Table 6 Results of the Numerical Simulations of the Correlation Intensities

	Input images	I_1	I_0	SD_1	SD_0	G	SNR
Without mask	Set 1	103.50	44.75	14.67	8.28	2.56	3691.02
	Set 2	126.77	55.82	40.33	17.69	1.22	
With mask	Set 1	155.67	57.05	15.75	11.04	3.68	8336.87
	Set 2	172.2	66.8	31.14	18.40	2.13	

Note: Set 1 corresponds to nonoverlapping noise, set 2 corresponds to spatially disjoint background clutters.

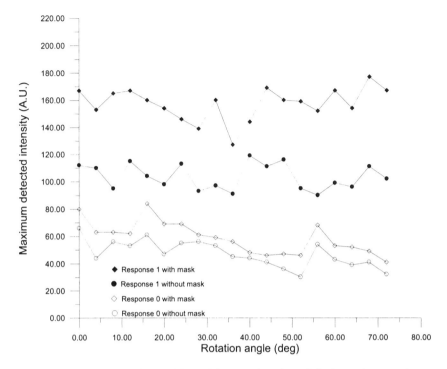

Figure 12 Maximum detected intensities as a function of the input image orientation for images with spatially disjoint random noise. Results are compared with and without the application of the mask. Images of set 1 (car) have to be recognized and images of set 2 (truck) have to be rejected.

This mask is computed on the basis of the reference images. The results of the numerical simulations show that the performances are improved by the application of the mask.

15.5 REDUCTION OF THE CORRELATION SENSITIVITY TO ILLUMINATION CHANGES

15.5.1 Introduction

When an input image suffers from illumination changes with respect to its reference model, the correlation peak amplitude decreases proportionally. This problem becomes more critical in intensity. In practical applications, Illumination changes are completely unpredictable and it is necessary to

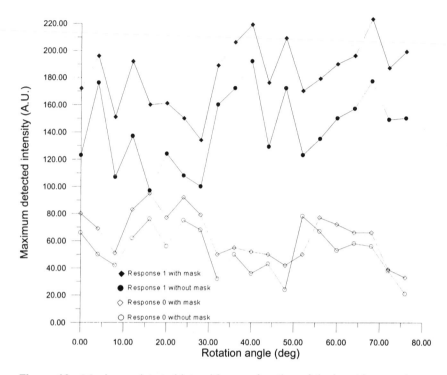

Figure 13   Maximum detected intensities as a function of the input image orientation for images with spatially disjoint background clutter. Results are compared with and without the application of the mask. Images of set 1 (car) have to be recognized and images of set 2 (truck) have to be rejected.

identify specific methods that allow one to automatically readjust the intensity levels of the correlation peak when illumination changes appear in the input images. Logarithmic methods have already been proposed by Arsenault et al. [50]. The method that we describe here is based on the statistical properties of the correlation intensities. It is based on the fact that the histogram shape of the correlation intensities is a stable process and depends very little on the presence or not on a recognized event.

15.5.2   Analysis

To understand the physical meaning of the process, we will consider a particular application where illumination changes and substitutive noise problem increase progressively during the recognition phase.

The goal of this actual application is to count the number of cancerous cell divisions in in vitro cultures (cell line U87 from brain cancerous tumors). The proliferation rate (number of cell divisions) is an important parameter to test the antiproliferation efficiency of an antitumor drug. The images are taken by a phase-contrast microscope (magnification ×10) each 4 min during 3 days, resulting in a sequence of 1000 images. A typical image of the culture is shown in Fig. 14.

With the phase-contrast microscope, at the beginning of a division, a cell gives rise to a bright ring. When the division is actually started, the cell takes the shape of a bright "height" and finally separates into two daughter cells. The identification of the object is performed when the cell appears as a "height." It has to be noted that the culture and the shapes of the cells strongly change with time. Examples of images at the beginning and at the end of the culture are shown in Fig. 15.

The objective of this application is to train a correlation process to automatically count the cells in division in a reliable way. For this purpose, several items have to be taken into account:

The orientation of the dividing cells is random. Therefore, the recognition process has to be completely rotation invariant.

As we are dealing with living cells, we can observe significant events with very varying shapes.

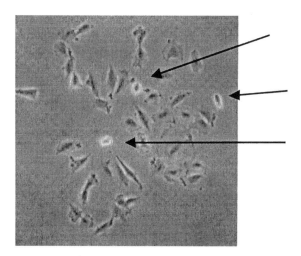

Figure 14 Example of in vitro culture. Objects to be detected (dividing cells) are indicated by black arrows.

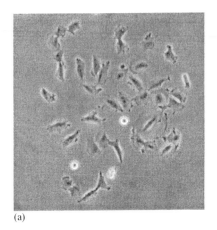
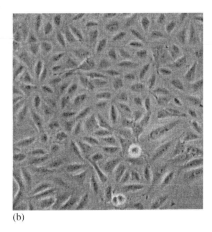

(a)   (b)

Figure 15   Examples of images at (a) the beginning and (b) end of the culture.

There is an important background and the illumination changes in time due to the fluctuations of the lighting.

In the following, the assumption is made that an ASFS filter has been computed to recognize the dividing cells (the specifications are described in Section 15.5.6). As the intensity changes in time, due to the cell proliferation and fluctuations of the background, it is mandatory to implement specific methods to reduce the influence of the light fluctuations. We have implemented two approaches. The first one consists of a preprocessing of the input images in order to suppress most of the background in the input images. The second one is based on a postprocessing of the correlation intensities based on the statistical properties of the correlation histograms. This last method is quite general and can be exported to a very wide range of applications.

15.5.3   Theoretical Aspects

First Method: Preprocessing of the Input Images

At the beginning of the culture, the number of cells is low and the background presents large regions of uniform background. This background can easily be removed by preprocessing the input images. This preprocess consists in the subtraction of the average intensity (computed for all the pixels of the input images) and a stretch of the resulting intensities between 0 and 255:

$$x'(s, t) = x(s, t) - \frac{1}{N * N} \sum_{s,t=0}^{N-1} x(s, t) \tag{55}$$

$$x''(s, t) = 255 \, \frac{x'(s, t) - \min\{x'(s, t)\}}{\max\{x'(s, t) - \min\{x'(s, t)\}} \tag{56}$$

where $x''(s, t)$ is the preprocessed image of the input image $x(s, t)$.

Because bright cells are small compared to the size of the input images (typically 32×32 pixels for a cell and 512×512 pixels for the input image), the preprocess has a small effect on the cells and suppresses most of the background. However, this is not true at the end of the culture, where the number of cells in the background is larger. This induces important fluctuations of the background, leading to a global change of the illumination. In this case, the subtraction of the mean value does not remove the background completely and introduces nonoverlapping noise in the input image.

Second Method: Postprocessing of the Correlation Intensities

Let us first investigate some statistical properties of the correlation intensities. For this purpose, we assume that the images $x(s, t)$ can be described as N^2 independent random variables filtered by a pupil function. This is corresponding to some colored noise. The corresponding Fourier transform $X(U, V)$ is defined as follows:

$$X(U, V) = N(U, V)P(U, V) \tag{57}$$

where $N(U, V)$ are N^2 independent random variables and $P(U, V)$ is a symmetric filtering pupil function. The corresponding $n(s, t)$ is defined positive. Because $n(s, t)$ is defined positive, the expected value is not null; we can thus write $N(U, V)$ as

$$N(U, V) = B(U, V) + B_0\delta(U, V) \tag{58}$$

where $B(U, V)$ is the Fourier transform of a white noise of zero mean $b(s, t)$ and B_0 is a constant value. We have

$$\begin{aligned} E\{N(U, V)\} &= B_0\delta(U, V) \\ E\{B(U, V)\} &= 0 \end{aligned} \tag{59}$$

For the second-order moment, we have

$$E\{B(U, V)B(U', V')\} = \sigma^2\delta(U - U', V - V') \tag{60}$$

with

$$\begin{aligned} E\{x(s, t)\} &= \mu \\ E\{x^2(s, t)\} &= \sigma^2 \end{aligned} \tag{61}$$

The relationship between μ and B_0 is obtained by considering

$$E\{x(s, t)\} = \frac{1}{N} \sum_{U,V=0}^{N-1} \exp\left\{+\frac{2i\pi}{N}(Us + Vt)\right\} P(U, V)E\{N(U, V)$$

$$\mu = \frac{1}{N} \sum_{U,V=0}^{N-1} \exp\left\{+\frac{2i\pi}{N}(Us + Vt)\right\} P(U, V)B_0\delta(U, V) \qquad (62)$$

$$\mu = \frac{1}{N} P(0, 0)B_0$$

Let us consider a filter $h(s, t)$ and an image $x(s, t)$. The correlation product is written

$$c(s, t) = (x \otimes h)(s, t) = \frac{1}{N} \sum_{U,V=0}^{N-1} \exp\left\{\frac{2i\pi}{N}(Us + Vt)\right\}$$
$$H^*(U, V)[N(U, V)P(U, V)]$$

$$c(s, t) = \frac{1}{N} \sum_{U,V=0}^{N-1} \exp\left\{+\frac{2i\pi}{N}(Us + Vt)\right\}$$
$$H^*(U, V)P(U, V)\{B(U, V) + B_0\delta(U, V)\}$$

$$c(s, t) = \frac{1}{N} H^*(0, 0)P(0, 0)B_0 + \frac{1}{N} \sum_{U,V=0}^{N-1} \exp\left\{+\frac{2i\pi}{N}(Us + Vt)\right\}$$
$$H^*(U, V)P(U, V)B(U, V)$$

$$c(s, t) = \frac{1}{N} B_0 H^*(0, 0)P(0, 0) + \frac{1}{N} \sum_{U,V=0}^{N-1} C(s, t, U, V)B(U, V)$$

where $C(s, t, U, V)$ is given by

$$C(s, t, U, V) = \exp\left\{+\frac{2i\pi}{N}(Us + Vt)\right\} H^*(U, V)P(U, V) \qquad (64)$$

Combining Eqs. (62) and (63) gives

$$c(s, t) = \mu H^*(0, 0) + \frac{1}{N} \sum_{U,V=0}^{N-1} C(s, t, U, V)B(U, V) \qquad (65)$$

The second term on the right-hand side of Eq. (65) states that the correlation product can be seen as a sum of random weighted variables $B(U, V)$. We assume that the number of random variables is large enough to invoke the central limit theorem, which stipulates that the sum of a large number of independent random variables is distributed as a normal distribution. Applying this theorem, we find a correlation distribution:

$$P[x(s, t) = a] \approx + \frac{1}{\Sigma\sqrt{2\pi}} \exp\left(-\frac{(a - \mu H^*(0, 0))^2}{2\Sigma^2}\right) \tag{66}$$

where $P[c(s, t) = a]$ represents the probability density to find a correlation amplitude equal to a.

Let us determine the value of Σ:

$$E\{c^2(s, t)\} = E\left\{\frac{1}{N^2} \sum_{U_1, V_1=0}^{N-1} W(s, t, U_1, V_1)B(U_1, V_1)\right.$$
$$\left. \sum_{U_2, V_2=0}^{N-1} W^*(s, t, U_2, V_2)B^*(U_2, V_2)\right\} \tag{67}$$

$$E\{c^2(s, t)\} = \frac{1}{N^2} \sum_{U_1, U_2, V_1, V_2=0}^{N-1} W(s, t, U_1, V_1)W^*(s, t, U_2, V_2)$$
$$\sigma^2\delta(U_1 - U_2, V_1 - V_2)$$

$$\Sigma^2 = \frac{\sigma^2}{N} \sum_{U, V=0}^{N-1} |H(U, V)P(U, V)|^2 \tag{68}$$

Let us now consider the properties of the correlation intensities. Starting from Eq. (66), we would like to determine $P[x^2(s, t) = j]$. For this purpose, we use a specific theorem [51]:

$$P[x^2(s, t) = j; j > 0] \approx + \frac{1}{j\Sigma\sqrt{2\pi}} \exp\left(\frac{(j - \mu H^*(0, 0))^2}{2\Sigma^2}\right) \tag{69}$$

This simple model of the images gives the general shape of the correlation intensity distribution. In practical applications, we never meet this model. Nevertheless, we observe, in practical applications, that the shape of the correlation histograms obtained with input images are very similar to the curve described by Eq. (69).

The result of Eq. (69) is very important: It states that we can consider that the correlation intensity distributions have a defined statistical distribution that can be more easily handled than the original input image.

Let us analyze the correlation histograms obtained in the particular application of the automatic detection of cancerous dividing cells and let us consider the effect of illumination changes on the correlation histograms.

We assume that the input images are preprocessed before to be correlated with the filter (the preprocess is presented in a previous section). At the beginning of the culture, the background presents large regions of constant value. In this case, the preprocess removes almost the background and we are close to the reference image illumination.

Examples of input images, correlation intensities, and correlation histogram (which represents the population of each intensity level) in the beginning of the culture are presented in Fig. 16. The correlation image presents a very large number of black pixels. The value of the histogram in 0 is very large (62.6%) with respect to the other intensity levels. The median (the value of the gray level, which leaves 50% of the population for the lowest levels and 50% of the population for the highest levels) of the histogram is close to zero.

Let us now consider an input image at the end of the culture where the background has become nonuniform. Examples of images are presented in Fig. 17. We observe that the pixel population of the zero intensity has decreased with respect to the beginning of the culture. Indeed, we found a population of 31% for the zero level and a median equal to 5. However, it can be seen in Figs. 16 and 17 that the global shape of both histograms are very close.

Let us now analyse what happens in the case of a response 0 correlation image. An example of image is shown in Fig. 18. This image is recorded 15 min (in the same sequence) after the recording of the image of Fig. 17. In this image, no event has to be recognized.

We found, in this case, a comparable value of the zero-level population (31%) and a median value of 5. This observation shows that the statistical properties (width, mean, median, etc.) of the correlation histograms depend very little on the presence of a recognized event. The main difference between the correlation histograms with recognized event or not lies in the population of the highest intensity levels. However, the higher-level

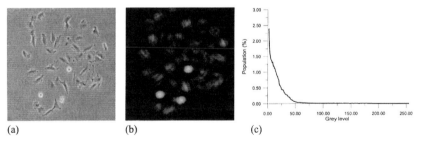

(a)   (b)   (c)

Figure 16   Example of (a) image, (b) correlation intensity, and (c) histogram in the case of a recognition at the beginning of the culture.

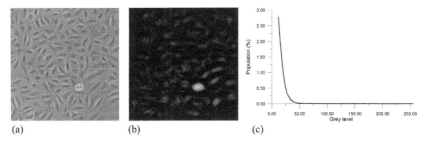

(a)   (b)   (c)

Figure 17 Example of (a) image, (b) correlation intensity, and (c) histogram in the case of a recognition at the end of the culture.

populations are very low because the number of pixels of higher intensities (pixels corresponding to a recognition peak) is very low in comparison to the total number of pixels in the correlation plane.

From those observations, we can conclude that the median of the correlation histogram remains very low and that the intensity levels lower than the median do not contribute to a recognized event. Therefore, those intensities are removed by applying the following process:

If the correlation intensity $I(s, t)$ is superior to the median of $I(s, t)$, then $I'(s, t) = I(s, t)-$ median $\{I(s, t)\}$.

If the correlation intensity $I(s, t)$ is less than the median $I(s, t)$, then $I'(s, t) = 0$.

As the correlation histograms are very similar for response 1 and response 0, we will apply adjustments based on the statistical properties of the correlation

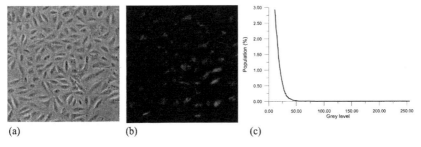

(a)   (b)   (c)

Figure 18 Example of (a) image, (b) correlation intensity, and (c) histogram in the case of a reject at the end of the culture.

histogram that will not depend on the present of a recognized event. It is a very important point because the adjustment can be applied automatically.

To correct the intensity fluctuations appearing when the illumination of the input image changes with respect to the reference model, we apply the following process:

At the initialization stage, the correlation intensity between the filter and a reference image to be recognized is computed. In this correlation intensity, we measure the level that is higher than the intensity of 95% of the pixels. This quantity is a measure of the spread of the correlation histogram and the width of the correlation peak. This value is noted *Per_Ref* (percentile 95 of reference image).

During the recognition process, we measure for each input image the correlation width under which 95% of the pixels have a smaller intensity correlation. This value is noted *Per_Obj* (percentile 95 of the correlation with an actual input image).

The readjustment of the correlation intensity is achieved by multiplying each gray level of the correlation image obtained with an input image by the ratio *Per_Ref/Per_Obj*:

$$I''(s, t) = \frac{Per\_Ref}{Per\_Obj} I'(s, t) \tag{70}$$

where $I'(s, t)$ and $I''(s, t)$ are the correlation intensities before and after the readjustment of the dynamics, respectively.

Let us now analyze the effect of the dynamics readjustment. If the input image $x'(s, t)$ undergoes a global illumination reduction with respect to its reference model $x(s, t)$, we have

$$x'(s, t) = \alpha x(s, t) \tag{71}$$

where α is a constant ($\alpha < 1$). Because the correlation product is a linear operation, the effect of the illumination change in the input image is directly reflected in the correlation amplitude:

$$c'(s, t) = (x' \otimes h)(s, t) = \alpha(x \otimes h)(s, t) \tag{72}$$

The width of the correlation histogram will be decreased. This decrease is measured by the quantity *Per_Obj*. As the value of *Per_Obj* will be less than the value of *Per_Ref*, the ratio *Per_Ref/Per_Obj* will be greater than 1. We will, thus, have an increase of the correlation intensities. On the contrary, if we have an increase of the illumination in the input image, the ratio *Per_Ref/Per_Obj* will be less than 1 and the correlation intensities will decrease to a comparable value obtained with the reference images. In both cases, we will have an automatic readjustment of the correlation intensities to overcome the illumination change of the input image.

15.5.4 Application to the Substitutive Noise

As has been outlined in the previous subsection, the substitutive noise has a negative contribution on the correlation peak intensity. The consequence is a decrease of the maximum correlation intensity. The intensity decrease is reflected in the value of *Per_Obj* of the correlation image. With the above-described methods, it is possible to readjust this decrease in order to obtain a final correlation intensity comparable to those obtained with the reference images.

15.5.5 Performances Evaluations on a Simple Test Problem

The problem considered here consists in the classification between a car and a truck independently of their out-of-plane orientation. Performances have also been studied in the case of substitutive noise and background.

In this test, we modified the illumination of the input image numerically by multiplying all the pixels of the input image $x(s, t)$ by a constant factor $\alpha(\alpha < 1)$. An ASFS filter has been computed on the basis of 20 reference images per class. Class 1 (of response 1) includes the car images with an incremental angle of $4°$. Class 2 (response 0) includes the truck images with an incremental angle of $4°$. As in the previous subsections, for each reference image, a distorted image is associated corresponding to an intermediate rotation with respect to the corresponding reference image. The filter has been tested on 80 images corresponding to the reference images and the distorted images. We found a value of the *reference percentile* of 50 (*Per_Ref* = 50).

As a first test, we lower the illumination of the input images by a factor of 2 ($\alpha = 0.5$). By readjustment of the correlation histograms, we obtain correlation intensities similar to those obtained with the reference images. Figures 19 and 20 present the results of the correlation images for the reference images and for the low-illumination input images, respectively. The statistic estimations are reported in Table 7.

The fluctuations of the correlation intensities are relatively important when the correlation intensities have been postprocessed. Nevertheless, even if the G ratio is decreased with postprocessing, the correlation intensities of response 1 are strongly improved by the postprocessing while intensities of response 0 keep low values. Comparing columns 2 and 3

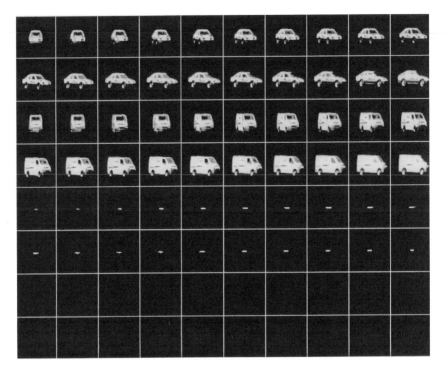

Figure 19   Input images and corresponding correlation images (with a threshold of 150).

of Table 7 clearly shows a drastic improvement of the correlation intensities, allowing one to use the value of the reference threshold when input images have different illumination.

For the second test, we added surrounding backgrounds on the input images. Two different kinds of background were used. The first considered was uniformly distributed noise in the interval [0, 255] (mean $\mu = 127$ and a standard deviation $\sigma^2 = 73$). The second one was a set of different textures representing general landscape views (of the same maximal intensity of the objects). A different texture has been used for each input image. The input images and the corresponding correlation peaks are presented in Figs. 21 and 22 in the case of texture background and random noise background, respectively. We used a threshold value of 150 (threshold used for the reference images). The statistical estimations are reported in Table 8. Results with and without postprocessing of the correlation intensities are compared. As in the case of low-illumination input images, we observe a larger differ-

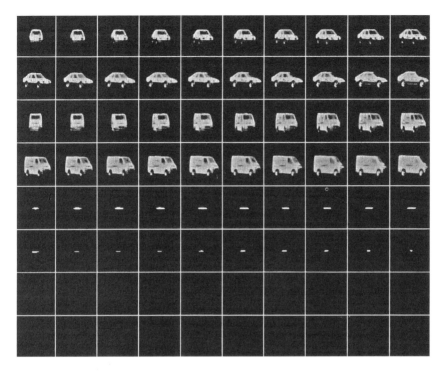

Figure 20 Input images (low illumination) and corresponding correlation intensities after postprocessing (with a threshold of 150).

Table 7 Statistic Estimations of the Numerical Correlation Intensities for Reference and Low-Illumination Input Images

	I_1	I_0	SD_1	SD_0	G
Reference images	246.40	141.8	2.99	17.50	5.10
Low-illumination images without illumination compensation	128.35	73.75	2.16	8.96	4.91
Low-illumination input images with illumination compensation	219.9	113.05	28.14	20.02	2.22

Note: I_1 is the mean maximum detected intensity of response 1; I_0 is the mean maximum detected intensity of response 0. SD_1 and SD_0 are the corresponding deviations. G is the discrimination ratio.

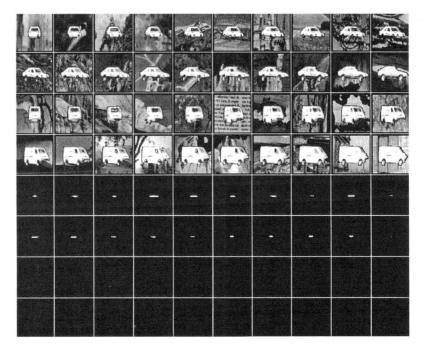

Figure 21   Input images (with nonoverlapping background clutter) and corresponding correlation intensities after postprocessing (with a threshold of 150).

ence between response 1 and response 0 correlation intensities after postprocessing but with a decrease of the G ratio due to the fluctuations of the correlation intensities after postprocessing.

The reason for those correlation intensity fluctuations originates in the size of the input images (64×64 pixels). Because the correlation peaks is relatively extended, the number of pixels having high intensities is relatively large with respect to the total number of pixels in the correlation image. This induces fluctuations of the Per_Obj values, leading to intensity fluctuations of the correlation intensities. Nevertheless, simulation results show a valuable improvement of the correlation intensities after postprocessing (compared to results obtained without postprocessing), adjusting intensity levels comparable to those obtained with the reference images.

On the other hand, it is important to note that the car–truck recognition problem presents high intraclass similarities. We found a mean interclass similarity of 0.669 (the mean similarity is defined as the average normalized scalar products of the reference images). As it has been outlined earlier, the ASFS filter presents better robustness to distortions if the intra-

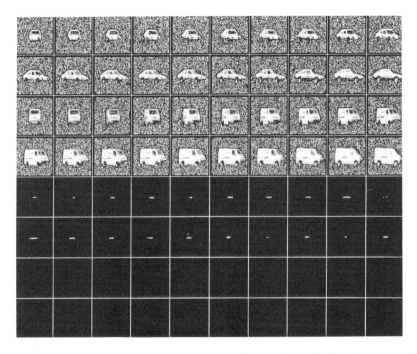

Figure 22 Input images (with nonoverlapping random noise) and corresponding correlation intensities after postprocessing (with a threshold of 150).

Table 8 Statistic Estimation of the Numerical Correlation Intensities for Reference and Input Images Surrounded by Spatially Disjoint Backgrounds.

	I_1	I_0	SD_1	SD_0	G
Reference images	246.40	141.80	2.99	17.50	5.10
Input images with background textures and without illumination compensation	189.10	111.60	18.82	19.37	2.03
Input images with background textures and with illumination compensation	204.95	113.95	24.95	21.72	1.95
Input images with background noise and without illumination compensation	153.15	103.40	13.48	9.67	2.15
Input images with background noise and with illumination compensation	195.50	125.10	17.34	13.49	2.28

Note: I_1 is the mean maximum detected intensity of response 1; I_0 is the mean maximum detected intensity of response 0. SD_1 and SD_0 are the corresponding deviations. G is the discrimination ratio.

class similarities are close to 1 and the interclass similarities are close to 0. This means that the filter may have difficulties in handling distortions, even if the input images are presented on a black background. As it can be seen from the results of the numerical simulations, the postprocessing based on the statistical properties of the correlation histograms allows one to correctly handle illumination changes and the presence of spatially disjoint backgrounds in the input images.

15.5.6 Automatic Detection of Cancerous Cells Divisions

In this subsection, we describe the computation of the ASFS filter for the detection of cancerous dividing cell application. The performance of the two methods on this actual problem are presented next.

The relative position of the daughter cells in formation is randomly oriented. Therefore, it is required to have a complete in-plane rotation invariance for the recognition. Rotations of the reference images are included in the reference image set with an incremental step of 22.5°. For response 0 images, we took several background zones representative of the culture. The reference image set is shown in Fig. 23. Reference images of 64×64 pixels are cleaned by hand in order to completely remove the background. For each reference image and according to the

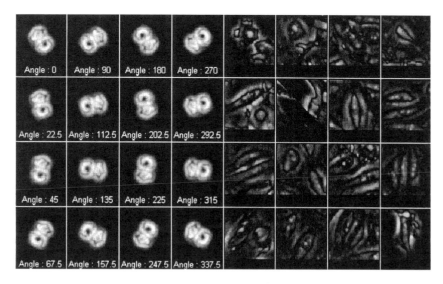

Figure 23   Reference image set. Columns 1–4 correspond to response 1 images and columns 5–8 correspond to response 0 images.

ASFS algorithm, a distorted image is associated, corresponding to an intermediate rotation of 12°. As explained in previous subsections, in order to remove most of the background, the input images are preprocessed before the correlation process. Results will be presented with and without preprocessing in order to show the efficiency of the illumination compensation postprocessing on the correlation intensities.

Input images, coming from a coupled-change device (CCD) camera placed on the microscope are digitalized in a resolution of 512×512 pixels.

During the recognition phase, the correlation intensities are postprocessed with the algorithm presented. Three tests have been performed:

Test 1: No preprocessing of the input images and no postprocessing of the correlation intensities

Test 2: No preprocessing of the input images; postprocessing of the correlation intensities

Test 3: Preprocessing of the input images; no postprocessing of the correlation intensities

The results of the simulated correlation intensities are presented in Table 9 and Fig. 24.

Comparing the results obtained for test 2 and test 3 clearly show that the intensities obtained with the postprocessing of the correlation intensities are similar to those obtained with preprocessing of the input images (especially at the beginning of the culture where the preprocessing of the input images removes almost the entire background). Furthermore, intensities of

Table 9 Results of the Maximum Detected Correlation Intensities of Response 1 and Response 0 in the Case of Input Images at the Beginning and at the End of the Culture

	Test 1	Test 2	Test 3
Average max intensity of response 1 at the beginning of the culture	135	189	190
Average max intensity of response 0 at the beginning of the culture	108	122	152
G	0.11	0.22	0.11
Average max intensity of response 1 at the end of the culture	122	181	203
Average max intensity of response 0 at the end of the culture	84	74	114
G	0.18	0.42	0.28

G is the discrimination ratio defined as $G = (I_1 - I_0/I_1 + I_0)$.

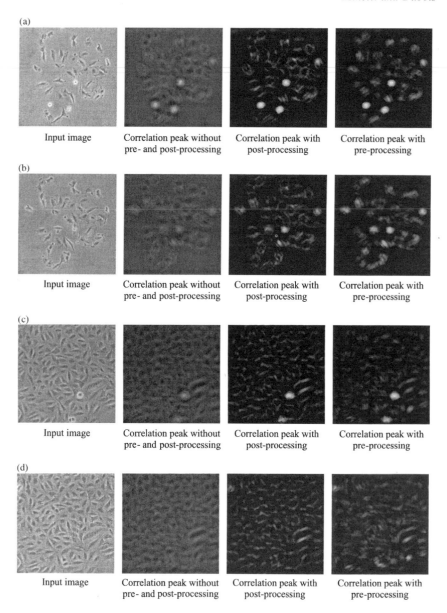

Figure 24 (a) Input image and correlation peaks in the case of a response 1 at the beginning of the culture; (b) input image and correlation peaks in the case of a response 0 at the beginning of the culture; (c) input image and correlation peaks in the case of a response 1 at the end of the culture; (d) input image and correlation peaks in the case of a response 0 at the end of the culture.

response 0 are decreased for test 2 compared to test 3 while keeping similar values for the response 1. An increase of the *G* ratio follows. On the contrary, results obtained with test 1 show that the intensity levels of response 1 and response 0 are too close to choose a suitable threshold that discriminates clearly events to be recognized and rejected.

Examples of input images and their corresponding correlation peaks for the different tests are presented in Fig. 24 at the beginning and at the end of the culture for response 1 and response 0 sample images. As expected by the results presented in Table 9, the correlation intensities obtained with test 2 and test 3 are similar. Results clearly show a strong decrease of the low-intensity levels and an increase of the high correlation intensities. Therefore, postprocessing allows one to readjust the correlation intensities to those obtained with the reference images, increasing the robustness of the correlation to illumination changes.

15.5.7 Application of the Postprocessing on the Nonlinear Cascaded Correlation System

Because the cascaded correlation architecture involves a nonlinearity, the choice of the parameters θ_0 and θ_i [see Eq. (18)] which allows to adjust the strength of the nonlinearity, may be critical, when the correlation intensity of the first stage suffers a too large loss of intensity. Indeed, the role of the nonlinearity is to perform a smooth threshold between the low and the high intensities. If the nonlinearity parameters are not adequately set, a loss of intensity at the output of the first stage will be increased by the nonlinearity, leading to a detection error at the output of the second stage.

By applying the postprocessing of the correlation intensities at the output of the first stage, before the nonlinear stage, allows one to avoid this negative effect of the nonlinearity while keeping the advantages of the cascaded correlation architecture.

15.5.8 Concluding Remarks

We have shown that it was possible to handle the problem of the illumination in the input image by a postprocessing of the correlation intensities. It is based on the fact that the correlation histograms have stable statistical properties and depend very little on the presence of a recognized event. The postprocessing is performed in two steps. The first step consists of suppressing the intensity levels that do not contribute to a recognized event. This is achieved by subtracting the median of the correlation histogram in order to recalibrate the histogram to the reference correlation histogram. The second step consists of stretching the resulting correlation histogram to make it correspond to the reference correlation histogram (obtained with a reference image). This is performed by multiplying the correlation intensity levels by a factor based on the statistics of the reference and the actual correlation populations.

The postprocessing has been applied for the detection of cells divisions in in vitro cell cultures, where it is used in actual applications. In this application, the illumination problems appear naturally in time. Results of the numerical simulations show that the influence of the changing illumination is strongly reduced by the application of the postprocessing. Concluding results have also been provided in the problem of the car–truck recognition where the illumination was changed numerically. Results were also improved in the case of the nonoverlapping noise problem.

15.6 UNSUPERVISED LEARNING AND POSTSEGMENTATION

15.6.1 Introduction

In pattern recognition problems, the user defines the recognition problem a priori. One can include several patterns and variable distortions, but the selection of the reference and distorted images is performed before the computation of the filter and before the matching. This scheme is not appropriate for practical applications, where the shapes of the objects to be recognized may change with time (see, for example, the identification of cancerous cells in division presented in the previous section). For such applications, it is mandatory to consider systems capable of dynamic self-learning. Such a system must be able to integrate automatically, during the matching phase, all of the recognized events in its database and to integrate them in the further recognition process. In this way, when an object has been recognized by the system, this object must be extracted from the input image where it has been detected in such a way that it can be used for an update of the recognition filters. This extraction is called an a posteriori segmentation process, which differs mainly from an a priori segmentation process because the position of the object to be segmented is known. This position, given by the position of the correlation peak, is called the *seed*.

The automatic integration of recognized events in the learning process considerably extends the application field of the pattern recognition by correlation. Indeed, such a system has the capacity to follow objects whose shapes may strongly vary in time without any human operator intervention. Furthermore, the constraints applied to the initial filter are strongly reduced because we can restrict the initial reference set to a limited number of images. The need of a posteriori segmentation in view to achieve an automatic update of the correlation filters has already been investigated.

15.6.2 Automatic Self-Learning System

To establish the a posteriori segmentation applied to our system, we process the data delivered by the correlation system with a segmentation system that profits from the seed (correlation peak position) inside the recognized object. The recognized and a posteriori segmented events are then used to dynamically compute new recognition filters. Such a system, combining the recognition, the postsegmentation and the automatic learning, is described in Fig. 25.

Because the postsegmentation benefits from the seed delivered by the correlation peak, it is possible to use specific segmentation techniques such as region-growing techniques. Research in this field has been performed by Réfrégier et al. in the case of image sequences where the high degree of correlation between successive frames is taken into account while designing the filter [52].

15.6.3 Segmentation Algorithm Based on the Snakes Technique

The Snakes technique is based on a active connected contour centered on the seed and iteratively distorted to match the external contour of the object to be segmented [53].

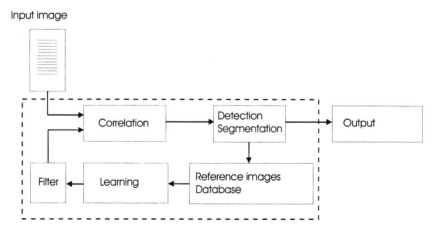

Figure 25 Global schema of the self-learning system.

Let us consider an image $x(s, t)$ (of dimension $N \times N$) in which an event has been recognized at position (x_e, y_e) (position of the maximum detected correlation intensity). We crop, around the point (x_e, y_e), a window $w_x(s, t)$ of dimension $n \times n$ including the object to be segmented and the background surrounding the object. The goal of the segmentation is to extract the object and to remove the surrounding background automatically. For this purpose, we choose an initial contour (initial Snake) that will be modified iteratively in order to match the external contour of the object. The initial contour is defined by a function $S_0(s, t)$ equal to 1 $[S_0(s, t) = 1]$ inside the contour and 0 $[S_0(s, t) = 0]$ outside the contour.

To extract the segmented zone, we multiply the image $w_x(s, t)$ by the initial Snake $S_0(s, t)$:

$$w'_x(s, t) = w_x(s, t) S_0(s, t) \tag{73}$$

It is then necessary to define a function that allows one to distort the shape of the initial Snake in order to extract correctly the object from its background. For this purpose, we use an optimization process based on the decrease of a cost function. The choice of the cost function is very important and depends on the actual application. In the particular application of the recognition of dividing cells (see previous section), we have chosen a cost function based on the histogram of the segmented zone because, in this particular application, the gray-level histograms of the cells are very significant. Other approaches have been investigated [54–57]. The choice of the cost function is very application dependent. We consider that the objects to be segmented (the dividing cells) present a stable histogram that is different from the background one. We define the initial cost function as follows:

$$F_0 = \sum_{i=0}^{Nn-1} h_{w'}(i) - h_{\text{ref}}(i))^2 \tag{74}$$

where

$h_{w'}(i)$ is the histogram of the image $w'_x(s, t)$ (i represent the gray-level intensity)

$h_{\text{ref}}(i)$ is the reference histogram computed on a reference image containing a segmented object (object segmented, for example, by a human operator)

Nn is the number of gray levels in the images (typically 256).

The optimization of the cost function F is achieved by simulated annealing algorithm. At each iteration, the Snake is modified and the cost function is computed. If the cost function of the modified Snake is less than the previous one, we keep the modified Snake as the initial state for the next iteration. Otherwise, the Snake is rejected and the previous Snake is kept as initial state for the next iteration. In this way, the optimization finds iteratively configurations, where the histogram of the segmented object is closer and closer to the reference histogram.

The Snake is defined as an active contour; this means a series of nodes $S(x_k, y_k)$ $(k = 1, \ldots, K)$, where K is the total number of nodes of the Snake related to the line segments. The contour is connected, which means that the first and the last node of the contour are identical.

The modifications $(\delta(x_k), \delta(y_k))$ applied to each node of the Snake are random values chosen in an interval $[-A_m, +A_m]$. This interval is parameterized initially by the user. The modified Snake at the $j + 1$st iteration is written

$$S^{j+1}(x_k, y_k) = S^j(x_k, y_k) + \delta^j(x_k, y_k) \tag{75}$$

Furthermore, we imposed the modifications $\delta(x_k), \delta(y_k)$ to be perpendicular to the contour. It is observed experimentally with such displacements of the nodes a faster convergence of the cost function F. On the other hand, we observe that this specific segmentation method is very sensitive to the shape of the initial Snake. To reduce this sensitivity, we have introduced a multiscale method [56] that allows one to increase the number of nodes during the iteration process.

It is important to note that the cost function F that allows one to optimize the Snake does not depend on the specific shape of the objects to be segmented. It only depends on the difference of the gray-level population of the segmented object and a reference image. It is a very important point because it is not necessary to introduce a distortion model in the system. However, it can cause problems when the illumination changes significantly in a short time delay.

15.6.4 Application to the Cell Division Problem

The segmentation method based on Snakes has been used in the problem of the automatic detection of cancerous cells. As described in the previous section, this problem is particularly interesting for unsupervised learning systems because the shape of the cells can vary significantly with time. The different parameters used for the test are described in Table 10. Examples of segmentation results are presented in Figs. 26 and 27.

We observe a drastic decrease of the cost function F from an initial value of 0.001948 to a final value of 0.000008. The mean difference between the 256 gray-level densities of the reference histogram and the histogram obtained with the segmented object is less than 0.01%. This preliminary

Table 10 Parameters Used for the Segmentation Test

Input image size	256×256 pixels
Window w_x size	128×128 pixels
Initial number of nodes	6
Final number of nodes	48
Maximum displacement amplitude (A_m)	3 pixels
Total number of iterations	1500

result show that the Snake segmentation method based on the histogram cost function minimization is well adapted to the segmentation of the cells.

15.6.5 Self-Learning Recognition System for the Cells Division Recognition Problem

During the recognition phase, the input images of the cellular culture are correlated with an initial filter. This initial filter is computed on the basis of a reference image set defined by the user. When an event is recognized in the input image, the correlation image presents a bright peak at the position of the detected object. This position provides the *seed* to the segmentation process, which is then capable of extracting the object from its background. This procedure is repeated for each image of the sequence (a sequence is made of 1080 images). Each recognized event is segmented and stocked in a reference image database. In this way, the system is capable of automatically determining if the shape of the cells significantly changed with respect to a reference model. This shape difference is measured by the normalized scalar

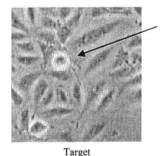

Target   Reference image

Figure 26 Target and reference image used for the preliminary segmentation test. The cell to be segmented is indicated by a black arrow.

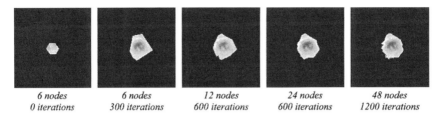

| 6 nodes | 6 nodes | 12 nodes | 24 nodes | 48 nodes |
| 0 iterations | 300 iterations | 600 iterations | 600 iterations | 1200 iterations |

Figure 27 Different steps of the segmentation process.

products of the reference and the segmented image for different rotations. Examples of segmented cells during the recognition phase are presented in Fig. 28. Note that the shapes of the different segmented cells are very different. This is a very important point because it means that the segmentation system is capable of segmenting new shapes that are very different from the original ones.

When the shape of the segmented cells becomes too different from the reference model, a new learning phase takes place. A minimum of two new segmented images (sufficiently different from the reference model) is required to perform a new learning phase. A new set of reference images is built, including those two new reference images. As in the case of the initial filter, the in-plane rotation invariance is required. Fourteen images are computed for each reference image, corresponding to rotations of $24°$. Intermediate rotations of $12°$ are computed for the distorted images (one

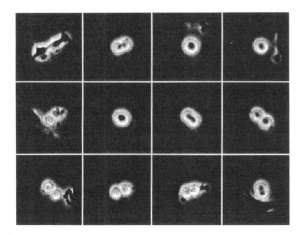

Figure 28 Examples of segmented cells during the recognition phase.

distorted image per reference image). A new ASFS filter is computed and the recognition phase goes on.

The unsupervised learning system has been tested on a complete sequence (containing 1082 input images). Four new filters have been computed during the entire sequence. We observed experimentally an increase of the correlation intensities with the new computed filters for events that differ strongly from the reference model. This intensity increase allows one to detect a larger number of cells, especially at the end of the culture. Examples of cells detected by new computed filters that were not detected by the initial filter are shown in Fig. 29 (a threshold value of 140 was used). We see that, with the initial filter, no cell has been detected. However, with the unsupervised learning system, the three cells are correctly detected. For the complete sequence, a normal ASFS (based on the reference set described in the previous section) detects correctly 164 cells.

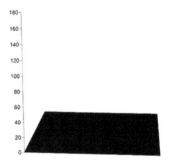

Correlation peak obtained with the initial filter

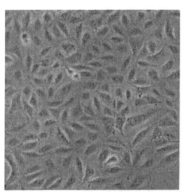

Input image at the end of the culture

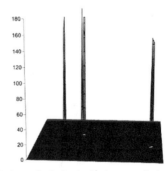

Correlation peak obtained with the new calculated filter

Figure 29   Example of input image and the corresponding correlation peaks in the case of the initial filter (top) and a dynamically computed filter (bottom)

The unsupervised learning system detects 308 cells correctly, which demonstrates the efficiency of adaptive systems. Furthermore, it has been observed that the number of false detections (the presence of a correlation peak for a cell that should not be detected) has similar values to a normal ASFS filter and with the unsupervised learning system.

15.6.6 Concluding Remarks and Perspectives

An unsupervised learning system has been studied and implemented with the ASFS algorithm. In this configuration, the user defines the initial filter with an initial reference set. During the matching phase, when an event is recognized in the input image, the position of the detected event (the *seed*) is transmitted to the segmentation stage, which is then capable of extracting the object from its background and to stock the segmented image in a database. Once the detected objects become too different from a reference model, a new learning phase starts with the new reference images coming from the segmented image database. In this way, once the recognition system has been correctly initialized by the user, the system is capable of self-adaptive learning and is capable of following objects whose shape may vary considerably with time.

We have used a segmentation method based on the Snakes algorithm. The optimization of the active contour is based on the minimization of the mean histogram difference between the segmented image and a reference model. This minimization criterion has the great advantage of being insensitive to the shape of the objects and well adapted to the particular problem of the cells. However, this choice could be problematic in other applications and it should then be necessary to use other criteria [54–57].

15.7 GENERAL CONCLUSIONS

Actual applications require specific methods to reduce the influence of the correlation to distortions that appear when input images differ from their reference model. When the distortion model can be described analytically (such as in-plane rotation, scaling, etc.), specific methods give invariant outputs to those particular distortions but reduces significantly the field of practical applications. To overcome those limitations, we have first proposed the ASFS algorithm that allows to include a set of distorted images in order to decrease the sensitivity of the correlation to nonanalytically described distortions. Based on the ASFS algorithm, we have developed a nonlinear two-step architecture, where the first correlation plane is processed by a nonlinearity before being correlated by the second correlation stage. It has been shown that such an architecture increased the performances of the correlation appreciably and allows one to solve a specific recognition problem that linear systems cannot handle.

When input images are surrounded by noise or background clutter, the correlation amplitude decreases because nonoverlapping noise has a negative contribution to the central correlation amplitude. To remove the contribution of the nonoverlapping background, two methods have been proposed and successfully tested. The first one includes, in the reference set, the window functions of the reference images. The method presents an improvement of the performances but requires twice the number of reference images and is restricted to nonbinary objects. The second method consists of multiplying the numerical filters by a mask computed on the basis of the reference image set in order to limit the spatial extension of the filters to the mean spatial extension of the reference images.

To overcome the sensitivity to illumination changes in the input images, we developed a postcorrelation method based on the statistical properties of the correlation histograms. Correlation histograms depend very little on the presence of a recognized event. Therefore, it is possible to apply a postprocessing of the correlation intensities independently of the presence of a correlation peak. The postprocessing allows one to readjust the correlation values obtained with low-illumination images to those obtained with the reference images. We emphasize the fact that, depending on particular pattern recognition problems, most of those techniques can be combined in a flexible way.

To extend the application field of the correlation to living objects whose shape may vary strongly in time, we proposed an unsupervised learning system with a postsegmentation and a learning stage. In this configuration, all of the recognized events are segmented and stocked in an image database. When the shape of the objects becomes too different with respect to a reference model, a new correlation filter is computed and the matching phase continues with this new filter.

ACKNOWLEDGMENTS

This work has been supported by the Walloon Region (Roidimac project) and the IUPA 4-06 supported by the Belgian Federal Office for Scientific, Technical and Cultural Affairs.

REFERENCES

1. AB VanderLugt. Signal detection by complex spatial filtering. IEEE Trans Inform Theory IT-10:139–145, 1964.
2. DP Casasent, D Psaltis. Position, rotation and scale invariant optical correlation. Appl Opt 15:1795–1799, 1976.
3. Y Hsu, HH Arsenault. Optical pattern recognition using circular harmonic expansion. Appl Opt 21:4016–4019, 1982.
4. GF Schils, DW Sweeney. Rotationally invariant correlation filtering. J Opt Soc Am A 2:1411–1418, 1985.
5. T Szoplik, HH Arsenault. Shift and scale invariant anamorphic Fourier correlator using multiple circular filters. Appl Opt 24:3179–3183, 1985.
6. AS Jensen, L Lindvold, E Rasmussen. Transformation of image positions, rotations and sizes into shift parameters. Appl Opt 26:1755–1781, 1987.
7. K Mersereau, GM Morris. Scale, rotation, shift invariant image recognition. Appl Opt 25:2338–2342, 1986.
8. J Rosen, J Shamir. Circular harmonic phase filters for efficient rotation-invariant pattern recognition. Appl Opt 27:2895–2899, 1988.
9. J Rosen, J Shamir. Scale invariant pattern recognition with logarithmic radial harmonic filters. Appl Opt 28:240–244, 1989.
10. D Casasent, A Lyer, G Ravichandran. Circular harmonic function, minimum correlation energy filters. Appl Opt 35:5169–5175, 1991.
11. HJ Caulfield, WT Maloney. Improved discrimination in optical character recognition. Appl Opt 8:2345–2356, 1969.
12. DP Casasent. Unified synthetic discriminant function computational formulation. Appl Opt 23:1620–1627, 1984.
13. DP Casasent, W-T Chang. Correlation synthetic discriminant functions. Appl Opt 25:2343–2350, 1986.
14. JR Leger, SH Lee. Image classification by an optical implementation of the Fukunaga–Koontz transform. J Opt Soc Am 72:556–564, 1982.
15. B Braunecker, R Hauck, AW Lohmann. Optical character recognition based on the nonredundant correlation measurements. Appl Opt 18:2746–2753, 1979.
16. GF Schils, DW Sweeney. Rotationally correlation filtering for multiple images. J Opt Soc Am 3:902–908, 1986.
17. RR Kallman. Construction of low noise optical correlation filters. Appl Opt
18. DP Casasent, WA Rozzi. Modified MSF synthesis by Fisher and mean-square-error techniques. Appl Opt 25:184–186, 1986.
19. BVK Vijaya Kumar. Minimum variance synthetic discriminant functions. J Opt Soc Am 3:1579–1584, 1986.
20. A Mahalanobis, BVK Vijaya Kumar, D Casasent. Minimum average correlation energy filters. Appl Opt 26:3633–3640, 1987.
21. M Fleisher, U Mahlab, J Shamir. Entropy optimized filter for pattern recognition. Appl Opt 29:2091–2098, 1990.
22. A Mahalanobis, D Casasent. Performance evaluation of minimum average correlation energy filters. Appl Opt 30:561–572, 1991.

23. SI Sudharsanan, A Mahalanobis, MK Sundareshan. Unified framework for synthesis of discriminant functions with reduced noise variance and sharp correlation structure. Opt Eng 29:1021–1028, 1990.
24. D Casasent, G Ravichandran, S Bollapragada. Gaussian minimum correlation energy filters. Appl Opt 30:5176–5181, 1991.
25. Ph Refregier, JP Huignard. Phase selection of synthetic discriminant function filters. Appl Opt 29:4772–4778, 1990.
26. Ph Refregier. Optimal trade-off filters for noise robustness, sharpness of the correlation peak, and Horner efficiency. Opt Lett 16:829–831, 1991.
27. Ph Refregier. Filter design for optical pattern recognition: multicriteria optimization approach. Opt Lett 15:854–856, 1990.
28. DL Flannery. Optimal trade-off distortion-tolerant constrainted modulation correlation filters. J Opt Soc Am A 12:66–72, 1995.
29. B Javidi. Generalization of the linear matched filter concept to nonlinear matched filters. Appl Opt 29:1215–1224, 1990.
30. Ph Réfrégier, V Laude, B Javidi. Basic properties of nonlinear global filtering techniques and optimal discriminant solutions. Appl OPt 34:3915–3923, 1995.
31. B Javidi, D Painchaud. Distortion-invariant pattern recognition with Fourier-plane nonlinear filters. Appl Opt 35:318–331, 1996.
32. P Willet, B Javidi, M Lops. Analysis of image detection based on Fourier plane nonlinear filtering in a joint transform correlator. Appl Opt 37:1329–1341, 1998.
33. Ph Réfrégier, F Goudail. Decision theory approach to nonlinear joint-transform correlation. J Opt Soc Am A 15:61–67, 1998.
34. F Dubois. Automatic spatial frequency selection algorithm for pattern recogntion by correlation. Appl Opt 32:4365–4371, 1993.
35. BVK Vijaya Kumar, L Hassebrook. Performance measures for correlation filters. Appl Opt 29:2997–3006, 1990.
36. JL Horner. Clarification of Horner efficiency. Appl Opt 31:4629, 1992.
37. S Kirkpatrick, CD Gelatt Jr, MP Vecchi. Optimization by simulated annealing. Sciences 220:671–679, 1983.
38. MS Kim, CC Guest. Simulated annealing algorithm for binary phase only filters in pattern classifications. Appl Opt 29:1203–1208, 1989.
39. F Dubois. Nonlinear cascaded correlation processes to improve the performances of automatic spatial-frequency-selective filters in pattern recognition. Appl Opt 35:4589–4597, 1996.
40. B Javidi, J Wang. Limitation of the classical definition of the correlation signal-to-noise ratio in optical pattern recognition with disjoint signal and scene noise. Appl OPt 31:6826–6829, 1992.
41. Ph Refregier, B Javidi, G Zhang. Minimum mean-square-error filter for pattern recognition with spatially disjoint signal and scene noise. Opt Lett 18:1453–1455, 1993.
42. B Javidi, F Parchekani, G Zhang. Minimum-mean-square-error filters for detecting a noisy target in background noise. Appl Opt 35:6964–6975, 1996.

43. B Javidi, G Zhang, F Pachekani, Ph Refregier. Performance of minimum mean-square-error filters for spatially nonoverlapping target and input noise. Appl Opt 33:8197–8209, 1994.

44. B Javidi, J Wang. Optimum filter for detection of a target in nonoverlapping scene noise. Appl Opt 33:4454–4458, 1994.

45. H Sjöberg, B Noharet. Distortion invariant filter for nonoverlapping noise. Appl OPt 37:6922–6930, 1998.

46. AH Fazlollahi, B Javidi. Error probability of an optimum receiver designed for nonoverlapping target and scene noise. J Opt Soc Am A 14:1024–1032, 1997.

47. C Minetti, F Dubois. Reduction in correlation sensitivity to background clutter by the automatic spatial frequency selection algorithm. Appl Opt 35:1900–1903, 1996.

48. F Dubois, C Minetti, J-C Legros. Nonlinear cascaded correlation process for recognition of target in nonoverlapping noise. II Euro-American Workshop, Optoelectronic Information Processing. Bellingham, WA: SPIE Optical Engineering Press, 1997, pp 195–220.

49. H Sjöberg, F Goudail, Ph Réfrégier. Optimal algorithms for target location in nonhomogeneous binary images. J Opt Soc Am A 15:2976–2985, 1998.

50. HH Arsenault, D Lefebvre. Homomorphic cameo filter for pattern recognition that is invariant with changes in illumination. Opt Lett 25:1567–1569, 2000.

51. A Papoulis. Probability, Random Variable and Stochastic Processes. New York: McGraw-Hill, 1965, pp 126–127.

52. F Goudail, Ph Réfrégier. Optimal target tracking on image sequences with a deterministic background. J Opt Soc Am A 14:3197–3207, 1997.

53. M Kass, A Witkin, D Terzopoulos. Snakes: Active contour models. Int J Computer Vision, 321–33, 1998.

54. O Germain, Ph Réfrégier. Optimal Snake-based segmentation of random luminance target on a spatially disjoint background. Opt Lett 21:1845–1847, 1996.

55. F Goudail, Ph Réfrégier. Optimal detection of a target with random gray levels on a spatially disjoint background noise. Opt Lett 21:495–497, 1996.

56. Ch Chesnaud, V Pagé, Ph Réfrégier. Improvement in robustness of the statistically independent region Snake-based segmentation method of a target-shape tracking. Opt Lett 23:488-490, 1998.

57. Ch Chesnaud, Ph Réfrégier, V Boulet. Statistical region Snake-based segmentation adapted to different physical noise models. IEEE Trans Pattern Anal Machine Intell PAMI-21:1145–1157, 1999.

Index